D0934485

PROTEINS

LAB F AX

The LABFAX series

Series Editors:

B.D. HAMES Department of Biochemistry and Molecular Biology, University of Leeds, Leeds LS2 9JT, UK

D. RICKWOOD Department of Biology, University of Essex, Wivenhoe Park, Colchester CO4 3SQ, UK

MOLECULAR BIOLOGY LABFAX

CELL BIOLOGY LABFAX

CELL CULTURE LABFAX

BIOCHEMISTRY LABFAX

VIROLOGY LABFAX

PLANT MOLECULAR BIOLOGY LABFAX

IMMUNOCHEMISTRY LABFAX

CELLULAR IMMUNOLOGY LABFAX

ENZYMOLOGY LABFAX

PROTEINS LABFAX

Forthcoming titles

NEUROSCIENCE LABFAX

MOLECULAR BIOLOGY: GENE CLONING LABFAX

MOLECULAR BIOLOGY: GENE ANALYSIS LABFAX

PROTEINS
LABFAX

EDITED BY

N.C. PRICE

Department of Biological and Molecular Sciences,
University of Stirling,
Stirling FK9 4LA, UK

*β*IOS
SCIENTIFIC
PUBLISHERS

ACADEMIC PRESS

© **BIOS Scientific Publishers Limited, 1996**

Published jointly by Academic Press, Inc., 525 B Street, San Diego, CA 92101-4495, USA and BIOS Scientific Publishers Limited, 9 Newtec Place, Magdalen Road, Oxford OX4 1RE, UK.

A CIP catalogue record for this book is available from the British Library.

ISBN 0 12 564710 7

Distributed exclusively throughout the world by Academic Press, Inc., 525 B Street, San Diego, CA 92101-4495, USA pursuant to agreement with BIOS Scientific Publishers Limited.

Typeset by APEK Typesetters, Bristol, UK.
Printed by Biddles Ltd, Guildford, UK.

Front cover: *stylized version of Roxindole molecule in the 5-HT1A model. Original photograph courtesy of Dr Friedrich Rippmann, E. Merck, Darmstadt, Germany.*

CONTENTS

Contributors   xvii
Abbreviations   xxi
Preface   xxv

PURIFICATION

1. CHOICE OF ASSAY OR DETECTION METHOD   1

Introduction   1
Continuous versus discontinuous assays   2
    Common techniques for the assay and detection of proteins (Table 1)   3
Photometric assays   2
  Absorption   2
    The coupled assay for fructose-1,6-bisphosphate aldolase (Figure 1)   4
  Fluorescence   5
  Turbidimetry   5
Radiometric assays   5
HPLC-based assays   6
Electrochemical assays   7
  The oxygen electrode   7
  The pH-stat (stationary pH)   7
Assays involving gel electrophoresis   8
  Visualization of protein bands by protein staining   8
  Visualization of protein bands by activity staining   8
  Other electrophoresis-based assays   9
Immunological detection   9
  Immunoblotting   9
  Enzyme-linked immunosorbent assay   10
  Radioimmunoassay and immunoradiometric assay   10
Conclusion   10
References   11

2. PREPARATION OF CELL EXTRACTS   13

Introduction   13
Disruption of cells and tissues   13
  Choice of tissue   13
  Disruption of tissue and separation of cells   14
  Disruption of cells   14
    Methods for disruption of cells (Table 1)   15
Protection of protein integrity   17
  Control of pH   17
  Control of temperature   17
  Control of proteolysis   17
    Some inhibitors used to contol proteolysis (Table 2)   18
  Protection of labile thiol groups   19
  Control of mechanical stress   19

Control of free radical formation	19
Control of the effects of dilution	19
Assays of proteins in unfractionated cell extracts	20
References	20

3. BUFFERS   21

Introduction	21
Factors determining the choice of buffer	21
Selected properties of commonly used buffers (Table 1)	22
Volatile buffer mixtures (Table 2)	24
Preparation of buffers	24
Metal ion buffers	25
References	26

4. THE DETERMINATION OF PROTEIN CONCENTRATION   27

Introduction	27
Principal methods for determination of protein (Table 1)	28
Recommendations	27
References	30

5. SALTING OUT: AMMONIUM SULFATE PRECIPITATION   31

Calculation of the amount of ammonium sulfate needed	31
Ammonium sulfate, grams to be added to 1 l of solution at $S1\%$ saturation, to take the saturation to $S2\%$ (Table 1)	32

6. CHROMATOGRAPHIC PROCEDURES   33

Basic chromatographic theory	33
Progress of protein bands down an adsorption column under isocratic conditions (Figure 1)	34
Typical appearance of a chromatogram of a protein mixture on an ion exchange column (Figure 2)	34
Chromatographic materials	35
Ion-exchange chromatography	35
Ion exchanger types and nomenclature (Table 1)	36
Titration curves for a variety of proteins (Figure 3)	36
Buffers for ion exchange	37
Buffers for use in ion-exchange chromatography (Table 2)	38
Operation of an ion-exchange column	37
Chromatofocusing — general principles	39
Progress of a chromatofocusing column (Figure 4)	40
Operation of chromatofocusing	39
Hydrophobic chromatography	40
Sketch of the surface of a typical protein (Figure 5)	41
Hydrophobic interaction chromatography — practical aspects	41
Operation of a hydrophobic interaction column	42
Typical appearance of a chromatogram of a hydrophobic column (Figure 6)	43
Reverse-phase chromatography	42
Affinity chromatography	43
Affinity chromatography theory	43
Stages in affinity chromatography (Figure 7)	44

A selection of some commercially available group affinity adsorbents (Table 3) 45
Activation of the matrix 44
Some activation chemistries for affinity chromatography and for protein immobilization (Table 4) 45
Operation of affinity chromatography 45
Recombinant proteins 46
Some recombinant-fusion proteins designed for affinity purification (Table 5) 47
Affinity elution techniques 46
Principle of affinity elution from a cation exchanger (Figure 8) 47
Principle of affinity elution from an affinity adsorbent (Figure 9) 48
Dye–ligand chromatography 48
Selection of dye adsorbent 48
Batchwise screening of dye–ligand adsorbents (Figure 10) 49
Some dyes commonly used in dye–ligand chromatography (Table 6) 50
Two-adsorbent procedures — differential chromatography 50
Principle of differential tandem column chromatography (Figure 11) 51
Buffers for use in dye–ligand chromatography 50
Buffers for use in dye–ligand chromatography (Table 7) 52
Immobilized metal affinity chromatography 52
IMAC principles and practice 52
Buffer systems for IMAC (Table 8) 53
Gel filtration 53
Diagram to represent the separation of proteins on the basis of size by a gel filtration bead (Figure 12) 54
Gel filtration: operating conditions for separating proteins 53
Gel filtration to separate proteins from low molecular mass compounds 54
Media for gel filtration 55
Gel filtration media (Table 9) 55
Methods for concentrating proteins 56
Concentration by precipitation 56
Concentration by adsorption 56
Concentration by removal of water 57
Concentration by centrifugal force 57
Ultrafiltration to concentrate proteins (Figure 13) 58
Concentration by gas pressure 57
Concentration by water pressure 58
Concentration by osmotic pressure 58
Purification of small peptides from biological sources 58
Introduction 58
Practical considerations 59
Parameters affecting RP-HPLC seperation of peptides 60
Isolation of guinea-pig adrenocorticotrophin (ACTH) 61
Purification of guinea-pig ACTH from anterior pituitary extracts by sequential HPLC steps (Figure 14) 62
Conclusion 62
References 62
Further reading 64

7. COVALENT CHROMATOGRAPHY BY THIOL–DISULFIDE INTERCHANGE USING SOLID-PHASE ALKYL 2-PYRIDYL DISULFIDES **65**

Introduction 65
Covalent chromatography by thiol–disulfide interchange (TDCC) 65
Other applications of 2-pyridyl disulfides and the origins of TDCC 66
Sequential elution TDCC 66
Covalent affinity chromatography 67
Covalent chromatography and salt-promoted thiophilic adsorption 67
Some examples of the applications of TDCC 67
Isolation of thiol-containing proteins 67

CONTENTS **vii**

Illustration of the range of thiol-containing proteins that have been isolated by covalent
chromatography (Table 1)   68
Isolation and sequencing of thiol-containing peptides   67
Examples of the application of covalent chromatography in the isolation and sequencing of peptides
(Table 2)   68
Removal of prematurely terminated peptides in solid-phase peptide synthesis   68
Enzyme immobilization with associated purification   69
Thiol-specific covalent chromatography by methods other than thiol–disulfide interchange   69
Selectivity in covalent chromatography   69
References   70

8. CRITERIA OF PROTEIN PURITY   73

Introduction   73
Methods for assessing the purity of protein preparations (Table 1)   74
Detection of nonprotein contaminants   73
The quantification of nucleic acid contaminants in protein samples (Figure 1)   75
Detection of protein contaminants   73
Determination of purity   76
Specific activity measurement   76
Electrophoretic methods   76
The choice of electrophoretic method for the determination of protein purity (Table 2)   77
Some common problems encountered in electrophoresis and their remedy (Table 3)   78
The uses of capillary electrophoresis (Table 4)   79
CE can be used in a variety of modes (Figure 2)   80
Chromatographic techniques   78
Centrifugation methods   81
Mass spectrometry   81
Amino acid analysis and sequence analysis   82
Measurements of binding sites or active sites   83
Summary   83
References   83

9. ELECTROPHORESIS METHODS   85

Introduction   85
Recipes for gels and buffers   85
One-dimensional polyacrylamide gel electrophoresis   85
Recipe for gel preparation using the SDS–PAGE discontinuous buffer system (Table 1)   86
Gel mixtures for a 5–20% SDS–PAGE gradient gel (Table 2)   86
Electrolytes used in IEF (Table 3)   87
Recipes for preparation of IEF gels (Table 4)   87
Two-dimensional gel electrophoresis   90
Recipes for first-dimensional IEF gel and second-dimensional SDS–PAGE gel (Table 5)   88
Immunoelectrophoresis   90
Reagents for immunoelectrophoresis (Table 6)   88
Sample preparation   91
Standard marker proteins used in either denaturing or IEF gels (Table 7)   89
Analysis of gels   91
In situ direct polypeptide detection methods in SDS gels without staining (Table 8)   90
Protein staining and detection methods used in IEF (Table 9)   91
Staining procedures for two-dimensional polyacrylamide gels (Table 10)   92
Staining procedures used after immunoelectrophoresis (Table 11)   92
In situ polypeptide detection methods in gels using fluorophore labeling (Table 12)   93
In situ detection of specific classes of proteins in gels (Table 13)   93

Common blotting membranes (Table 14)   94
Blocking solutions to prevent nonspecific binding of the probe to the matric in protein blotting
    (Table 15)   94
Transfer buffers used in blotting (Table 16)   95
General protein stains for blot transfers (Table 17)   95
In situ detection of radioactive proteins in gels (Table 18)   95
Sensitivities of methods for radioisotope detection in polyacrylamide gels (Table 19)   96
Manufacturers and suppliers   96
References   97

10. INTEGRAL MEMBRANE PROTEINS   101

Introduction   101
Problems of solubilization, purification and assay of membrane proteins   101
    Choice of detergent   101
        General properties of a range of detergents used with membrane proteins (Table 1)   102
    Methods of purification   101
    Assay: use of reconstitution   105
Structural analysis of membrane proteins   105
References   106
Further reading   107
    Liposomes and reconstitution methods   107
    General   107

11. EXPRESSION SYSTEMS AND FUSION PROTEINS   109

Introduction   109
    References for the production of fusion proteins (Table 1)   109
Proteins commonly used in gene fusion technology   110
    Physical properties of the most common proteins used in gene fusion technology (Table 2) 110
    Gene fusion vectors (Table 3)   111
Induction of expression   111
    Conditions that influence induction of a recombinant fusion protein (Table 4)   112
    Promoter and strain relationships (Table 5)   113
Proteolytic cleavage of fusion proteins   112
    Properties of some of those proteases frequently used to liberate recombinant proteins from a
        fusion partner (Table 6)   114
Purification of fusion proteins   112
    Purification of fusion proteins (Table 7)   114
Detection of fusion proteins   113
    Spectrophometric assays for fusion proteins (Table 8)   115
References   116

12. INCLUSION BODIES AND REFOLDING   119

Introduction   119
    Origin of inclusion bodies and their advantage in the purification of recombinant proteins 119
    Structure of inclusion bodies   119
    Factors affecting inclusion body formation   120
    When should the inclusion body route be chosen?   120
Methods for the isolation and solubilization of inclusion bodies   121
    Recovery of proteins from inclusion bodies (Figure 1)   121
    Methods for the isolation of inclusion bodies   122

CONTENTS   **ix**

Solubilization 122
Renaturation of solubilized inclusion bodies 123
  Refolding 123
    Refolding strategies (Figure 2) 124
  Disulfide bond formation 125
    Proteins purified from inclusion bodies (Table 1) 126
Designing a procedure *de novo* for refolding and purification of an inclusion body protein 126
    Generic purification approach for inclusion body protein when no or limited information on the
    properties of the protein is available (Figure 3) 127
References 128

13. INDUSTRIAL SCALE PURIFICATION OF PROTEINS   131

Introduction 131
  Scope 131
  Process scale up 131
Developing a scale down process 132
Block diagrams and process flow sheeting 132
    Block diagrams and process flow sheeting (Figure 1) 133
Selection of unit operations 133
    Downstream processing technologies used in the industrial production of proteins (Figure 2) 134
  Primary recovery 133
  High-resolution purification steps 135
References 137

14. PEPTIDE SYNTHESIS   139

Introduction 139
General aspects of solid-phase peptide synthesis 140
Solid-phase supports, modes of synthesis and related apparatus 140
    Materials for peptide synthesis supports (Table 1) 142
Peptide–resin linkage and protecting group strategy 141
    Protection strategies in solid-phase peptide synthesis (Table 2) 144
    Deprotection/scavenger mixtures suitable for the Fmoc strategy (Table 3) 144
Carboxyl activation and coupling: peptide bond formation 143
    Active species and coupling reagents (Table 4) 145
Synthesis monitoring and control 146
    Qualitative color tests on resins (Table 5) 147
    Continuous flow monitoring systems (Table 6) 148
    Other methods for monitoring peptide synthesis (Table 7) 149
Synthetic peptide analysis, purification and handling 146
    Problems in handling and use of peptides (Table 8) 150
Synthetic peptides for antibody production 149
    Prediction of immunogenic peptide sequences (Table 9) 151
    Carriers for synthetic peptide immunogens (Table 10) 151
    Conjugation methods to link peptide carriers (Table 11) 152
References 152

15. CRYSTALLIZATION PROCEDURES   155

Introduction 155
Overview 155
    Some of the factors which affect the crystallization of a protein (Table 1) 156
    Methods for crystallizing proteins (Table 2) 157

Some specialist supplies and suppliers (Table 3)   158
Growth of large crystals   156
General scheme   157
A simple procedure for hanging drop crystallization   158
A simple procedure for crystallization by small-scale dialysis   159
  Methods for 'buttons'   159
Conclusion   159
References   159
Further reading   160
  Sources of information on protein crystallization   160

PHYSICAL CHARACTERIZATION

16. SIZE DETERMINATION OF PROTEINS   161

A. HYDRODYNAMIC METHODS   161

Introduction   161
  Data from the analytical ultracentrifuge   161
    Parameters measurable in the analytical ultracentrifuge (Table 1)   161
  Complementary techniques   162
    Complementary techniques (Table 2)   162
Theory   162
  Sedimentation equilibrium   162
    The diagnostic plot of ln (c) versus r^2 (Figure 1)   163
    A plot of $M^c_{w,app}$ versus total solute loading concentration for two hypothetical systems, both with a
      monomer molecular mass of 100 kDa (Figure 2)   164
  Sedimentation velocity   164
    Schematic diagram of the traces recorded with absorption optics in a typical sedimentation velocity
      experiment (Figure 3)   165
    Absorption optics traces for a sedimentation velocity experiment with a heterogeneous sample in
      which the sedimentation coefficients of the two components differ sufficiently to give a
      characteristic stepped boundary (Figure 4)   167
Advantages and disadvantages   167
    The advantages and disadvantages of analytical centrifuge experiments (Table 3)   168
Essential additional information   167
  Partial specific volume   167
    Approximate partial specific volumes for common biological macromolecular constituents (Table 4)   168
  Buffer density   169
  Buffer viscosity   169
Instrumentation   169
  Optics   169
Materials and methods   169
  Sample preparation   169
  Sedimentation equilibrium   170
  Sedimentation velocity   171
Data analysis   171
  Sedimentation equilibrium   171
    Models embodied in Beckman Optima XL-A analysis software for equilibrium data (Table 5)   171
  Sedimentation velocity   172
    Models embodied in Beckman Optima XL-A analysis software for velocity data (Table 6)   172
National analytical ultracentrifuge facilities   173
  RASMB   174

CONTENTS   **xi**

B. MASS SPECTROMETRIC METHODS   174

Introduction   174
    Mass spectrometric techniques for the determination of protein molecular mass (Table 7)   175
MALDITOF   176
    Applications of MALDITOF and ESIMS (Table 8)   177
ESIMS   176
    ESIMS of a mixture of horse heart myoglobin and human hemoglobin α and β subunits (Figure 5)   179
Common problems and limitations   180
  MALDITOF   180
  ESIMS   181
    Some do's and don'ts for ESIMS (Table 9)   180
Calculation of molecular masses   183
    Average isotopic masses of amino acids and protein end groups (Table 10)   181
    Common post-translational modifications (excluding glycosylations) (Table 11)   182
    Common sugars occuring in glycosylated proteins (Table 12)   183
References   183

17. FTIR AND RAMAN SPECTROSCOPIES IN THE STUDY OF PROTEINS AND OTHER BIOLOGICAL MOLECULES   187

Elementary theory of molecular vibrations   187
  General considerations   187
  Infrared (IR) spectroscopy   187
  Raman spectroscopy   187
Instrumentation   188
  IR spectroscopy   188
  Raman spectroscopy   188
Sampling   188
  IR spectroscopy   188
    The most widely used methods for sampling biological material (Table 1)   189
  Raman spectroscopy   190
  Microsampling   190
Vibrational frequencies   190
  Strongly absorbing bands in biological materials   190
    IR bands of primary importance in the study of biological material and the molecules that show these bands (Table 2)   191
Assignment of bands in complex biological molecules   190
  Resonance Raman spectroscopy   191
  Difference spectroscopy   191
  Heavy atom isotope editing   191
Some selected applications of IR and Raman spectroscopies   192
  Lipids/phospholipids   192
  Bacteriorhodopsin   192
  Heme proteins   192
  DNA and nucleoprotein complexes   192
Determination of the secondary structure of proteins   192
  Band fitting methods   192
  Frequencies of secondary structural elements in proteins   193
    Typical values of frequencies relating to specific secondary structures (Table 3)   193
  Factor analysis methods   193
Enzymes   193
  Enzyme–substrate interaction   193
References   194

18. CIRCULAR DICHROISM   195

Introduction   195
Applications   195
    Applications of circular dichroism in the study of proteins and peptides (Table 1)   195
  Examination of secondary and tertiary structure of proteins   195
    CD characteristics of different secondary structures found in proteins in solution (Table 2)   196
    Identifying features for residues contributing to the near-UV CD spectra of proteins (Table 3)   197
  Studying protein denaturation   197
  Detecting altered protein conformation   197
  Studying interactions between molecules   198
  Kinetics   199
Practical considerations   199
  Instrumentation   199
  Presentation of data   200
  Amounts of material   200
  Choice of solvent   200
    Low wavelength cut-off (A=1.0) in a 1 mm pathlength cuvette for different solvents and for common
      salts and buffers dissolved in water (Table 4)   201
  Concentration   201
  Sources of error in CD measurements   201
  Low-temperature measurements   203
  Membrane proteins   203
References   204

19. NMR STRUCTURE DETERMINATION OF PROTEINS   205

What is NMR?   205
    Properties of NMR sensitive nuclei that are commonly used in the study of proteins (Table 1)   206
Protein NMR assignment techniques   207
  Homonuclear assignment techniques   207
    A list of two-dimensional NMR experiments commonly used in the study of proteins without isotopic
      labels (Table 2)   207
    A list of H–H NOEs and their relative intensities normally observed in regular secondary structure
      elements (Table 3)   208
  Heteronuclear assignment techniques   208
    A list of multi-dimensional NMR experiments commonly used in the study of proteins with nitrogen,
      carbon or both nuclei isotopically labeled (Table 4)   210
Protein structure determination   209
    The backbone conformation of 17 solution structures of human stefin A showing regions of high
      definition of conformation and regions showing considerable disorder (Figure 1)   212
    A list of protein structures determined by NMR spectroscopy the coordinates of which been deposited
      at the Brookhaven Protein Data Bank by April 1995 (Table 5)   212
    The NMR parameters that give information on dynamic processes within proteins and the timescale
      of the processes than can be probed (Table 6)   215
References   215

20. EPR SPECTROSCOPY OF PROTEINS   217

Introduction   217
    EPR-detectable species in proteins (Table 1)   218
Principles   217
    Principles of EPR spectroscopy (Figure 1)   220
  How the EPR spectrometer works   221
  Characteristics of EPR spectra   221

CONTENTS   **xiii**

Examples of EPR spectra of transition metal ions (Figure 2) 221
Hyperfine (electron–nuclear) interactions 222
    Some nuclei with magnetic moments (Table 2) 223
    Types of EPR spectroscopy (Table 3) 224
Electron–electron interactions 222
Specialist EPR techniques 223
Sample requirements 225
Applications in the study of proteins 225
Measurements in whole-cell and tissue systems 225
    Types of information obtained by EPR spectroscopy (Table 4) 226
Information on paramagnetic centers in proteins 225
Spin labels and spin probes 227
    Examples of EPR spectra of free radicals (Figure 3) 228
    Applications of spin labels to studies of proteins (Table 5) 229
Measurement of rates of motion 229
Interactions between paramagnetic centers and between macromolecules 230
References 230

21. PROTEIN STABILITY

233

Introduction 233
Measuring protein stability 233
Selecting a technique to monitor unfolding 233
    Techniques used to monitor protein unfolding (Table 1) 234
    Criteria for selecting a technique to monitor unfolding (Table 2) 235
Determining an unfolding curve 233
    Urea and GdnHCl solutions (Table 3) 235
Equilibrium and reversibility 235
Data analysis 235
    Urea denaturation of RNase T1 monitored by fluorescence (Figure 1) 236
    Thermal denaturation of RNase T1 monitored by CD at 244 nm (Figure 2) 237
    Regions of an unfolding curve (Table 4) 237
    Methods used to measure ΔC_p (Table 5) 238
The conformational stability of globular proteins 238
References 238

22. COMPUTER ANALYSIS OF PROTEIN STRUCTURE

241

General biocomputing 241
Hardware and software requirements 241
World wide web 241
Gopher 241
    Some key URLs (Table 1) 242
ftp 241
Telnet 242
Network news 242
How to access, view and and analyze protein 3D structures 242
Finding and transferring protein structures 242
    A small section of a PDB file (Table 2) 243
Viewing proteins 243
    RASMOL instructions (Table 3) 244
    Mouse button and key combinations for RASMOL on PCs and UNIX (Table 4) 246
Structural analysis 245
    A sample of a DSSP file (Table 5) 247
Making evolutionary comparisons 246

A sample of an HSSP file (Table 6)   248
Assigning a protein to categories   249
Conformational changes   249
How to retrieve and interpret protein sequences   249
    Finding sequences   249
       Protein databases (Table 7)   250
    Comparing sequences   250
       Some alternative ways of finding matching sequences (Table 8)   251
    Multiple alignment   251
References   252

PROTEIN CHEMISTRY

23. PROTEIN CHEMISTRY METHODS, POST-TRANSLATIONAL MODIFICATION, CONSENSUS SEQUENCES   253

Introduction   253
    Consensus sequences and functional motifs (Table 1)   254
    Co- and post-translational modifications in proteins (Table 2)   262
Post-translational modifications   269
Consensus sequences   269
    One and three letter abbreviations for the amino acids (Table 3)   270
    Computer analysis on the internet/WWW   269
Cleavage and hydrolysis of proteins and peptides   270
    Hydrolysis of proteins for amino acid analysis   270
       Protein hydrolysis methods (Table 4)   271
       Elution order of amino acids and derivatives from ion-exchange analyzers (Table 5)   272
    Enzymic and chemical cleavage of proteins   272
       Digestion and cleavage of proteins and peptides (Table 6)   273
Identification and purification of modified peptides by HPLC   277
    Prediction of peptide retention times on RP-HPLC   278
       Altered retention of post-translationally modified peptides on RP-HPLC (Table 7)   279
       Relative retention coefficients on RP-HPLC (Table 8)   280
Structure elucidation of modified peptides   280
    N-terminal modifications of proteins   280
       N-terminal modifications of proteins (Table 9)   281
    Mass spectrometry   281
    Interfering salts and buffers   282
       Elution order of PTH-derivatives on HPLC (Table 10)   283
References   284

24. CHEMICAL MODIFICATION OF AMINO ACID SIDE-CHAINS   287

General aspects of the chemical modification of proteins   287
Characterization of site-specific modified proteins   287
    Establishment of reaction stoichiometry   287
    Correlation of the extent of modification with concomitant changes in biological activity   288
    Identification of residue(s) modified   288
    Comparison of the conformation of the modified protein to that of the native protein   289
Modification of specific amino acid residues   289
    Cysteine   289
       Reagents for the modification of cysteine in proteins (Table 1)   289
    Cystine   290
       Reagents for the modification of cystine in proteins (Table 2)   290

Methionine   290
   Reagents for the modification of methionine in proteins (Table 3)   290
Histidine   290
   Reagents for the modification of histidine in proteins (Table 4)   291
Lysine   291
   Reagents for the modification of lysine in proteins (Table 5)   291
Arginine   292
   Reagents for the modification of arginine in proteins (Table 6)   292
Tryptophan   292
   Reagents for the modification of tryptophan in proteins (Table 7)   292
Tyrosine   293
   Reagents for the modification of tyrosine in proteins (Table 8)   293
Carboxyl groups   293
   Reagents for the modification of carboxyl groups in proteins (Table 9)   293
References   293

25. AFFINITY-BASED COVALENT MODIFICATION

299

Introduction   299
  Scope and general principles of the technique   299
  Applications: an overview   299
Classical affinity labels   300
Quiescent affinity labels   301
Photoaffinity labels   301
Transition state affinity ($K^{\ddagger}_{s}$) labels   302
Mechanism-based irreversible inhibitors   302
Substrate-dependent nonaffinity labels   303
Substrate-derived site-specific reactivity probes   303
References   304

26. CROSS-LINKING REAGENTS FOR PROTEINS

307

Introduction   307
Reagents   307
   Some amino group-directed homobifunctional cross-linking reagents (Table 1)   308
   Some sulfhydryl group-directed homobifunctional cross-linking reagents (Table 2) 309
   Some representative heterobifunctional cross-linking reagents (Table 3)   310
  Control of pH   308
  Lysine-directed cross-linking reagents   310
  Cysteine-directed cross-linking reagents   311
  Other types of cross-linking reagent   311
Applications   311
  Estimation of the native molecular mass of oligomeric proteins   312
  Nearest neighbor analysis in oligomeric proteins   312
  Studies on membrane proteins   313
  Detection of conformational change   313
  Detection of subunit association/dissociation   313
  Conjugation of proteins   313
References   314

INDEX   315

CONTRIBUTORS

A. AITKEN
Laboratory of Protein Structure, National Institute for Medical Research, The Ridgeway, Mill Hill, London NW7 1AA, UK

A. BERRY
Department of Biochemistry and Molecular Biology, University of Leeds, Leeds LS2 9JT, UK

K. BROCKLEHURST
Laboratory of Structural and Mechanistic Enzymology, Department of Biochemistry, Queen Mary and Westfield College, University of London, Mile End Road, London E1 4NS, UK

O. BYRON
NCMH, Department of Biochemistry, University of Leicester, University Road, Leicester LE1 7RH, UK

R. CAMMACK
Centre for the Study of Metals in Biology and Medicine, Division of Life Sciences, King's College London, Campden Hill Road, London W8 7AH, UK

J. CHAUDHURI
School of Chemical Engineering, University of Bath, Claverton Down, Bath BA2 7AY, UK

R.J. COGDELL
Division of Biochemistry and Molecular Biology, Institute of Biomedical and Life Sciences, University of Glasgow, Glasgow G12 8QQ, UK

J.R. COGGINS
Division of Biochemistry and Molecular Biology, Institute of Biomedical and Life Sciences, University of Glasgow, Glasgow G12 8QQ, UK

A. GILETTO
Department of Medical Biochemistry and Genetics, Texas A & M University, College Station, TX 77843-1114, USA

D.P. HORNBY
Krebs Institute, Department of Molecular Biology and Biotechnology, University of Sheffield, PO Box 594, Firth Court, Western Bank, Sheffield S10 2UH, UK

P.J. HURD
Krebs Institute, Department of Molecular Biology and Biotechnology, University of Sheffield, PO Box 594, Firth Court, Western Bank, Sheffield S10 2UH, UK

J.G. LINDSAY
Division of Biochemistry and Molecular Biology, Institute of Biomedical and Life Sciences, University of Glasgow, Glasgow G12 8QQ, UK

R. LUNDBLAD
Baxter-Biotech-Hyland Division, 1720 Flower Avenue, Duarte, CA 91010, USA and Dental Research Center, University of North Carolina, Chapel Hill, NC 27599, USA

S.R. MARTIN
Division of Physical Biochemistry, National Institute for Medical Research, The Ridgeway, Mill Hill, London NW7 1AA, UK

B.J. MCGINN
Division of Biochemistry and Molecular Biology, Institute of Biomedical and Life Sciences, University of Glasgow, Glasgow G12 8QQ, UK

E.J. MILNER-WHITE
Division of Biochemistry and Molecular Biology, Institute of Biomedical and Life Sciences, University of Glasgow, Glasgow G12 8QQ, UK

R.L. MORITZ
Ludwig Institute for Cancer Research, PO Box 2008, Royal Melbourne Hospital, Parkville, Victoria 3050, Australia

C.N. PACE
Department of Biochemistry and Biophysics, Texas A & M University, College Station, TX 77843-1114, USA

D. PATEL
Department of Biology, University of Essex, Wivenhoe Park, Colchester, Essex CO4 3SQ, UK

A.R. PITT
Department of Pure and Applied Chemistry, University of Strathclyde, 295 Cathedral Street, Glasgow G1 1XL, UK

N.C. PRICE
Department of Biological and Molecular Sciences, University of Stirling, Stirling FK9 4LA, UK

D. RICKWOOD
Department of Biology, University of Essex, Wivenhoe Park, Colchester, Essex CO4 3SQ, UK

L. SAWYER
Department of Biochemistry, University of Edinburgh, George Square, Edinburgh EH8 9XD, UK

R.K. SCOPES
School of Biochemistry, La Trobe University, Bundoora, Melbourne, Victoria 3083, Australia

R.J. SIMPSON
Ludwig Institute for Cancer Research, PO Box 2008, Royal Melbourne Hospital, Parkville, Victoria 3050, Australia

J.K. SHERGILL
Centre for the Study of Metals in Biology and Medicine, Division of Life Sciences, King's College London, Campden Hill Road, London W8 7AH, UK

L. STEVENS
Department of Biological and Molecular Sciences, University of Stirling, Stirling FK9 4LA, UK

D.R. THATCHER
Therexsys, The Science Park, University of Keele, Keele, Staffordshire ST5 5SP, UK

G.J. THOMSON
Department of Biochemistry and Molecular Biology, University of Leeds, Leeds LS2 9JT, UK

J.P. WALTHO
Krebs Institute for Biomolecular Research, Department of Molecular Biology and Biotechnology, University of Sheffield, PO Box 594, Firth Court, Western Bank, Sheffield S10 2UH, UK

C.W. WHARTON
School of Biochemistry, University of Birmingham, Edgbaston, Birmingham B15 2TT, UK

B. WHITEHEAD
Krebs Institute for Biomolecular Research, Department of Molecular Biology and Biotechnology, University of Sheffield, PO Box 594, Firth Court, Western Bank, Sheffield S10 2UH, UK

P. WILKS
Therexsys, The Science Park, University of Keele, Keele, Staffordshire ST5 5SP, UK

ABBREVIATIONS

ACTH	adrenocorticotrophin
AMAS	analyze multiple sequence alignments
ANS	anilinonaphthalene sulfone
APS	ammonium persulfate
APTG	p-aminophenyl-β-D-thiogalactoside
AS	ammonium sulfate
BCA	bicinchoninic acid
Bis-Tris	bis-(2-hydroxyethyl)imino-tris(hydroxymethyl)methane
BNL	Brookhaven National Laboratory
BSA	bovine serum albumin
CCD	charge coupled device
CD	circular dichroism
CDNB	1-chloro-2, 4-dinitrobenzene
CE	capillary electrophoresis
CHAPS	3-(3-cholamidopropyl)-dimethylammonio-1-propanesulfonate
CHAPSO	3-(3-cholamidopropyl)-dimethylammonio-2-hydroxypropane-1-sulfonate
CIF	crystallographic information file
CMC	critical micelle concentration
CMTD	carboximidomethyltartaramide
COSY	correlated spectroscopy
CTAB	cetyltrimethylammonium bromide
cw	continuous wave
DACM	N-(dimethylamino-4-methylcoumarinyl) maleimide
DCC	dicyclohexylcarbodiimide
DCI	3, 4-dichloroisocoumarin
DEAE	diethylaminoethyl
DMS	dimethyl suberimidate.2HCl
DMSO	dimethylsulfoxide
DSC	differential scanning calorimetry
DSSP	dictionary of secondary structure information in proteins
DTBB	dithiobisbutyrimidate
DTT	dithiothreitol
EBI	European Biomolecular Institute at Cambridge
EDTA	ethylenediaminetetraacetic acid
ELISA	enzyme-linked immunosorbent assay
ELP	entrapment of liberated phosphate
e.m.f.	electromotive force
EMR	electronic magnetic resonance
ENDOR	electron-nuclear double resonance
EPR	electron paramagnetic resonance
ES	enzyme–substrate
ESEEM	electron spin-echo envelope modulation
ESI	electrospray ionization
ESIMS	electrospray ionization mass spectrometry

ESR	electron spin resonance
FAB	fast atom bombardment
FCS	fetal calf serum
Fmoc	9-fluorenylmethoxycarbonyl
FPLC	fast protein liquid chromatography
FSCE	free solution capillary electrophoresis
FT	Fourier transform
FTICR	Fourier transform ion cyclotron resonance
FTIR	Fourier transform infrared
ftp	file transfer protocol
FWHM	full width at half maximum
GCG	genetics computer group
GDE	genetic data environment
GdnHCl	guanidine hydrochloride
HFBA	heptafluorobutyric acid
HIC	hydrophobic interaction chromatography
HMQC	heteronuclear multiple-quantum correlation
HPLC	high-pressure liquid chromatography
HSQC	heteronuclear single-quantum correlation
HSSE	high speed sedimentation equilibrium
HSSP	homology-derived secondary structure of proteins
HRP	horseradish peroxidase
IEF	isoelectric focusing
IPG	immobilized pH gradient
IMAC	immobilized metal affinity chromatography
Ir	immunoradioactive
IR	infrared
IRMA	immunoradiometric assay
LGT	low gelling temperature
LSSE	low speed sedimentation equilibrium
MALDI	matrix-assisted laser desorption ionization
MALDITOF-MS	matrix-assisted laser desorption ionization-time of flight mass spectrometry
Mes	morpholinoethane sulfonate
Mops	morpholinopropane sulfonate
MS	mass spectrometry
NCBI	US National Center for Biotechnology Information
NCS	newborn calf serum
ncsa	National Center for Supercomputer Applications
NMR	nuclear magnetic resonance
NOE	nuclear Overhauser effect
NOESY	nuclear Overhauser effect spectroscopy
NP-40	Nonidet P-40
OPA	*ortho*-phthalaldehyde
PAS	periodic acid-Schiff
PDB	protein databank
PEG	polyethylene glycol
PEO	polyethyleneoxide
PFPA	pentafluoropropionic acid
PMSF	phenylmethanesulfonyl fluoride
PPO	2,5-diphenyloxazole
PVDF	polyvinyldifluoride

rf	radiofrequency
RIA	radioimmunoassay
rmsd	root-mean-squared distance
ROESY	rotating frame nuclear Overhauser effect spectroscopy
RP	reverse-phase
SCOP	structural classification of proteins
SDS–PAGE	sodium dodecyl sulfate–polyacrylamide gel electrophoresis
SEC	size-exclusion chromatography
S/N ratio	signal-to-noise ratio
TBAP	tetrabutylammonium phosphate
TCA	trichloroacetic acid
TDCC	covalent chromatography by thiol–disulfide interchange
TEA	triethylamine acetate
TEAE	triethylaminoethyl
TEAP	triethylamine phosphate
TEMED	N, N, N', N',-tetramethylethylenediamine
TFA	trifluoroacetic acid
TFE	trifluoroethanol
TLCK	Tos-LysCH$_2$Cl
TNBS	2,4,6-trinitrobenzoic acid
TOCSY	total correlation spectroscopy
TPCK	tos-pheCH$_2$Cl
URL	uniform resource loader
UV	ultraviolet
WS	working solution
WWW	world wide web

PREFACE

The study of proteins and enzymes has been almost inseparable from that of biochemistry itself; during the last century, important developments in structural and functional characterization of these molecules have led to the award of a succession of Nobel prizes. In the late 1970s and early 1980s, however, it seemed as though interest in proteins might be swamped by the spectacular advances in molecular genetics. In fact, the application of recombinant DNA technology over the last 10–15 years has ushered in a new 'golden age' for proteins, with increasing quantities of material available for detailed structural study and mutagenesis providing novel insights into the function of individual amino acid side chains.

This volume in the *Labfax* series aims to provide the reader with an approach to the theory and practice of techniques which are widely used in the study of proteins. Given the constraints of space, coverage cannot be encyclopedic; the topics have been chosen to be complementary to those in the companion volume *Enzymology Labfax*. The first section of this book deals with protein purification and incorporates recent developments such as recovery of proteins from inclusion bodies, fusion proteins and special methods for isolating membrane proteins. Crystallization of proteins (a prerequisite for X-ray structural determination) requires highly homogeneous preparations; these topics are included in this section. In the second part of the book, accounts are given of a number of spectroscopic techniques which can be used to characterize proteins. Recent advances in mass spectrometry and in computer-based analysis of protein structures are also featured. The third and final section of the book deals with protein chemistry, with particular emphasis on the analysis of post-translational modifications and the design of specific reagents for modifying or cross-linking proteins.

I owe a great deal to the contributors to the book, who have not only produced excellent chapters and have shown remarkable tolerance of my persistent attempts to keep the production of the book roughly on schedule, but have also been generous in accepting suggestions for changes or condensation. I would also like to thank Jonathan Ray and Priscilla Goldby at BIOS who have helped enormously in piloting the project through some rough passages. My final thanks are to the creators of the Internet, who have allowed me to keep in touch with authors around the globe during the production of the book. It is hard to imagine how the project could have succeeded without this electronic marvel.

Nicholas C. Price

CHAPTER 1
CHOICE OF ASSAY OR DETECTION METHOD

G.J. Thomson and A. Berry

1. INTRODUCTION

The purification and characterization of proteins is of great importance to understand fully their function at a molecular level, and therefore to gain insights into the molecular mechanisms of cellular events. Before one can begin to consider purifying a particular protein it is essential to have a method of detecting that protein within the complex cellular milieu. The aim of this chapter is, therefore, to provide a general guideline to techniques currently available for the detection and assay of proteins. The scope of this chapter does not permit detailed discussion of various assay procedures, nor of the practical protocols involved. It is hoped, however, that by illustrating each method with one or two examples the reader will gain a better feel for the techniques available. More exhaustive reviews are published elsewhere (1, 2). Similarly, assays for the measurement of protein concentration will not be considered in this chapter (see Chapter 4) but rather, more specific methods that rely on the protein's biological activity or structural features will be discussed. Relatively more assays are available for enzymes than for proteins, which do not have a catalytic activity, and much of this chapter, therefore, deals with enzymes. However, other proteins often have some specific function that may be measured (e.g. binding) and these methods for their assay are also described.

Ideally, assays should be simple, convenient, highly specific and rapid. One consideration in choosing an assay is to decide whether it is sufficient simply to detect the presence of the target protein, or whether a more specific assay is required to both detect the protein and quantify precisely its presence or determine its kinetic characteristics. During protein purification, perhaps where no specific enzymatic assay is available, a target protein may be detected by Western blotting (if an antibody has been raised against it) or in the case of an overproduced protein simply by observing a band at the appropriate molecular mass by sodium dodecyl sulfate–polyacrylamide gel electrophoresis (SDS–PAGE). The development of rapid gel electrophoresis systems, such as the Pharmacia PhastGel apparatus, has made the latter approach viable in a short period of time. Generally speaking, however, a more specific assay based on the biological activity of the protein will be employed whenever available. One unusual assay is that for the taste-modifying protein miraculin which can be detected simply by taste through its sweet-inducing activity (3). The potential problems of such assays, especially when assaying crude mixtures of protein, will be briefly considered. During such early stages of protein purification the assay may be complicated by factors such as endogenous inhibitors or interfering, competing pathways. Further purification may, therefore, be required to remove such contaminants in order to get a 'true' measure of the rate of reaction, although inhibitors of low molecular mass may be easily removed by dialysis or gel filtration. It may also be possible to inhibit specifically any competing pathways (taking care that the pathway under study is unaffected). Complications may also arise in photometric assays of crude samples due to turbidity of the solution or the presence of highly absorbing compounds, particularly in the far ultraviolet (UV) range, since these may interfere with the assay. Careful controls should always be carried out to determine the validity of an assay method.

2. CONTINUOUS VERSUS DISCONTINUOUS ASSAYS

In many cases, the protein under investigation will be an enzyme and will have a well known biological activity. Enzymes may be monitored either continuously or discontinuously. Continuous assays allow the monitoring of the reaction directly and, as such, are able to detect any deviations from the linearity. Continuous assays may be measured spectrophotometrically (see Section 3) or by changes in O_2 concentration (see Section 6.1), ionic composition or pH (see Section 6.2), depending on the reaction. The ease and convenience of continuous assays is such that many reactions, the substrates or products of which cannot be directly measured in a continuous assay, may be monitored continuously by the use of linked enzyme reactions. Such coupled assays then provide all the benefits of continuous assays to the reaction under study. If it is not possible to measure a reaction continuously, then assays may be performed discontinuously. In many cases the protein under study will either not have a suitable continuous assay, or will have unknown or no enzymatic activity. Proteins in the first of these groups would include proteins expressed from unidentified genes and it would therefore be impossible to choose substrates for any reaction; whereas proteins in the second group would include structural proteins, transport proteins or binding proteins. Discontinuous assays are generally more time-consuming and involve the stopping of a reaction at a given time, followed by, either, separating the components to be measured, as in the case of many radiometric or high-pressure liquid chromatography (HPLC) based assays, or further treating the products or substrates in a separate reaction in order to develop a specific color reaction, as in the assay for inorganic phosphate (4). In these cases one must take care to ensure that the reaction is stopped immediately, and that the reaction proceeds at a constant rate over the time period of the discontinuous assay in order to be able to equate it with the initial rate of the reaction. *Table 1* summarizes the various continuous and discontinuous methods which may be used for the assay of proteins, and these will now be discussed in turn.

3. PHOTOMETRIC ASSAYS

3.1. Absorption

Direct continuous assays
The changes in absorption properties of natural chromophores have routinely been employed as the basis for assay procedures. Commonly used examples are the nicotinamide coenzymes and the flavin coenzymes. The reduced forms of the nicotinamide adenine dinucleotide coenzymes, NADPH and NADH, absorb radiation at 340 nm, whilst the oxidized versions, $NADP^+$ and NAD^+, do not. Oxidation or reduction of the nicotinamide coenzymes is, therefore, accompanied by a large change in absorbance at 340 nm making it a simple, convenient method for the direct continuous assay for the various dehydrogenase enzymes. Because of the large number of dehydrogenases this is probably the most widely used assay of its type, and many other enzymes can be assayed continuously by coupling to a suitable dehydrogenase reaction (see following sub-section). A number of other enzymes may also be assayed directly by virtue of the absorbing characteristics of their natural substrates or products including fumarase (5), citrate synthase (6) and xanthine oxidase (7). These make use of the absorption of fumarate (λ_{max} = 240 nm), acetyl CoA (λ_{max} =232 nm) and uric acid (λ_{max} = 232 nm), respectively.

Indirect continuous assays
The reactions of a large number of enzymes do not yield products that are directly measurable by changes in absorbance. It is possible, however, to employ one or more additional enzymes to couple the reaction to be measured with other chemical reactions in order to yield a final

Table 1. Common techniques for the assay and detection of proteins

Technique	Assay type	Comments
Photometric Absorption	Continuous/discontinuous	Commonly used assay, simply and rapidly performed. Natural or synthetic chromophores may be used. May be affected by color or turbidity
Fluorescence	Continuous/discontinuous	More sensitive than absorption but more limited choice of fluorescent substrates available. Quenching may be a problem
Radiometric	Discontinuous (generally)	Highly sensitive, especially for detection of low amounts of activity in crude extracts. Problems associated with radioactivity, e.g. health hazard, quenching, multiple-steps involved
HPLC	Discontinuous	Separations performed rapidly (5–30 min) and with high sensitivity. Conditions to separate most metabolites can be devised. Use of toxic solvents a disadvantage
Electrochemical	Continuous/discontinuous	Inexpensive and convenient, unaffected by color or turbidity
Electrophoresis	Discontinuous	Simple procedure using standard techniques. Limited types of proteins can be assayed this way
Immunological	Discontinuous	Specific and highly sensitive, multiple samples can be screened easily. Time-consuming, does not detect inactive enzymes

identifiable product. Many coupled assays often employ dehydrogenases for the final step of the reaction, for the reasons described above; for example, the assay of fructose-1,6-bisphosphate aldolase, which catalyzes the reversible cleavage of fructose-1,6-bisphosphate to dihydroxy-

Figure 1. The coupled assay for fructose-1,6-bisphosphate aldolase (8).

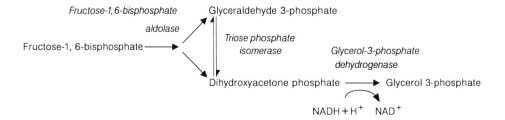

acetone phosphate and glyceraldehyde 3-phosphate, involves two additional auxiliary enzymes, namely triose phosphate isomerase and glycerol-3-phosphate dehydrogenase, and decline in absorbance at 340 nm provides a measure of the aldolase activity (8) (*Figure 1*).

This type of linked reaction is a very common assay procedure for a wide variety of enzymes. It is important to ensure that it is the reaction under study that is responsible for any changes in absorbance observed, and not other compounds in the system which may be utilized by the coupling enzymes. In addition it should be noted that the concentration of coupling enzymes present in the assay mixture should always be in excess so as not to become rate-limiting for the assay being measured.

The use of synthetic chromophores
Where natural substrates do not show useful absorbance properties and where coupled enzyme assays are not convenient, then the use of synthetic chromogenic substrates may provide a solution. The highly absorbent compounds nitrophenol and nitroaniline have been widely used as the basis for a number of artificial substrates for the assay of esterases and peptidases. Synthetic peptides containing linked chromogenic groups are often used in the assay of serine proteases (9). A variation on this theme is the coupled assay of prolyl isomerase, an enzyme involved in the *cis/ trans* isomerization of Xaa-Pro peptide bonds during polypeptide refolding. This assay is based on the ability of chymotrypsin rapidly to cleave the 4-nitroanilide moiety from the synthetic oligopeptide N-succinyl-Ala-Ala-Pro-Phe-4-nitroanilde only when the preceding Ala-Pro peptide bond is in the *trans* conformation. Hydrolysis of the ester is followed by increase in absorbance at 390 nm (10). Other uses of synthetic chromogenic substrates include the assay of β-galactosidase by the synthetic analog nitrophenyl-β-D-galactoside (11). This type of reaction forms the basis of the well-known blue/white selection procedure during cloning of genes. In this case, disruption of the gene by insertion of DNA results in a loss of β-galactosidase activity and colonies showing this phenotype can be selected by growth on plates containing the chromogenic substrate 5-bromo-4-chloro-3-indolyl-β-D-galactoside (X-gal). Another example of the use of chromogenic substrates comes from the spectrophotometric assay of protein phosphatases by following the hydrolysis of 4-nitrophenylphosphate at 400 nm (12).

Discontinuous absorption methods
Despite the use of the variety of methods described above, there are enzymes for which continuous assays are not easily devised. In these cases discontinuous assays are the only possibility. These discontinuous methods may still be based on absorbance, although they are more likely to use separation methods (e.g. see Section 5, below) or radiochemical methods (see Section 4). One example of a discontinuous absorption assay is that used for porphobilinogen deaminase. This enzyme converts the pyrrole porphobilinogen into a linear tetrapyrrole, preuroporphyrinogen. This reaction has no directly usable absorption and no direct coupled

method is available. The assay of porphobilinogen deaminase is based on the fact that preuroporphyrinogen is spontaneously converted into uroporphyrinogen I. Oxidation of uroporphyrinogen I by benzoquinone or iodine results in the formation of uroporphyrin I, which may be detected by its strong absorbance at 405 nm (13).

3.2. Fluorescence

A potentially more sensitive alternative to the use of absorption of radiation is to measure the fluorescence of a compound as the basis of an assay technique. Fluorescence emission is observed on the return to the ground state of an excited electron. Nonfluorescent compounds lose the energy of the excited molecule as heat, whereas fluorescent compounds emit part of this energy as light. Due to energy loss the fluorescence emission is shifted to a longer wavelength compared to the absorption of the chromophore. The reduced forms of the nicotinamide nucleotides again provide excellent examples of this type of assay. They can be excited by light at 340 nm and the fluorescent emission maximum observed at 465 nm can be used for the development of assays for dehydrogenases. Other naturally fluorescent compounds which exist include anthranilate, an intermediate in tryptophan biosynthesis, which can be can be used to assay the enzyme anthranilate synthase and other enzymes in a coupled assay procedure (14); similarly porphyrins may be assayed by their fluorescence (15). A limiting factor in the use of fluorescence assays is that few natural compounds fluoresce sufficiently to be of use, although, in a manner similar to that described above, synthetic fluorogenic substrates may be employed to assay for activity; for example, the development of a functional active site during the refolding of hen lysozyme has been monitored through the binding of a fluorescent substrate analog, 4-methylumbelliferyl-N, N'-diacetyl-β-chitobiose (16).

3.3. Turbidimetry

In a limited number of cases, the degree of light-scattering of a solution may also be used to determine the amount of an enzyme present. One such example is the assay of lysozyme, an enzyme which degrades bacterial cell walls. This leads to rupture of the cell and therefore results in a clarification of a bacterial suspension. An assay has been developed based on this principle by simply measuring the decrease in turbidity of a suspension of bacterial cells on addition of the enzyme (17).

4. RADIOMETRIC ASSAYS

One of the most frequently used discontinuous types of assay involves the use of radioactively labeled compounds. The large number of radiolabeled substrates commercially available allows the possibility of performing radioactive-based assays on many proteins. Radiometric assays are highly sensitive and offer several advantages over nonradiometric assays, particularly when assaying low levels of activity in crude extracts. These assays are unaffected by highly absorbing compounds that may be present in the extract, and that often complicate photometric assays, and can also be easily used to study the effects of inhibitors or activators on enzymes. These latter effects can be difficult to interpret when using coupled enzyme assays.

Radiometric assays are generally performed discontinuously and the procedure for most enzymes is similar, involving the conversion of radiolabeled substrate to labeled product, the stopping of the reaction at defined time-points and then measuring the radioactivity of the product or residual substrate after quantitative separation; for example, the assay of protein kinases is most widely studied using radiometric techniques. These assays involve the transfer of ^{32}P from [γ^{32}P]ATP to a protein or peptide substrate, the separation of phosphorylated substrate from unincorporated

$[\gamma^{32}P]$ATP and the quantification of radioactively labeled substrate. In this example the choice of substrate is important depending on how specific an assay is required. Substrates which act as phosphoryl acceptors without necessarily having any physiological significance, such as casein and histones, have been used to assay protein kinases. However, since both casein and histones lack substrate specificity and may be phosphorylated by a number of different kinases (e.g. as may be present in crude extracts), their use is generally more suited to the assay of purified enzymes. The use of the physiological substrate is, therefore, infinitely preferable when one wishes to develop a specific assay for the detection of a particular kinase. Should obtaining the large quantities required for routine assays be a problem, the recent advances in the use of synthetic peptide substrates, based on the known phosphorylation sites of the physiological substrates, may provide the solution. A number of protein kinases can now be assayed highly specifically using synthetic peptide substrates (18, 19).

The separation procedures used to remove unincorporated radiolabel or unreacted substrate are obviously important in the optimization of the assay. Separation techniques are mostly based on differences in the solubility, size and charge between the substrate and product. These may involve the use of ion-exchange resin or paper, trichloroacetic acid (TCA) precipitation, SDS–PAGE or charcoal adsorption.

Radioactively labeled substrates are also of crucial importance in assays of proteins whose only easily measured biological activity is to bind substrate. Examples in this area include DNA binding proteins. Such proteins are often assayed by band-shift assays. These make use of electrophoresis to separate bound from unbound substrate and are more fully described in Section 7.3.

5. HPLC-BASED ASSAYS

When no continuous assay is available for a protein, but where a catalytic activity is known, the assay often involves the separation of the individual components of the reaction by HPLC. The use of HPLC may also find an important use in conjunction with radiolabeled material (see Section 4). HPLC is performed under high pressure and separation is often achieved very rapidly (a matter of minutes in most cases). In brief the process is this. The mixture of solutes is introduced via a mobile phase and then retained or retarded on the stationary phase, the HPLC column. Desorption and elution is finally achieved by altering the mobile phase, if necessary, usually via a linear gradient. The retention and elution of the sample will depend on its physical properties, and those of the column packing. Thus, for optimal separation, the choice of both column and mobile phase is critical. A large number of HPLC columns are available, although they can be categorized into three main groups: reverse-phase columns for the separation of nonpolar or hydrophobic compounds; ion-exchange columns to separate ionic, charged compounds; and size-exclusion columns to separate on the basis of size. Eluted compounds may be detected spectrophotometrically (by absorption or fluorescence) or by radioactive or electrochemical means if suitable monitors are on-line. HPLC is a highly sensitive technique and can be used for the separation of amino acids, peptides, proteins, nucleotides, steroids, vitamins, sugars and fats. In particular, the use of HPLC is an extremely powerful method of separating structurally homologous compounds, which may be indistinguishable using other methods.

A particular case in point is the assay of glutamine synthetase, a key enzyme in the assimilation of nitrogen into organic compounds. This enzyme catalyzes the amidation of glutamate to glutamine and is assayed by the fluorimetric detection of o-phthalaldehyde derivatives of the metabolites after reverse-phase HPLC separation. Glutamate synthase is assayed using the same method (20). The assay of several other enzymes may be performed by HPLC, including a novel aldolase from *Streptomyces amakusaensis*, which catalyzes the conversion of β-hydroxy-α-amino acids to the corresponding aldehyde and glycine (21). This assay follows the formation of the aldehyde using reverse-phase HPLC separation and UV detection. In all such cases it is crucially important that pure standards of the substrate, product or intermediates are available for use as standards in HPLC separation.

6. ELECTROCHEMICAL ASSAYS

6.1. The oxygen electrode

Enzymes that utilize or produce molecular oxygen as part of their reaction mechanism may be assayed directly using an oxygen electrode, which measures changes in the oxygen content of any given solution. The oxygen electrode works on the following principle: by applying a potential difference of 0.6–0.9 V across a platinum–Ag/KCl electrode system (such that the platinum electrode is negative with respect to the Ag electrode) a current will only flow on the electrolytic reduction of oxygen to water at the surface of the platinum electrode. The current produced is proportional to the oxygen content of the solution and is measured using a suitable amplifier and chart recorder.

The oxygen electrode has been widely used to examine oxygen consumption in tissues, cells and organelles maintained under physiological conditions, such as in the study of the respiratory chain in mitochondrial extracts. The study of individual enzymes has also been assisted by the advent of the oxygen electrode; for example, the quantitative assay of catalase, which breaks down hydrogen peroxide produced by anaerobic metabolism to water and oxygen, proved difficult until the development of an oxygen electrode based assay (22). Other enzymes which can be assayed using an oxygen electrode include L-amino acid oxidase (23), cytochrome oxidase (24), xanthine oxidase (25) and glucose oxidase (26). The latter enzyme catalyzes the oxidation of D-glucose to gluconolactone and since the oxygen depletion during this reaction can be directly related to both enzyme activity and the concentration of glucose, glucose oxidase impregnated membranes stretched over an oxygen electrode have been adapted as glucose sensors (27). Similarly other oxidative enzymes have been utilized in an analogous manner and have been used to assay biochemical intermediates ranging from cholesterol to xanthine (27).

6.2. The pH-stat (stationary pH)

Chemical reactions in which protons are either generated or consumed can be measured using a pH-stat system. Essentially this titration technique measures the amount of acid or base, of known concentration, that has to be added to keep the pH of a non-buffered or weakly buffered solution constant. Such assays can therefore be performed using relatively basic equipment (i.e. burette, pH meter, magnetic stirrer and stopwatch), although, for ease of operation and accuracy, fully automated, computer-driven pH-stat devices are commercially available.

The wide variety of chemical reactions in which there is a net change in hydrogen ion concentration has meant that the pH-stat has proved useful for the assay of a wide variety of enzymes and numerous examples are available in the literature (e.g. 28, 29). The reaction catalyzed by dihydrofolate reductase,

$$\text{dihydrofolate} + \text{NADPH} + \text{H}^+ \longrightarrow \text{tetrahydrofolate} + \text{NADP}^+, \qquad \text{Equation 1}$$

is an example of an assay that can be performed both by spectrophotometric means (by the

CHOICE OF ASSAY OR DETECTION METHOD   7 decrease in absorbance at 340 nm) or by using a pH-stat (measuring the addition of sulfuric acid to maintain pH) (30), thus offering a useful comparison of the two techniques. In this particular case the pH-stat assay is approximately 10 times more sensitive than the spectrophotometric assay, can be performed rapidly and can also be used to assay crude samples including those which may be turbid or contain colored compounds (the latter being a general advantage of pH-stat assays over photometric assays). Electrochemical assays may also be applicable to a variety of other reactions and the use of other ion-specific electrodes or gas electrodes can be used to monitor the changes in concentration of other reactants; for example, those reactions involving uptake or release of metal ions.

7. ASSAYS INVOLVING GEL ELECTROPHORESIS

7.1. Visualization of protein bands by protein staining
The general availability of equipment for gel electrophoresis has meant that many protein assays involve the use of electrophoresis in one form or another. In this section we deal with a variety of these methods. In a number of cases, for example when a gene of unknown function is expressed or where a protein or fragment of a protein with no known function is assayed, the only method of detection and assay may be the visualization of a stained protein band of the expected molecular mass on an SDS-gel. The practical details of electrophoresis and gel staining are described elsewhere (see Chapter 9).

7.2. Visualization of protein bands by activity staining
The preceding sections may have given the misleading impression that the assay and detection of enzymatic activity can only be carried out when the protein of interest is free in solution. A refinement of the electrophoretic assays is to use a specific biological stain for the protein of interest. These methods also include Western blotting or immunological detection and these will be discussed in Section 8. It is possible to stain for activity when the protein is present within a polyacrylamide gel matrix or after subsequent blotting onto a nitrocellulose or polyvinylidene difluoride (PVDF) membrane. Such an approach may be useful, for example, for the detection of isoenzymes or to locate the position of a protein, perhaps to be further identified by N-terminal amino acid sequencing. Naturally it is important that separation should take place under conditions whereby the biological activity of the protein is preserved and that normally involves nondenaturing PAGE, although some proteins maintain their activity in the presence of detergents such as SDS or may be renatured following SDS–PAGE. Several protocols are available for the renaturation of proteins after SDS–PAGE and these vary depending on the particular protein to be renatured (31–33).

In general, detection of activity *in situ* may be performed either by following changes in the optical properties of either the substrate or product by visualization under UV or visible light (34), or by converting the product of the enzymatic reaction under study to an insoluble colored product which may be directly visualized on the gel. A common procedure, to illustrate the latter point, is the staining for activity of nicotinamide coenzyme utilizing enzymes. This reaction depends on the formation of NAD(P)H which, via the electron carrier phenazine methosulfate, can reduce tetrazolium compounds to produce a strongly colored, insoluble formazan which indicates the position of the enzyme of interest (35). This procedure is useful for the staining of both dehydrogenases and any other enzyme, the activity of which can be linked to dehydrogenases in a manner analogous to coupled enzyme assays, as described earlier. The staining method for both pyruvate carboxylase and phosphoenolpyruvate carboxylase provides an alternative example. The product of the enzymatic reaction, oxaloacetate, can simply be coupled with Fast Violet B (6-benzamido-4-methoxy-*m*-toluidine diazonium chloride) to produce a strong red band at the site of oxaloacetate synthesis (36).

7.3. Other electrophoresis-based assays

Apart from detecting the protein of interest directly, as described above, electrophoresis plays a major role in other protein assays; for example, the detection of sequence-specific DNA binding proteins may be achieved by one of two electrophoresis-based assays, namely band-shift assays or footprinting assays. Band-shift assays compare the electrophoretic mobilities of DNA–protein complexes and unbound DNA (DNA–protein complexes migrate more slowly than free DNA) (37). Footprinting assays are based on the protection of DNA from mild enzymatic cleavage or chemical hydrolysis when complexed to the DNA-binding protein. In this case protection is revealed by an almost complete absence of DNA fragments in the region of the protein binding site following analysis on nondenaturing polyacrylamide gels (38). Both assays require a prior knowledge of the DNA-binding region to which the protein complexes, but are very sensitive, particularly with radiolabeled oligonucleotides and are of great use in monitoring DNA-binding proteins during purification.

The differential mobility of synthetic peptides on an agarose gel may also be used as an assay procedure for several groups of proteins. As already mentioned, synthetic peptides have proved useful in the radiometric-based assay for kinase activity (see Section 4), however the assay of protein kinase C and cAMP-dependent protein kinase activity may be determined by nonradioactive means. Phosphorylation of fluorescent tagged peptides with an overall (+1) positive charge changes the net charge of the peptide to negative (−1) and thus can be detected by rapidly running an agarose gel (39), providing a simple, convenient assay for these kinases. Serine protease activity may be assayed in a similar manner, whereby the proteolytic cleavage of a fluorescent-tagged peptide changes its net charge and size and thus can be detected by its mobility on an agarose gel (40). The latter assay may be useful for the detection of contaminating protease activity within a protein preparation or for designing protease inhibition cocktails to use with crude extracts.

8. IMMUNOLOGICAL DETECTION

8.1. Immunoblotting

Sensitive detection of proteins after electrophoresis may also be achieved by immunoblotting if a suitable antibody has been raised against the protein of interest. Immunoblotting may be particularly useful if it is difficult to develop an activity-based assay for a specific protein. The technique basically involves the transfer of the separated proteins onto a blotting membrane such as nitrocellulose, 'blocking' the membrane to prevent non-specific binding of the reagents before treating with the specific primary antibody, raised against the target antigen, and developing with a secondary antibody. Detection by the secondary antibody is mediated by a conjugated reporter group which may be a radiolabel, a colloidal gold label or an enzyme label. Secondary antibodies coupled to enzymes, usually horseradish peroxidase or alkaline phosphatase, are most commonly used. In such cases the presence of the secondary antibody may be revealed by the conversion of a chromogenic substrate to an insoluble product, denoting where the enzyme label is present on the membrane (41), or, as is becoming increasingly popular, by light emission when using a chemiluminescent substrate (42).

8.2. Enzyme-linked immunosorbent assay

Enzyme-linked immunosorbent assay (ELISA) may be used for the detection of a wide range of antibodies or antigens. The assay works on a similar principle to that already described for immunoblotting, whereby an antibody or antigen is immobilized onto a specially coated surface (microtitration plates are commonly used), non-specific sites are then 'blocked' before a solution containing the protein or polypeptide of interest is added. Unbound protein is removed by washing and bound protein/polypeptide is quantified using a secondary enzyme-linked antibody, which, on the addition of substrate, produces a colored product which may be detected visually or photometrically. Fluorometric and chemiluminescent detection systems are also available. Commercial ELISA-based assay kits are available for a wide range of antigens such as immunoglobulins, hormones, cytokines and serum proteins (43).

8.3. Radioimmunoassay and immunoradiometric assay

The classical immunoassay techniques are based on the radiometric detection of antigen (44). Although many radioimmunoassays are now being superseded by enzyme immunoassays, of the type described above (thus avoiding the potential hazards of working with radioactivity), radiometric-based immunoassays offer high sensitivity and in many cases are the standardized procedure. These assays will now be briefly discussed. Radioimmunoassay (RIA) is based on the competition between radiolabeled and unlabeled antigen for a fixed, but limiting, number of binding sites on antibody molecules. Radioactively labeled antigen is added in excess to ensure saturation of the binding sites, and displacement of radioactivity is achieved in the presence of unlabeled antigen. Bound and free labeled and unlabeled ligand are then separated and quantified. Using (unlabeled) antigen standards of known concentration a calibration curve is determined, from which the concentration of test samples to be assayed can be measured. RIA has proved useful for the assay of protein hormones, hepatitis B antigen and smaller molecules such as steroids, prostaglandins and morphine-related drugs. In contrast to RIA, immunoradiometric assay (IRMA) involves the reaction of an excess of labeled antibody with the ligand of interest. The two-step or sandwich IRMA is most commonly used and requires two specific antibodies, raised against distinct epitopes of the target molecule. As such, IRMA is a more sensitive technique than RIA. The first of the antibodies is bound to a solid phase and is used to capture the ligand, whilst the second antibody is radiolabeled and acts as the 'detector' molecule. After a washing step the amount of ligand can be quantified. One limitation of the two-step IRMA is that small molecules, such as steroids or small peptides, will not be recognized, since the technique requires the molecule of interest to have two distinct antigenic determinants which can be recognized simultaneously by both antibodies. Although this is a highly sensitive technique it should be noted that IRMA, like RIA and ELISA, only recognizes immunological activity and therefore may still detect denatured and thus inactive protein.

9. CONCLUSION

The assays described above are examples of those most commonly used for the detection of proteins, but by no means constitute a definitive or exclusive list. Numerous other assay procedures are available, which range from rapid luminescence-based assays for the detection of ATP production or disappearance by firefly luciferase (45), to more time-consuming growth-factor assays, where the triggering of growth of cultured cells is the basis of the assay (46), and the more unusual ones such as that for miraculin, already mentioned (3).

This chapter aims to point out the general possibilities available to the biochemist interested in assaying a specific protein. In the end, the choice of the most convenient and successful assay or detection procedure will depend on the particular protein to be studied, its degree of purity and the purpose of the assay. The use of more than one assay procedure may prove necessary to provide a complete analysis of the kinetic behavior of a particular enzyme. In each case the time and cost involved should also be carefully considered when designing or selecting an assay procedure.

10. REFERENCES

1. Colowick, S.P., and Kaplan, N.O. (1955 et seq.) *Methods Enzymol.*, **1**, 1–763; **2**, 1–870; **5**, 1–992; **6**, 1–450.

2. Eisenthal, R. and Danson, M.J. (eds) (1992) *Enzyme Assays: A Practical Approach*. IRL Press, Oxford.

3. Theerasilp, S. and Kurihara, Y. (1988) *J. Biol. Chem.*, **263**, 11536.

4. LeBel, D., Poirier, G.G. and Beaudoin, A.R. (1978) *Anal. Biochem.*, **85**, 86.

5. Kanarek, L. and Hill, R.L. (1964) *J. Biol. Chem.*, **239**, 4202.

6. Srere, P.A. and Kosicki, G.W. (1961) *J. Biol. Chem.*, **236**, 2557.

7. DeGroot, H., DeGroot, H. and Noll, T. (1985) *Biochem. J.*, **229**, 255.

8. Blostein, R. and Rutter, W.J. (1963) *J. Biol. Chem.*, **238**, 3280.

9. Chase, T. Jr. and Shaw, E. (1970) *Methods Enzymol.*, **19**, 20.

10. Lang, K. and Schmid, F.X. (1988) *Nature*, **331**, 453.

11. Tenu, J.-P., Viratelle, O.M., Garnier, J. and Yon, J. (1971) *Eur. J. Biochem.*, **20**, 363.

12. Takai, A. and Mieskes, G. (1991) *Biochem. J.*, **275**, 233.

13. Rimington, C. and Sveinsson, S.L. (1950) *Scand. J. Clin. Lab. Invest.*, **2**, 209.

14. Gozo, Y., Zalkin, H., Kein, P.S. and Heinrisburg, R.L. (1976) *J. Biol. Chem.*, **251**, 941.

15. Bishop, D.F. and Desnick, R.J. (1986) *Methods Enzymol.*, **123**, 339.

16. Itzhaki, L.S., Evans, P.A., Dobson, C.M. and Radford, S.E. (1994) *Biochemistry*, **33**, 5212.

17. Terzaghi, E., Okada, Y., Streisinger, G., Emrich, J., Inouye, M. and Tsugita, A. (1966) *Proc. Natl Acad. Sci. USA*, **56**, 500.

18. Kemp, B.E., Graves, D.J., Benjaming, E. and Krebs, E.G. (1977) *J. Biol. Chem.*, **252**, 4888.

19. House, C., Wettenhall, R.E.H. and Kemp, B.E. (1987) *J. Biol. Chem.*, **262**, 772.

20. Marques, S., Florencio, F.J. and Candau, P. (1989) *Anal. Biochem.*, **180**, 152

21. Herbert, R.B., Wilkinson, B., Ellames, G.J. and Kunec, E.K. (1993) *J. Chem. Soc. Chem. Comm.*, **2**, 205.

22. Halbach, S. (1977) *Anal. Biochem.*, **80**, 383.

23. Massey, V. and Curti, B. (1967) *J. Biol. Chem.*, **242**, 1259.

24. Ferguson-Miller, S., Brautigan, D.L. and Margoliash, E. (1976) *J. Biol. Chem.*, **251**, 1104.

25. Zakakis, J.P. and Treece, J.M. (1971) *J. Dairy Sci.*, **54**, 648.

26. Hertz, R. and Barenholz, Y. (1973) *Biochim. Biophys. Acta*, **330**, 1.

27. Lesser, M.A. (1982) *Methods Biochem. Anal.*, **28**, 175.

28. Hammes, G.G. and Kochavi, D. (1962) *J. Am. Chem. Soc.*, **84**, 2076.

29. Aguirre, R., Hatchikian, E.C. and Monson, P. (1983) *Anal. Biochem.*, **131**, 525.

30. Gilli, R., Sari, J.C., Sica, L., Bourdeaux, M. and Briand, C. (1986) *Anal. Biochem.*, **152**, 1.

31. Weber, K. and Kuter, D.J. (1971) *J. Biol. Chem.*, **246**, 4504.

32. Hagar, D.A. and Burgess, R.R. (1980) *Anal. Biochem.*, **109**, 76.

33. Chang, L.M.S., Plevani, P. and Bollum, F.J. (1982) *Proc. Natl Acad. Sci. USA*, **79**, 758.

34. Seymour, J.L., Lazarus, R.L. and May, J.W. (1989) *Anal. Biochem.*, **178**, 243.

35. Liebenguth, F. (1974) *Biochem. Genet.*, **13**, 263.

36. Scrutton, M.C. and Fatabene, F. (1975) *Anal. Biochem.*, **69**, 247.

37. Fried, M. and Crothers, D.M. (1981) *Nucleic Acids Res.*, **9**, 6505.

38. Galas, D.J. and Schmitz, A. (1978) *Nucleic Acids Res.*, **5**, 3157.

39. White, D. and Shultz, J. (1992) *Promega Notes*, **35**, 2.

40. White, D., Stevens, J. and Shultz, J. (1993) *Promega Notes*, **44**, 2.

41. Turner, B.M. (1983) *J. Immunol. Methods*, **63**, 1.

42. Hauber, R. and Geiger, R. (1987) *J. Clin. Chem. Clin. Biochem.*, **25**, 511.

43. Thorpe, S.J. and Kerr, M.A. (1994) in *Immunochemistry Labfax* (M.A. Kerr and S.J. Thorpe, eds). BIOS Scientific Publishers/Blackwell Scientific Publications, Oxford, p. 175.

44. Thorpe, R. and Rafferty, B. (1994) in *Immunochemistry Labfax* (M.A. Kerr and S.J. Thorpe, eds). BIOS Scientific Publishers/Blackwell Scientific Publications, Oxford, p. 115.

45. DeLuca, M. and McElroy, W.D. (1978) *Methods Enzymol.*, **57**, 3.

46. Shing, Y., Folkman, J., Sullivan, R., Butterfield, J., Murray, J. and Klagsbrun, M. (1984) *Science*, **223**, 1296.

CHAPTER 2
PREPARATION OF CELL EXTRACTS
N.C. Price

1. INTRODUCTION

The purpose of this chapter is to outline the principal techniques used to extract proteins from cells. The principal aim of such procedures is to obtain the protein *in as high a yield as possible*, consistent with *retention of maximal biological activity*. In fulfilling these aims it is necessary to consider the choice of a suitable assay method for the protein of interest and appropriate methods for the determination of total protein; these aspects are covered in Chapters 1 and 4 respectively of this volume. The various procedures that can be applied to extracts to purify any particular protein are described in Chapters 5, 6 and 7; the success of these procedures can be judged by the criteria given in Chapters 8 and 9.

It is not always necessary to break open cells in order to extract proteins. Several proteins such as immunoglobulins and a number of hydrolytic enzymes are secreted from cells or tissues and these can be purified directly from culture filtrates or supernatants. A different problem can occur with the high level expression of recombinant proteins (Chapter 11) in hosts such as *Escherichia coli,* where the expressed protein often appears in an insoluble form as inclusion bodies within the cells. The recovery of active protein often involves solubilization of the inclusion body in a strong denaturant and subsequent refolding; these aspects are dealt with in Chapter 12.

In Section 2 of this chapter, the factors underlying the choice of tissue and the method of extraction of that tissue are described. Section 3 outlines some of the factors involved in the protection of the integrity of the protein during the extraction procedures. Section 4 mentions some of the points to bear in mind when proteins are assayed in crude (i.e. unfractionated) extracts.

2. DISRUPTION OF CELLS AND TISSUES

2.1. Choice of tissue
The choice of tissue for extraction of proteins depends on a number of factors.

Availability, cost or abundance of protein
If a readily available tissue is rich in the protein of interest, it is sensible to use that tissue; for instance, large quantities of heart or brain tissue are available from meat animals; these would be suitable sources for enzymes of the tricarboxylic acid cycle or myelin proteins respectively.

Comparative studies
Comparative information might be sought on a protein which has been previously studied in another species or another tissue from the same species.

Use of microorganisms
The levels of proteins in microorganisms can often be altered substantially by changes in the

growth conditions or by genetic manipulation. Most recombinant proteins are expressed in prokaryotic (e.g. *E. coli*) or lower eukaryotic (e.g. yeast (*Saccharomyces cerevisiae*)) hosts. The disruption of these cells often requires fairly harsh procedures.

Minimization of proteolytic activity
Unless the proteolytic enzymes are themselves the object of study it may well be possible to avoid potential problems by a suitable choice of source. Thus certain animal tissues (e.g. liver, spleen, kidney and macrophages) are rich in lysosomal proteinases (notably cathepsins); this should be borne in mind when such tissues are used as sources of proteins. In the case of microorganisms, it may be possible to choose or construct mutant strains that are deficient in certain proteinases. This approach has been successfully employed in yeast and *E. coli* (1).

2.2. Disruption of tissue and separation of cells
In some cases it may be desirable to disrupt tissues and prepare homogeneous populations of intact cells, prior to disruption of these cells. The separation of the different types of cells may well allow comparative information to be obtained which may not be available if the whole tissue is studied.

Cell suspensions can be prepared from tissues by mechanical or enzymatic methods or a combination of the two (2). Enzymatic methods are generally preferred since there is less damage to the integrity of the cells. Collagenase from *Clostridium histolyticum* at a concentration of 0.01–0.1% (w/v) is normally added for 15 min–1 h; other proteinases such as trypsin, elastase and pronase have also been used. It is usual to add ethylenediaminetetraacetic acid (EDTA) to chelate Ca^{2+} ions which are often involved in cell adhesion. Separation of the cells obtained by these treatments is most often performed by centrifugation on the basis of cell size and density (3). The media used for such separations must be nontoxic, nonpermeable to cells and form iso-osmotic gradients of the appropriate density; solutions of Percoll, Ficoll and metrizamide have been widely used.

2.3. Disruption of cells
The principal procedures used to disrupt cells are listed in *Table 1*. The methods are classified rather broadly in terms of their harshness. In general it is advisable to use as gentle a method as possible, consistent with extraction of the protein of interest. This should avoid possible damage to the protein or the release of degradative enzymes from subcellular organelles such as vacuoles or lysosomes. Further details on the various methods can be found in refs 3, 4 and 5.

Mammalian tissues
The tissue should be cut into small pieces and trimmed as free as possible of fat and connective tissue. Soft tissue, such as liver, can be homogenized in a Potter–Elvehjem homogenizer in which a rotating pestle (a Teflon piston attached to a metal shaft) fits into an outer glass vessel, with a clearance in the range 0.05–0.6 mm, depending on the type of homogenizer. In the case of tougher tissues such as skeletal muscle, it is advisable to mince the chopped tissue prior to homogenization in a Waring blender. Three or four bursts, each of 15 sec, are normally sufficient to give a smooth extract, which is then stirred for about 30 min to allow further extraction of proteins. Centrifugation at 10 000 *g* for 20 min will then give a clear extract. The choice of the solution used for homogenization will depend on the nature of the extract required. If the isolation of subcellular organelles is important, iso-osmotic sucrose or mannitol (0.25 M) lightly buffered with Tris, Hepes or Tes (5–20 mM at pH 7.4) is generally used. If it is not important to isolate the intact organelles, the solution should be chosen to give a good yield of the desired protein. Thus the globular (G) form of actin can be extracted from muscle using low ionic strength solutions (0.01 M KCl). If the ionic strength is raised (e.g. to 0.1 M KCl), actin aggregates to form fibres (F-actin). Myosin can be selectively extracted from muscle in solutions of high ionic strength (0.3 M KCl, 0.15 M potassium phosphate).

Table 1. Methods for disruption of cells

Method	Underlying basis
Gentle	
Cell lysis	Osmotic disruption of cell membranes
Enzyme digestion	Digestion of cell wall; contents released by osmotic disruption
Potter–Elvehjem homogenizer	Cells forced through narrow gap. Cell membrane disrupted by shear forces
Moderately harsh	
Waring blender	Cells broken by rotating blades
Grinding (with sand, alumina or glass beads)	Cell walls broken by abrasive action of particles
Vigorous	
French pressure cell	Cells forced through small orifice at high pressure and disrupted by shear forces
Explosive decompression	Cells equilibrated with inert gas (e.g. N_2) at high pressure; on exposure of cells to 1 atm., disruption occurs
Bead mill	Rapid vibrations with glass beads remove cell wall
Ultrasonication	High-pressure sound waves cause cell rupture by cavitation and shear forces

For further details see ref. 4.

Plant tissues
Plant tissues pose a number of difficulties as far as the preparation of extracts is concerned:

(i)   moderately harsh methods such as the Waring blender (*Table 1*) are required to break the cellulose cell wall;
(ii)   disruption of vacuoles would lead to the release of proteinases and a lowering of the pH of the extract if there is inadequate buffering capacity;
(iii)   in the presence of oxygen the phenolic compounds present in the cells are converted by the action of phenol oxidases to polymeric pigments which can adsorb and damage proteins in the extract. It is usual to add a reducing agent (see Section 3.4) such as 2-mercaptoethanol (10 mM) to inhibit the phenol oxidases and a polymer such as polyvinylpolypyrollidone (2% (w/v)) to adsorb the phenolic polymers.

Yeasts
Yeasts and other fungi possess a tough cell wall requiring harsh methods for disruption. In addition these organisms contain large amounts of proteinases which could damage the protein of interest. It is therefore important to use methods of cell breakage where such proteolysis is minimized (see Section 3.3). The following methods are commonly used.

(i)   Autolysis with toluene (6). Incubation of yeast cake with toluene and 2-mercaptoethanol at 37°C followed by incubation in the presence of EDTA leads to degradation of the cell wall. However, the high temperature and presence of toluene may damage sensitive proteins.
(ii)   Lysis of spheroplasts (5,7). Treatment of yeast with glucanases (from snail gut or from microorganisms e.g. *Arthrobacter luteus*) or cellulases (from plant tissues) in the presence of

sucrose or mannitol leads to the formation of spheroplasts by digestion of the cell wall. These are then readily lysed by incubation with diethylaminoethyl-dextran under iso-osmotic conditions (7), allowing the integrity of subcellular organelles to be maintained.

(iii) Shaking with glass beads (8). This technique, which involves shaking a yeast suspension with small glass beads (1 mm diameter), is especially suitable for small scale extraction procedures, e.g. when an overexpressed protein is being purified. Maximum degrees of extraction (usually over 80%) are obtained after about 20 min shaking.

(iv) Freezing and thawing (9). Yeast cake is added in small amounts to toluene at $-20°C$ and then allowed to melt at $4°C$ after addition of extraction buffer. The low temperatures in this procedure may well be important in minimizing damage caused by proteinases (9).

Bacterial cells

Bacteria possess very tough cell walls; it is usually necessary to employ very vigorous mechanical methods (French pressure cell, ultrasonication, bead milling, etc.) to cause disruption. These can damage the cellular contents and are not easy to scale up. Methods based on enzymatic breakdown of the wall are gentler and are more easily applied to large amounts of cells. Gram-positive bacteria (such as *Bacillus*, *Micrococcus*, *Streptococcus*) are highly susceptible to the action of lysozyme. Typically the bacterial suspension would be incubated with hen egg-white lysozyme (0.2 mg ml^{-1}) at 37°C for 15 min (4). Gram-negative bacteria (such as *E. coli*, *Klebsiella* spp., *Pseudomonas* spp.) require additional pre-treatments to render them susceptible to lysozyme. A typical treatment would include washing the cells in a dilute detergent (0.1% (v/v) *N*-lauroyl-sarcosine) followed by an osmotic shock in which a cell suspension in sucrose (0.7 M), Tris (0.2 M), EDTA (0.4 M) is diluted with four volumes of distilled water. The effectiveness of these treatments is related to disruptions of the outer and cytoplasmic membranes and the consequent drawing of lysozyme molecules into the murein layer of the cell wall, promoting its digestion (10).

The release of DNA on cell lysis makes the resulting extract highly viscous, which can lead to severe problems in subsequent steps in the purification procedures. The DNA can be degraded by the addition of deoxyribonuclease 1 (10 $\mu g\ ml^{-1}$) or can be precipitated along with other nucleic acids (and some acidic proteins) by the addition of a neutralized solution of protamine sulfate (4). The amount of protamine sulfate to be used (which can be up to 5 mg g^{-1} tissue) should be determined in preliminary experiments.

Assessing the degree of cell breakage

It is important to be able to measure the degree of cell breakage in order to judge the success of the particular extraction method used. For suspensions of single-celled organisms, an estimate of the degree of cell breakage can be obtained by analysis of the extract using a hemocytometer (8). A quantitative estimate of the release of cell constituents can be obtained by measurements of the protein that is not sedimented by centrifugation (30 000 *g*. min) relative to the total protein in the organism (5).

Membrane-bound proteins

Many proteins within cells occur physically associated with membranes; such associations range in strength from the relatively weak, mainly electrostatic, interactions of peripheral proteins, to the much stronger, mainly hydrophobic, interactions of integral membrane proteins (11). Peripheral proteins can usually be extracted by an increase in ionic strength. However, extraction of integral proteins usually requires treatment with organic solvents (e.g. butanol), chaotropic agents (e.g. urea or $NaClO_4$) or detergents (e.g. Triton X-100, octyl glucoside or sodium deoxycholate), often in conjunction with enzymes such as phospholipases or proteinases which would disrupt the membrane structure. The extraction of membrane proteins is considered in more detail in Chapter 10.

3. PROTECTION OF PROTEIN INTEGRITY

During the process of tissue or cell disruption and during the subsequent steps of purification it is possible that the structural integrity (and hence activity) of the protein of interest may be lost. Some of the important factors which should be controlled are described in this section.

3.1. Control of pH

In general, there is only a limited range of pH values over which the structure of any particular protein is stable. It is therefore usual to include a buffer in the solution used for the extraction. This would be particularly important in cases where certain subcellular organelles such as vacuoles or lysosomes are ruptured, since the pH within these organelles is well below that of the cytosol. The principal factors involved in choosing a suitable buffer, which are analysed in more detail in Chapter 3, include:

(i) range of pH;
(ii) degree of buffering capacity;
(iii) influence of temperature and ionic strength on pH;
(iv) interaction with metal ions;
(v) compatibility with subsequent purification procedures.

Some proteins display unusual stability at extremes of pH and this can be taken advantage of during extraction procedures; for instance, a large degree of purification of adenylate kinase can be achieved by incubating muscle extracts at pH 2.0; under these conditions, the enzyme is stable and unwanted proteins are denatured and precipitated (12).

3.2. Control of temperature

During cell disruption, especially by the harsher mechanical methods listed in Table 1, the temperature of the extract can rise by at least 30°C. It is usual to attempt to minimize such temperature changes by using pre-cooled solutions and apparatus, and if necessary taking steps to ensure heat dissipation during the extraction. Although the general practice would be to keep temperature low in order to maintain protein structure and to minimize proteolytic activity, it should be remembered that there are some well-documented examples of cold-inactivation of proteins (13). In addition, the unusual heat stability of certain proteins can be used to advantage in their purification. Thus, the protein phosphatase inhibitor-1 can be purified to a considerable degree by heating muscle extracts at 90°C to precipitate unwanted proteins (14).

3.3. Control of proteolysis

The degradation of proteins by endogenous proteinases during extraction represents one of the more serious problems faced in this type of work. Indications that proteolysis is a potential problem include the findings that a protein is isolated in poor yield, that the biological activity of a protein is unstable, that the appearance of the protein on SDS–PAGE is heterogeneous, or that the M_r value and other properties differ from reported values. Although in some cases, the problem can be successfully controlled by the use of low temperatures or of proteinase-deficient strains or tissues, a much more usual approach is to include proteinase inhibitors during the extraction procedure and subsequent purification steps. The major types of proteinases present in various tissues and 'cocktails' of inhibitors to control them are listed in *Table 2*; further details are supplied in refs 1 and 15.

Particular points to note concerning the inhibitors are:

Safety. The inhibitor phenylmethanesulfonyl fluoride (PMSF) (or α-toluenesulfonyl fluoride) is reported to be highly toxic. Care (gloves, mask, etc.) should be taken when handling the solid and solutions. 3,4-Dichloroisocoumarin (DCI), which is more reactive towards many serine proteinases (16), could be used as a (considerably more expensive) alternative to PMSF (see *Table 2*). Suitable precautions should also be taken when using organic solvents used to prepare a number of the stock solutions.

Solubility. Several of the inhibitors in *Table 2* are of limited solubility in aqueous solvents and are thus made up as stock solutions in organic solvents, such as methanol or dimethyl sulfoxide (DMSO). The volume of the stock solution added should be kept to a minimum in order to avoid any potential damage to the protein of interest.

Stability. PMSF is unstable in aqueous solutions with a half-life of about 30 min at 25°C and pH 7.0 (1). The stability of DCI in aqueous solutions is reported to be similar (16). In particular cases it may be necessary to make repeated additions of the stock solutions of these inhibitors.

It is important to bear in mind that the 'cocktails' described in *Table 2* are of a general type, and in any particular case it is important to check that the problem is being controlled, for example by demonstrating that the protein is stable after extraction and that its behavior on SDS–PAGE is in accordance with expectations (see Chapter 9).

Table 2. Some inhibitors used to control proteolysis

Type of tissue and major proteinases present	Inhibitors added	Stock solution[a]
Animal tissues (serine, metallo, aspartic)	PMSF (1 mM)	0.2 M in methanol
	(*or* DCI (0.1 mM))	(10 mM in DMSO)
	Benzamidine (1 mM)	0.1 M
	Leupeptin (10 μg ml^{-1})	1 mg ml^{-1}
	Pepstatin (10 μg ml^{-1})	5 mg ml^{-1} in methanol
	Aprotinin (1 μg ml^{-1})	0.1 mg ml^{-1}
	Antipain (0.1 mM)	10 mM
Plant tissues (serine, cysteine)	PMSF (1 mM)	0.2 M in methanol
	(*or* DCI (0.1 mM))	(10 mM in DMSO)
	Chymostatin (20 μg ml^{-1})	1 mg ml^{-1} in DMSO
	EDTA (1 mM)	0.1 M
	E64 (10 μg ml^{-1})	1 mg ml^{-1}
Yeasts, fungi (serine, aspartic, metallo (?))	PMSF (1 mM)	0.2 M in methanol
	(*or* DCI (0.1 mM))	(10 mM in DMSO)
	Pepstatin (15 μg ml^{-1})	5 mg ml^{-1} in methanol
	1,10-Phenanthroline (5 mM)	1 M in ethanol
Bacteria (serine, metallo)	PMSF (1 mM)	0.2 M in methanol
	(*or* DCI (0.1 mM))	(10 mM in DMSO)
	EDTA (1 mM)	0.1 M

[a]Aqueous solution unless otherwise indicated.
Abbreviations: PMSF, phenylmethanesulfonyl fluoride; DCI, 3,4-dichloroisocoumarin; DMSO, dimethylsulfoxide; E64, L-*trans*-epoxysuccinyl-leucylamide-(4-guanidino)-butane; EDTA, ethylenediaminetetraacetic acid.
M_r values of inhibitors listed: PMSF, 174; DCI, 215; EDTA (disodium salt, dihydrate), 372; benzamidine (hydrochloride), 157; leupeptin, 427; pepstatin, 686; aprotinin, 6500; antipain (dihydrochloride), 678; chymostatin, 605; E64, 357; 1,10-phenanthroline, 198.
For further details see refs 1 and 15.

3.4. Protection of labile thiol groups

The thiol groups of cysteine side-chains can be damaged during extraction either by oxidation or by reaction with heavy metal ions. Removal of proteins from the reducing environment within cells and exposure to oxygen can lead to the formation of disulfide bonds or of oxidized species such as sulfinic acids. Heavy metal ions (such as those of Cu, Pb, Hg) could arise from the water, reagents or equipment used for the extraction procedure, and are likely to inactivate proteins by reaction with cysteine side-chains. These problems can be overcome by the inclusion of a reducing agent (such as 2-mercaptoethanol or dithiothreitol) and a chelating agent (such as EDTA) in the extraction medium. 2-Mercaptoethanol is a dense liquid (1.12 g ml^{-1}), with a most disagreeable odor and is toxic. It is usually necessary to add it to a final concentration of 10–20 mM to provide protection for thiol groups in a protein for up to 24 h (4). Dithiothreitol, because of its greater reducing power (17), can provide effective protection at lower concentrations (1 mM). It is also more convenient to handle, being a white (somewhat hygroscopic) solid with little odor. The high cost of dithiothreitol (it is some 20 times more expensive to make a solution of 1 mM dithiothreitol than the equivalent volume of 20 mM 2-mercaptoethanol) can be a serious problem when large volumes of a reducing solution are required for large-scale dialysis or chromatography.

The stability of solutions containing these reducing agents is markedly reduced at high pH and in the presence of oxygen or certain metal ions such as Cu^{2+} (18). Inclusion of EDTA (typically 1 mM) in the extraction medium would thus not only increase the stability of the reducing agents, but would also provide protection against a variety of heavy metal ions. In addition, EDTA would serve as an inhibitor of metalloproteinases (*Table 2*).

3.5. Control of mechanical stress

During cell disruption by harsh techniques such as the French pressure cell or ultrasonication, the cell contents are subjected to high pressure which can lead to the irreversible inactivation of some proteins. It is important to control the period of time and the pressure applied during such procedures so as to minimize the potential damage to the protein(s) of interest, provided that satisfactory degrees of extraction of the protein(s) are obtained.

3.6. Control of free radical formation

Cell extracts prepared by ultrasonic disintegration are susceptible to damage caused by free radicals, presumed to originate from the breakdown of H_2O molecules caused by local high temperatures in the solution. These effects can be minimized by the addition of certain sugars or other compounds which act as radical scavengers (19,20).

3.7. Control of the effects of dilution

The extraction of cells or tissues can lead to high degrees of dilution of the proteins within the cell. Many proteins appear to lose biological activity on storage in dilute solution, even when many of the precautions listed earlier in this section have been taken into account. This process can often be prevented by the inclusion of an 'inert' protein such as bovine serum albumin at a concentration of 1 mg ml^{-1}. It is possible that the added protein may help to prevent loss of the protein of interest by adsorption on the walls of the vessel, or may act as a sacrificial substrate for traces of proteinases or other reactive species in the extraction medium (4).

Alternative (or additional) protective agents are polyols such as glycerol or sucrose which stabilize the folded structures of proteins in aqueous solution by increasing the hydration of the protein (21). Glycerol at high concentrations (50% (v/v)) can be added to aqueous solutions for long-term storage of protein solutions at $-20°C$ (the normal temperature of domestic freezers) without freezing of the solution; this can be important since cycles of freezing and thawing damage a number of proteins.

Dilution could also lead to the loss of a cofactor (e.g. heme, flavin, pyridoxal-5′-phosphate) from a protein leading to a decrease in stability. If this appears to be the case, addition of the appropriate cofactor may be necessary during the extraction.

4. ASSAYS OF PROTEINS IN UNFRACTIONATED CELL EXTRACTS

The general considerations to be borne in mind when assays of proteins are performed are described in Chapter 1. It is important to realize, however, that complications to assays can be caused by the presence of the multitude of additional components in crude extracts. In the case of a binding protein, the presence of large amounts of components which bind non-specifically to the protein of interest may complicate the results of binding studies. The presence of endogenous inhibitors (22) or interference or competition from other related reactions (23) can make the interpretation of enzyme assays of crude extracts unreliable. The usual approach to the problem would be to attempt to remove such interfering substances or inhibit the competing related reactions; however, this is not always readily achieved. In such cases, the total amount of 'biological activity' present in the crude extract can only be determined approximately, and would not serve as a reliable basis on which to calculate the yield of the protein of interest during the purification procedure.

5. REFERENCES

1. North, M.J. (1989) in *Proteolytic Enzymes: A Practical Approach* (R.J. Beynon and J.S. Bond, eds). IRL Press, Oxford, p. 105.

2. Elliott, K.R.F. (1979) in *Techniques in Metabolic Research* (H.L. Kornberg, ed.). Elsevier/North Holland, Amsterdam, Vol. B204, p. 1.

3. Graham, J. (1984) in *Centrifugation: A Practical Approach (2nd edn)* (D. Rickwood, ed.). IRL Press, Oxford, p. 161.

4. Scopes, R.K. (1994) *Protein Purification: Principles and Practice (3rd edn)*. Springer Verlag, New York.

5. Lloyd, D. and Coakley, W.T. (1979) in *Techniques in Metabolic Research* (H.L. Kornberg, ed.). Elsevier/North Holland, Amsterdam, Vol. B201, p. 1.

6. Yun, S.-L., Aust, A.E. and Suelter, C.H. (1976) *J. Biol. Chem.*, **251**, 124.

7. Schwenke, J., Canut, H. and Flores, A. (1983) *FEBS Lett.*, **156**, 274.

8. Naganuma, T., Uzuka, Y. and Tanaka, K. (1984) *Anal. Biochem.*, **141**, 74.

9. Fell, D.A., Liddle, P.F., Peacocke, A.R. and Dwek, R.A. (1974) *Biochem. J.*, **139**, 665.

10. Schwinghamer, E.A. (1980) *FEMS Microbiol. Lett.*, **7**, 157.

11. Singer, S.J. (1974) *Annu. Rev. Biochem.*, **43**, 805.

12. Heil, A., Müller, G., Noda, L., Pinder, T., Schirmer, H., Schirmer, I. and von Zabern, I. (1974) *Eur. J. Biochem.*, **43**, 131.

13. Bock, P.E. and Frieden, C. (1978) *Trends Biochem. Sci.*, **3**, 100.

14. Nimmo, G.A. and Cohen, P. (1978) *Eur. J. Biochem.*, **87**, 341.

15. Beynon, R.J. and Salvesen G. (1989) in *Proteolytic Enzymes: A Practical Approach* (R.J. Beynon and J.S. Bond, eds). IRL Press, Oxford, p. 241.

16. Harper, J.W., Hemmi, K. and Powers, J.C. (1985) *Biochemistry*, **24**, 1831.

17. Cleland, W.W. (1964) *Biochemistry*, **3**, 480.

18. Stevens, R., Stevens, L. and Price, N.C. (1983) *Biochem. Educ.*, **11**, 70.

19. Coakley, W.T., Brown, R.C., James, C.J. and Gould, R.K. (1973) *Arch. Biochem. Biophys.*, **159**, 722.

20. McKee, J.R., Christman, C.L., O'Brien, W.D. Jr. and Wang, S.I. (1977) *Biochemistry*, **16**, 4651.

21. Arakawa, T. and Timasheff, S.N. (1982) *Biochemistry*, **21**, 6536.

22. Ansari, H. and Stevens, L. (1983) *J. Gen. Microbiol.*, **129**, 1637.

23. Crabtree, B. and Newsholme, E.A. (1972) *Biochem. J.*, **126**, 49.

CHAPTER 3
BUFFERS

L. Stevens

1. INTRODUCTION

The term 'buffer' is most frequently used in a biochemical context to refer to an agent that resists changes in hydrogen ion concentration. However it can be broadened to include agents which resist changes in ion concentration, generally metal ions, and also agents used to maintain a constant redox potential (1,2).

In almost all experimental procedures in which proteins have to be maintained in their native conformation, buffers are used to stabilize pH. The choice of buffer depends on the experimental procedure being carried out, and on the particular properties of the protein. In this chapter the factors determining the choice of buffer are first considered, the salient features of particular buffers are summarized in *Table 1*, and finally suggestions and hints are given on how best to make up buffers. The use of buffers specifically for enzyme work is described in the companion volume *Enzymology Labfax*.

2. FACTORS DETERMINING THE CHOICE OF BUFFER

The choice of buffer will be determined by the following general considerations:

(i)   the type of experimental procedures being performed;
(ii) the characteristics of the protein under study;
(iii) the range of buffers available.

The experimental procedures may involve purification, structural analysis or biological activity including binding studies. During protein purification the buffered media in which the proteins are suspended are required to give the maximum stability. In the early stages of purification when cells or tissue are disrupted, it is often necessary to stabilize a protein against acid released from lysosomes. A medium which provides osmotic stabilization for subcellular organelles is often used. Chromatographic procedures are amongst the most frequently used procedures for protein purification. For ion-exchange chromatography and affinity chromatography, the pH is chosen so that the protein is selectively bound or selectively released from the column. For anion-exchange chromatography the buffer chosen is generally such that the buffer provides the cation (e.g. Tris–HCl for diethylaminoethyl (DEAE)-cellulose), and for cation-exchange chromatography the buffer provides the anion (e.g. Mes or acetate for carboxymethyl (CM)-cellulose). For gel filtration, conditions are generally chosen so that the proteins being separated have least tendency to adsorb to the column. This is generally achieved by ensuring that the eluent has a minimum ionic strength of about 0.1, but the choice of buffer is not critical. For hydroxyapatite ($Ca_{10}(PO_4)_6(OH)_2$) and calcium phosphate gel chromatography, phosphate buffers are most frequently used for elution. They probably act by desorbing proteins off Ca^{2+}-binding sites on the gel.

Electrophoretic separation is sometimes used in the later stages of purification, most frequently with polyacrylamide as support. The pH chosen is generally that which gives best resolution from

BUFFERS   **21**

Table 1. Selected properties of commonly used buffers

pK_a[a] Compound[b]	d(pK_a)/dT	Saturated solution at 0°C (M)	Interaction with metal ions[c] w, m or s	Price[d]	Comments
4.64 Acetic acid	0.0002	>10		L	Interferes with Lowry
5.28 Succinic acid	0	0.36	Cu^{2+},m	L	Interferes with Lowry
5.80 Citric acid	0	>2	Mg^{2+},Ca^{2+},Mn^{2+}, and Cu^{2+} m	L	Interferes with Lowry
6.02 Mes	−0.011	0.65	Negligible	M	
6.32 Bis–Tris	−0.017			H	
6.32 Pyrophosphate	−0.01	0.1	Mg^{2+},Ca^{2+},m; Cu^{2+},s	L	Product of ligases and may inhibit
6.62 Ada	−0.011	v. sol.	Mg^{2+},w; Ca^{2+}, Mn^{2+}, m; Cu^{2+},s	H	Interferes with Lowry, absorbs <260 nm
6.67 Aces	−0.020	0.22	Cu^{2+},m	H	Interferes with Lowry, absorbs <260 nm
6.77 Mopso	−0.015	0.75	Negligible	M	Interferes with Lowry
6.84 Phosphate	−0.0028	0.2	Mg^{2+},Ca^{2+},Mn^{2+}, and Cu^{2+},m	L	Inhibits kinases, dehydrogenases, carboxypeptidase, fumarase, urease, aryl sulfatase, adenosine deaminase, phosphoglucomutase and enzymes involving phosphate esters. Stabilizes phosphoribosepyrophosphate synthase
6.86 Pipes	−0.0085	v. sol.	Negligible	H	Interferes with Lowry
6.97 Imidazole	−0.020	v. sol.	Mn^{2+}, Cu^{2+}, m	L	Reactive, unstable, Mops or Bes generally better alternative
6.98 Bes	−0.016	3.2	Cu^{2+},m	M	Interferes with Lowry
7.02 Mops	−0.015	3.09	Negligible	M	Interferes with Lowry
7.27 Tes	−0.020	2.6	Cu^{2+},m	H	Interferes with Lowry
7.39 Hepes	−0.014	2.25	Negligible	M	Interferes with Lowry
7.42 Dipso	−0.015	0.24		H	Interferes with Lowry

7.49 Tapso	−0.018	1.0		H	Interferes with Lowry
7.77 Heppso	−0.010	2.2		H	Interferes with Lowry
7.78 Triethanolamine	−0.020	v. sol.		L	
7.82 Popso	−0.013	v. sol.		H	
7.85 Hepps	−0.015	1.58	Negligible	M	Interferes with Lowry
7.92 Tricine	−0.021	0.8	Mg^{2+}, w; Ca^{2+}, Mn^{2+}, m	M	Interferes with Lowry
8.00 Tris	−0.031	2.4	Cu^{2+}, Zn^{2+}, m	L	Interferes with Lowry and Bradford, reacts with carbonyl groups
8.09 Glycylglycine	−0.028	1.1	Ca^{2+}, Mg^{2+}, w; Mn^{2+}, Cu^{2+}, m	M	Not suitable for peptidases
8.17 Bicine	−0.018	1.1	Mg^{2+}, Ca^{2+}, Mn^{2+}, m; Cu^{2+}, s	L	Interferes with Lowry
8.19 Taps	−0.018	v. sol.		H	
8.88 Diethanolamine	−0.024	0.05		L	
9.08 Borate	−0.008	0.05		L	Binds vic diols, not recommended with carbohydrates or ribonucleotides
9.23 Ches	−0.029	1.14		M	Interferes with Lowry
9.47 Ethanolamine	−0.029	v. sol.		L	Reacts with carbonyl groups
9.55 Glycine	−0.025	4.0	Mg^{2+}, Ca^{2+}, w; Mn^{2+}, m; Cu^{2+}, s	L	Interferes with Lowry and Bradford
9.96 Carbonate	−0.009	0.8		L	Requires closed system
10.05 Caps	−0.032	0.47		M	Interferes with Lowry

[a] pK_a, the practical dissociation constant defined as: $pH = pK_a + \log[(A^-)/(HA)]-(2z + 1)\{(0.5I^{0.5}/(1 + I^{0.5})) - 0.1I\}$, I = ionic strength and z = charge on the conjugate ion.

[b] The structures of compounds with abbreviated names are given in the companion volume *Enzymology Labfax* (p. 273).

[c] Interaction with metal ions; $\log K_M$ = 1–2 for weak binding (w), 2–5 for medium binding (m), and >5 for strong binding (s), where K_M is metal–buffer association constant.

[d] The price categories are based on 1995 manufacturers' catalogs. L = less than £20 per liter of 1 M buffer solution, M = between £20-50 per liter of 1 M buffer solution, and H = greater than £50 per liter of 1 M buffer solution.
v. sol., very soluble.

Table 2. Volatile buffer mixtures

pH range	Constituents
1.2–3.0	Acetic acid + formic acid
3.1–6.5	Pyridine + acetic acid
6.8–8.8	Triethanolamine + HCl
7.0–10.0	Ammonia + formic (or acetic) acid
7.9–8.9	NH_4HCO_3 + $(NH_4)_2CO_3$ solutions

Recipes are given for these buffers in refs 1 and 6.

other proteins. An important consideration with electrophoresis is to provide the maximum buffering capacity for a given ionic strength, and at the same time to minimize the concentrations of small highly mobile anions and cations (e.g. Na^+, K^+, Mg^{2+}, Cl^-, SO_4^{2-}, HPO_4^{2-}). These small mobile ions carry most of the current, and their presence will lead to heating effects, and also decrease the mobility of the proteins being separated. Minimizing the concentrations of small mobile ions is generally achieved by using two buffers differing by about 1 pH unit, one of which provides the anion and the other the cation; for example, for pH 4.5, β-alanine$^+$-acetate$^-$; pH 6.8–7.5 imidazole$^+$-Mops$^-$; pH 8.0–9.0 either Tris$^+$-borate$^-$ or Tris$^+$-glycine$^-$.

Certain stages in protein purification often require large volumes of buffered medium, either for elution, or for dialysis. The large volumes favor the choice of the less expensive buffers.

During purification procedures, structural analysis, measurement of biological activity or binding studies, it is often necessary to measure the absorbance of the protein without interference from the buffer, usually in the range 260–400 nm, but sometimes at shorter wavelengths. Most of the *N*-substituted glycine and taurine buffers listed in *Table 1* have low absorbance in this range (3,4). It may be necessary to determine the amount of protein present, using either the Lowry method or one of the Coomassie blue binding methods (see Chapter 4 for details). Both methods are subject to interference by certain buffers (see *Table 1*), but the range of interfering buffers is much greater with the Lowry method. Whether or not this is a serious problem will depend on the relative concentrations of the proteins and buffer. For concentrated protein solutions, which have to be diluted before assay, little interference may result, and this can be corrected satisfactorily by a buffer blank.

Proteins are often concentrated at various stages during purification procedures, either by ultrafiltration, dialysis against polyethylene glycol, precipitation with ammonium sulfate, or by freeze drying. For the last of these procedures the protein has first to be dialyzed to remove the solute, although this can be avoided if a volatile buffer is used (5). A list of volatile buffers is given in *Table 2*.

3. PREPARATION OF BUFFERS

When making up buffers for use with protein solutions, the pH of the buffer is generally more important than its concentration. Concentrations are generally chosen so that they are in the correct range to buffer adequately under the conditions being used (e.g. 0.01 M, 0.1 M, 0.2 M, etc.) but the exact concentration is generally not critical. Buffers are generally used within ±1 pH unit of the pK_a, where their buffering capacity is greatest. The most dilute buffer solution that will suffice is usually employed. The amounts of the two components can be calculated from the pK, using the Henderson–Hasselbalch equation:

$$pH = pK_a + \log[A^-]/[HA] \qquad\qquad \text{Equation 1}$$

and then weighed out and dissolved. However if this is done, it is necessary to check the final pH using a pH meter, since the components may contain moisture and this will affect concentration, or in some cases the free base may be liquid which is hygroscopic. More frequently, acidic buffers are often made up by taking the required amount of free acid (e.g. acetic acid) to give the required molarity and then adding concentrated base (e.g. NaOH) until the required pH is reached. In the case of an alkaline buffer the required amount of the base (e.g. Tris) to give the required molarity is taken and acid (e.g. HCl) is added until the required pH is reached. Sometimes the starting solution is the salt (e.g. triethanolamine HCl), especially if this is more readily obtained in purified crystalline form. In this case alkali is added to obtain the required pH, and the buffer will then also contain NaCl. If a low concentration of metal ions is required, tetramethylammonium hydroxide may be used as an alternative strong alkali.

Many manipulations with proteins, unless they are strongly hydrophobic, are carried out at 4°C in order to increase the stability of the protein preparation, but buffers are rarely made up in the cold room. At the other extreme, proteins from thermophilic organisms may be assayed at 50°C or higher. The d(pK_a)/dT for carboxylic acids is generally very low (*Table 1*), but for many other buffers it is sufficiently high to make a significant change. The primary standard buffers used to calibrate pH meters have very low d(pK_a)/dT; for example, potassium hydrogen phthalate which is most commonly used as the pH 4.0 standard, has a pH value of 3.999 at 15°C, 4.003 at 0°C and 4.060 at 50°C (6). The steps in the adjustment of the pH of a buffer for a particular temperature are:

(a) equilibrate the pH standard to the required temperature (for most purposes it will be sufficiently accurate to assume that the pH of the standard is unchanged);
(b) set the temperature compensation control on the meter to the required temperature (this usually compensates for the change in the relationship of electromotive force (e.m.f.) to pH with temperature);
(c) calibrate the pH reading on the meter to the standard;
(d) measure and adjust the pH of the buffer solution being made up only after it has also been equilibrated at the temperature required.

4. METAL ION BUFFERS

Proteins bind metal ions to varying extents. For some, the metal ions bind strongly (K_d 10^{-8} to 10^{-10} M) and are essential for the functioning of the protein, and this may also apply to less strongly bound ions. There are also examples where metal ions inhibit the function of a protein. It is important in a number of experimental procedures to control the metal ion concentration, for example to study the effects of the metal ion on activity, to measure its binding affinity for the protein, or to prevent its inactivation of the protein. Where the absence of monovalent ions such as Na^+ or K^+ is required, tetramethylammonium hydroxide can be used in place of NaOH or KOH to titrate the buffering acid, for divalent or trivalent metal ion chelating agents are used as metal ion buffers. The choice of chelating agents depends on the stability constants for complexation of the metal ion of interest, and as with pH buffers, buffering capacity is greatest when the free metal ion concentration is in the range of the K_d. (For a table of stability constants, see refs 1 and 2). Also the chelating agent should exert its effect by controlling the free metal ion concentration in solution and should not interact with the protein. Most metal ion buffers are carboxylic acid buffers (e.g. EDTA, EGTA, citric acid, 2-*N*-hydroxyethylethylenediaminetri-acetic acid and nitrilotriacetic acid), or polynuclear aromatic hydrocarbon buffers (e.g. 1,10-phenanthroline, 8-hydroxyquinoline-5-sulfonic acid, and 2,2′-dipyridyl) (7). There are also insoluble metal chelating agents such as Chelex 100, a cross-linked polystyrene resin having iminodiacetic acid functional groups. These have the advantage that in equilibrium dialysis experiments they can be kept separate from the protein on the opposite side of the dialysis membrane.

5. REFERENCES

1. Perrin, D.D. and Dempsey, B. (1974) *Buffers for pH and Metal Ion Control.* Chapman and Hall, London.

2. Blanchard, J.S. (1984) *Methods Enzymol.,* **87**, 405.

3. Good, N.E. and Izawa, S. (1972) *Methods Enzymol.,* **24**, 53.

4. Ferguson, W.J., Braunschweiger, K.I., Braunschweiger, W.R., Smith, J.R., McCormick, J., Wasmann, C.C., Jarvis, N.P., Bell, D.H. and Good, N.E. (1980) *Anal. Biochem.,* **104**, 300.

5. Stoll, V.S. and Blanchard, J.S. (1990) *Methods Enzymol.,* **182**, 243.

6. Dawson, R.M.C., Elliot, D.C., Elliot, W.M. and Jones, K.M. (1986) *Data for Biochemical Research (3rd edn).* Oxford University Press, Oxford, p. 417.

7. Baker, J.O. (1988) *Methods Enzymol.,* **158**, 33.

CHAPTER 4
THE DETERMINATION OF PROTEIN CONCENTRATION
N.C. Price

1. INTRODUCTION

The purpose of this short chapter is to set out in summary form (*Table 1*) the principal methods for the determination of protein concentration. Further details about each method are given in the companion volume *Enzymology Labfax* (17).

2. RECOMMENDATIONS

It is recommended that the absolute amount of protein in a sample is determined, if possible, using amino acid analysis (8,9). This result can then be used to calibrate a convenient routine procedure for example based on UV absorbance (4–6), dye binding (2,5,13–15), bicinchoninic acid (5,11,12) or Lowry (3,10). If amino acid analysis is not possible, use of a method based on peptide bond content, for example biuret (2,3) or far UV measurements (7) (more sensitive), will give a good calibration. Measurements of A_{280} are convenient and, for most pure proteins, can be used to determine the amount of protein reasonably precisely, provided that the A_{280} for a 1 mg ml^{-1} solution can be calculated from the amino acid composition (6).

Table 1. Principal methods for determination of protein

Method	Amount of protein required (μg)	Recovery of protein	Complexity of method[a]	Response of identical masses of different proteins	Reference protein used	Major sources of interference[b]	Equipment needed
1. Gravimetric (1)	5000–20 000	No	4	Identical	No	Other macromolecules	Microbalance
2. Biuret (2,3)	500–5000	No	2	Very similar	Yes	Tris, NH_4^+, sucrose, glycerol	Spectrophotometer
3. UV absorbance							
(i) 280 nm (4–6)	100–1000	Yes	1	Variable	No	Nucleic acids, chromophores such as heme	Spectrophotometer
(ii) far UV (7)	5–50	Yes	1	Very similar	No	Many buffers and other compounds absorb strongly in the far UV	Spectrophotometer
4. Amino acid analysis (8,9)	10–200	No	4	Variable	No	Other contaminating proteins	Amino acid analyzer

5. Lowry (3,10)	5–100	No	3	Variable	Yes	Some amino acids, NH_4^+, zwitterionic buffers, nonionic detergents, thiol compounds, sucrose	Spectrophotometer
6. Bicinchoninic acid (5,11,12)	5–100 (1–10 in microprotocol)	No	3	Variable	Yes	Glucose, NH_4^+, EDTA	Spectrophotometer
7. Coomassie blue binding (2,5,13–15)	5–50 (1–10 in microprotocol)	No	2	Variable	Yes	Triton, SDS	Spectrophotometer
8. Reaction with o-phthalaldehyde (4,16)	0.1–2.0	No	3	Variable	Yes	Tris, glycine, NH_4^+	Spectrofluorimeter

[a] Graded on a 1–4 scale. 1 Involves pipetting of the sample only. 2 Involves mixing the sample with one reagent solution. 3 Involves mixing the sample with more than one reagent solution. 4 Involves lengthy manipulation of the sample.
[b] In some cases the interference can be minimized or overcome.

3. REFERENCES

1. Blakeley, R.L. and Zerner, B. (1975) *Methods Enzymol.*, **35**, 221.

2. Darbre, A. (1986) in *Practical Protein Chemistry: A Handbook* (A. Darbre, ed.). John Wiley and Sons, Chichester, p. 227.

3. Stevens, L. (1992) in *Enzyme Assays: A Practical Approach* (R. Eisenthal and M.J. Danson, eds.). IRL Press, Oxford, p. 317.

4. Peterson, G.L. (1983) *Methods Enzymol.*, **91**, 95.

5. Stoscheck, C.M. (1990) *Methods Enzymol.*, **182**, 50.

6. Gill, S.C. and von Hippel, P.H. (1989) *Anal. Biochem.*, **182**, 319.

7. Scopes, R.K. (1974) *Anal Biochem.*, **59**, 277.

8. Chang, J.-Y., Knecht, R. and Braun, D.G. (1983) *Methods Enzymol.*, **91**, 41.

9. Ozols, J. (1990) *Methods Enzymol.*, **182**, 587.

10. Lowry, O.H., Rosebrough, N.J., Farr, A.L. and Randall, R.J. (1951) *J. Biol. Chem.*, **193**, 265.

11. Smith, P.K., Krohn, R.I., Hermanson, G.T., Mallia, A.K., Gartner, F.H., Provenzano, M.D., Fujimoto, E.K., Goeke, N.M., Olson, B.J. and Klenk, D.C. (1985) *Anal. Biochem.*, **150**, 76.

12. Harris, D.A. (1987) in *Spectrophotometry and Spectrofluorimetry: A Practical Approach* (D.A. Harris and C.L. Bashford, eds.). IRL Press, Oxford, p. 49.

13. Bradford, M.M. (1976) *Anal. Biochem.*, **72**, 248.

14. Sedmak, J.J. and Grossberg, S.E. (1977) *Anal. Biochem.*, **79**, 544.

15. Congdon, R.W., Muth, G.W. and Splittgerber, A.G. (1993) *Anal. Biochem.*, **213**, 407.

16. Butcher, E.C. and Lowry, O.H. (1976) *Anal. Biochem.*, **76**, 502.

17. Price, N.C. (1995) in *Enzymology Labfax* (P.C. Engel, ed.). BIOS Scientific Publishers, Oxford/Academic Press, San Diego, p. 34.

CHAPTER 5
SALTING OUT: AMMONIUM SULFATE PRECIPITATION
R. Scopes

One of the oldest methods of protein separation is by precipitation at high salt concentration. Many salts can be used, but rarely is any one better than ammonium sulfate, which has the advantages of high solubility, low heat of solution, low density of solution, and low cost.

The principle behind high salt precipitation ('salting out') is that the presence of salts greatly decreases the solvating power of water, by tying up many of the water molecules by interaction with the salt ions. Water molecules are removed from the protein's surface, and hydrophobic patches due to surface-located residues such as leucine, valine, phenylalanine, and so on, are exposed. These then interact with each other, causing aggregation of the proteins present. In any mixture of proteins, the aggregates will contain many different proteins interacting with each other, but some will aggregate more readily than others due to greater numbers of surface hydrophobic residues. As a result, some proteins precipitate at relatively low salt concentrations, while others require very high salt concentrations to form aggregates that can be centrifuged. Thus the method results in separation, with the more hydrophobic proteins precipitating first, and the least hydrophobic proteins precipitating last. To some extent there is an additional factor, that larger proteins precipitate earlier than small ones.

Salting out is a useful first step in a protein purification protocol. The quality of the starting material is not important; it can be cloudy, and have a variety of salts and buffers in it without greatly affecting the process. A purification factor of 3–5 should be achieved, with overall recovery of at least 70% of the protein you are aiming to purify. Alternatively, it can be used simply to concentrate a protein fraction; typically one would add ammonium sulfate at up to 600 g l^{-1} (about 85% saturation, see below) to precipitate out all the proteins present. Thus one can reduce one liter of a dilute solution to a few milliliters of concentrated protein. The redissolved precipitate then contains a good deal of ammonium sulfate, which may need to be removed before the next step in purification.

1. CALCULATION OF THE AMOUNT OF AMMONIUM SULFATE NEEDED

When adding any salt to an aqueous solution, there is likely to be a volume change. With ammonium sulfate this is approximately a 7.5% increase when the final salt concentration is 1 M. This must be taken into account when adding salt to a protein solution. Traditionally, ammonium sulfate concentrations have been expressed as '% saturation'. Saturated ammonium sulfate at 20°C is close enough to 4.0 M. Thus, 60% saturated ammonium sulfate is 2.4 M. It is *not* 60 g 100 ml^{-1}, or 60 g dissolved in 100 ml. In fact, to get from zero to 60%, you must add 39 g to every 100 ml of solution. The amount can be calculated from the following formula.

To take a solution from $S1\%$ saturation to $S2\%$ saturation, dissolve in solid ammonium sulfate, to every liter of solution:

$$\text{grams} = \frac{533 \times (S2 - S1)}{100 - 0.3 \times S2},$$ Equation 1

or, for V ml:

$$\text{grams} = \frac{0.533 \times V \times (S2 - S1)}{100 - 0.3 \times S2}.$$ Equation 2

It is easy to use a table to find how much ammonium sulfate to add, especially if the range is in round numbers, for example from 35 to 60%. *Table 1* gives the values in divisions of 5% from zero to 90% saturation. It should never be necessary to go above 90% because:

(i)   solutes already present will decrease the actual solubility of the ammonium sulfate;
(ii) there would most likely be some undissolved salt included in any protein precipitate formed;
(iii) the density of precipitated protein is so close to that of the solvent above 90% saturation that the precipitate may not be able to be centrifuged.

Ammonium sulfate fractionation should not normally be used on solutions that contain significant amounts of detergent (> 0.5%). The aggregated proteins interact with detergent molecules which also become insoluble, and because detergents mostly have a low density, the combined 'precipitate' will float rather than sink on centrifugation. In some circumstances this can be useful, especially if it removes the detergent while the target protein remains in solution, but the presence of detergent tends to reduce the differences between the behavior of the proteins, so there may not be much separation.

Table 1. Ammonium sulfate, grams to be added to 1 l of solution at $S1\%$ saturation, to take the saturation to $S2\%$

	Molarity																	
	0.5		1.0			1.5			2.0			2.5			3.0			3.5
S2:	5	10	15	20	25	30	35	40	45	50	55	60	65	70	75	80	85	90
S1: 0	27	55	84	113	144	176	208	242	277	314	351	390	430	472	516	561	608	657
5		27	56	85	115	146	179	212	246	282	319	357	397	439	481	526	572	621
10			28	57	86	117	149	182	216	251	287	325	364	405	447	491	537	584
15				28	58	88	119	151	185	219	255	292	331	371	413	456	501	548
20					29	59	89	121	154	188	223	260	298	337	378	421	465	511
25						29	60	91	123	157	191	227	265	304	344	386	429	475
30							30	61	92	126	160	195	232	270	309	351	393	438
35								30	62	94	128	163	199	236	275	316	358	402
40									31	63	96	130	166	202	241	281	322	365
45										31	64	97	132	169	206	245	286	329
50											32	65	99	135	172	210	250	292
55												33	66	101	138	175	215	256
60													33	67	103	140	179	219
65														34	69	105	143	183
70															34	70	107	146
75																35	72	110
80																	37	75
85																		37

CHAPTER 6
CHROMATOGRAPHIC PROCEDURES
R. Scopes

Chromatography

The two most frequently used techniques in protein chemistry are chromatography and electrophoresis. Whereas electrophoresis has been overwhelmingly used for analytical purposes, chromatography is used preparatively. The basic principle of column chromatography of proteins depends on the partitioning of the proteins between a solid phase (the adsorbent) and a liquid phase. In a column, the liquid passes around and diffuses into immobile solid particles of the adsorbent. Generally these are spherical particles, optimally of identical diameter; the chromatography takes place principally inside the particles, where all processes are governed by diffusion. Once outside the particle, the protein flows with the buffer down the column. Separation occurs because the individual protein components have differing adsorption behavior.

1. BASIC CHROMATOGRAPHIC THEORY

The chromatographic process consists of a succession of adsorption and desorption steps. After desorbing from the particle the protein moves down a small distance to the next adsorbing position (1). These individual steps are sometimes called 'plates', being compared with the successional plates in a fractional distillation column for volatile compounds. Each individual protein spends a certain proportion of its time adsorbed; the least adsorbed passes through the column quickly, and the most adsorbed will stay bound to the column. Thus only minor differences in adsorption behavior can result in complete separation because of the large number of steps (plates) that each protein is subjected to during chromatography (*Figure 1*).

Unfortunately most chromatography is not that simple, because in most protein mixtures the bulk of the components either do not adsorb at all, or adsorb so tightly that they cannot be removed under the starting conditions. Relatively few components can be eluted under isocratic conditions, that is without changing the buffer composition. However, quite slight changes to the buffer conditions can cause a totally adsorbed component to 'loosen', and become partially adsorbed, which means that it will start to move down the column. The normal practice is to apply a smooth gradient of conditions so that each component gradually becomes more weakly adsorbed, and starts to move down, at different stages of the gradient. Thus the commonest application of chromatography is first to choose buffer conditions that result in total adsorption of the protein component that you want, and then gradually release it by applying a gradient, causing separation from the other adsorbed components (*Figure 2*).

Theoretically ideal chromatography operates slowly enough for complete equilibration between adsorbing sites and the proteins. In practice this is not possible; diffusion effects must be minimized, and time of operation is important. In fact near theoretical results can be obtained at speeds which allow less than 90% equilibration, which with modern adsorbents means very rapid separation processes are possible. The optimum flow rate used depends on the bead diameter, in that the larger the bead, the slower the flow must be to allow for reasonable diffusion times. As bead size decreases, flow rate can be higher, but also flow resistance increases, as the inverse square of the bead diameter. Consequently the pressure required to operate at the optimum flow rate increases as the inverse cube of the bead diameter; pressure begins to become a significant

Figure 1. Progress of protein bands down an adsorption column under isocratic conditions. The mixture first binds to the top of the column (left), then one component slowly moves ahead of the other as they experience successive adsorption and desorption steps. Complete separation is achieved before they emerge from the column (right).

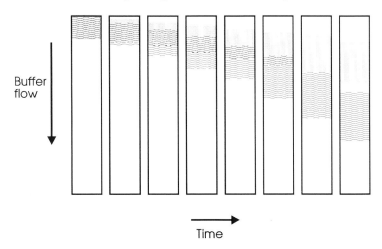

factor at below about 50 μm diameter beads. However, resolution is improved with smaller beads (the more beads that can be packed into a column, the better, up to around 10^9); this led to the development of 'High Performance' systems, generally known as HPLC (high-performance liquid chromatography), which can operate at many atmospheres of pressure. HPLC also introduced a new principle of 'reverse-phase' (RP) chromatography, which has been highly successful for the separation of small proteins and peptides (see Section 9).

Figure 2. Typical appearance of a chromatogram of a protein mixture on an ion exchange column. Unbound and partially retarded proteins are eluted before commencing the salt gradient.

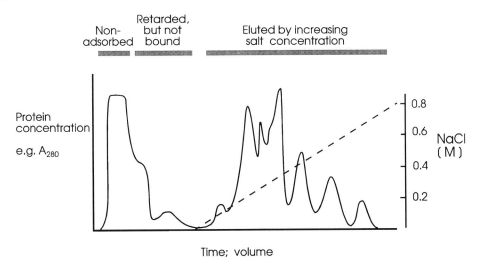

Apart from reverse-phase, there are a dozen or more adsorbent types. The main one is ion exchange, but many others are being used increasingly, and are discussed individually below. It should be appreciated that the principle of separation, the equipment used, and the scale of operation are not linked. HPLC systems can use the same principles of ion exchange, and so on, as are used in low-pressure systems. The latter are sometimes known as 'open columns', to indicate that samples can be applied to a column open to the atmosphere, under hydrostatic pressure only. HPLC does not necessarily achieve 'high performance' if the experimental design is inappropriate. On the other hand excellent results can be obtained with very simple apparatus, and the same basic principles can be used to purify one microgram or one gram of the same protein (though economics may dictate an alternative approach).

1.1. Chromatographic materials

The matrix from which the adsorbent is made has important requirements. First, it should preferably consist of spherical beads of relatively homogeneous size, readily porous for large proteins, yet sufficiently rigid to sustain pressure gradients, and have a high capacity for adsorbing proteins. Not all of the optimum parameters can be achieved in one material; a really open structure is not likely to be rigid enough to cope with high pressures; rigid materials have relatively low adsorptive capacity, and so on. There are many considerations to be taken into account, the main ones being the scale of the operation, and whether economics plays any part in the experimental design. For a one-off purification, the costs may not matter, but if it is to be a continuous commercial protein production, then obviously costs are vitally important. Small-scale column chromatography, using milligram to microgram amounts of protein, can make use of the most sophisticated high-performance systems to optimize resolution. Large-scale methods can achieve acceptable results with simpler and much cheaper adsorbents. The most commonly used adsorbent matrices for large-scale, low-pressure systems are agarose based; for RP-HPLC silica based; for high performance through the intermediate range, a wide variety of organic polymers and modified carbohydrate materials are used. Each manufacturer has a range for most operations.

Finally, some mention of alternative chromatographic-like procedures should be made. Several new materials are appearing; these include porous particles in which the liquid actually flows through the beads, and similar membrane structures (2,3). In both of these examples the adsorbing sites are adjacent to the flowing buffer, so diffusion times are negligible. This has enabled extremely fast separation times to be achieved, which can be advantageous in some circumstances.

2. ION-EXCHANGE CHROMATOGRAPHY

Although there is a wide variety of adsorbents to choose from, the ion-exchange principle depends on an electrostatic interaction between oppositely charged adsorbent and protein, and so there are only two basic types, the positive (anion-) and the negative (cation-) exchangers. Within these two types there are varieties which behave differently due to varying charge density, matrix type, accessibility of adsorption sites, and so on. There are 'weak' exchangers, containing groups which titrate (and so become uncharged) in a pH range that is likely to be used in the chromatography. The most notable of these are the DEAE- (diethylaminoethyl-) exchangers, which have in the past been the most used of all chromatographic media. More used these days are the 'strong' exchangers which remain fully charged throughout the pH range (*Table 1*).

The charge on a protein varies from positive at low pH to negative at high pH, and is zero at its isoelectric point which may be anything between pH 2.0 and 11.0, but is most commonly

Table 1. Ion exchanger types and nomenclature

Name	Ligand	pH range of titration of ligand
Anion exchangers		
DEAE-	Diethylaminoethyl-	6.0-9.5
TEAE-	Triethylaminoethyl-	Not applicable
QAE-	Trimethylaminoethyl- and	Not applicable
Q-	other quaternary amines	Not applicable
P-	Ampholytic polymers	Complete pH range available for chromatofocusing
Cation exchangers		
CM-	Carboxymethyl-	3.5–5.0
P-	Phospho-	5.5–7.5
SP-	Sulfopropyl-	Not applicable
S-	Sulfoethyl-	Not applicable

between 4.5 and 8.0 (*Figure 3*). With notable exceptions, the natural environment of most proteins is within the pH range 6.0 to 8.0, and proteins tend to be optimally stable at their normal physiological pH. Thus in the pH range 6.0 to 8.0, most proteins are stable, but may be either positively or negatively charged, depending on their isoelectric point. In fact, the majority of proteins tend to have isoelectric points below 7.0, especially proteins from bacteria and plants. In order to get a positive charge on a protein for cation exchange chromatography, it is necessary to go to a pH below the isoelectric point, and if this means going below about pH 5.5, there is a danger of inactivation in many cases. On the other hand, operating at pH 8.0 or slightly higher

Figure 3. Titration curves for a variety of proteins. Isoelectric points vary from 3.5 to 8.0. Note that there is relatively little change in charge between pH 7.0 and 9.0, due to the fact that few amino acids titrate in that pH range.

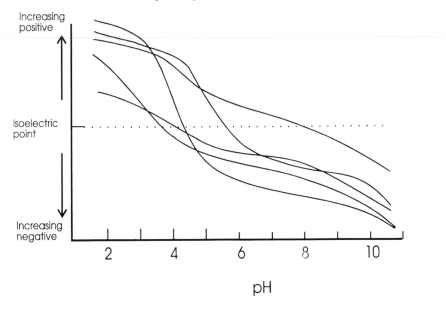

has been found to be relatively harmless in most cases; most proteins are negatively charged at this pH. Consequently most ion-exchange chromatography has been carried out at a slightly alkaline pH on anion exchangers such as DEAE- or the stronger quaternary nitrogen exchangers.

The size of column used depends on the amount of protein and the resolution required. Optimum separation is achieved with low loading, but if the component required behaves quite differently from other proteins present, then a much greater loading can be used without major cross-contamination. As a rough rule, a 100 ml ion-exchange column can easily accommodate up to 1 g of (adsorbed) protein, and will give optimum results with 100 mg. For smaller amounts than 100 mg, a smaller column should be used; indeed a *minimum* of 1 mg ml^{-1} of ion exchanger is a reasonable figure. Using much less than this can cause significant losses on the column, and highly dilute fractions which are eluted may be unstable.

2.1. Buffers for ion exchange
The buffer to be used should be selected carefully, as the behavior can critically depend on buffer conditions. This is particularly so when the required protein only binds weakly, for instance a protein of isoelectric point 6.5 on an anion exchanger in a pH 8.0 buffer. It is generally advisable to choose a buffer of the same charge characteristics as the exchanger. Thus for anion exchange, positively charged species should be used (e.g. diethanolamine, pH 9.0; Tris pH 8.0; imidazole pH 7.0; histidine pH 6.0), and for cation exchange negatively charged buffers (e.g. acetate pH 5.0; *N*-morpholinoethane sulfonate (Mes) pH 6.0; *N*-morpholinopropane sulfonate (Mops) pH 7.0; Tricine, pH 8.0). Preferably the acid and basic buffer species should have one and zero (or zero and one, respectively) charges, to maximize the buffering capacity for a given ionic strength (I). It is the ionic strength that determines how strong the interaction between protein and ion exchanger is: the lower the value of I the stronger the interaction. The buffering power, which is calculated from both the concentration of the buffer and the value of (pH$-$pK_a), should be *at least* 5 mM. That is, it would take 5 mM acid (or base) to titrate the buffer out of its range. This means a minimum of 10 mM buffer used at its pK_a, or more if the value (pH$-$pK_a) is not zero. The value of (pH$-$pK_a) should never be greater than 1.0; if it is 1.0, then the buffer should be at least 50 mM to control the pH satisfactorily. With a wide range of buffers available, it is possible to choose a buffer within 0.5 pH units of the value you wish to operate at.

It should be noted that the pH of certain buffers, especially those with an amine function, such as Tris, can be strongly affected by temperature. The pK_a decreases as the temperature rises. If it is critical that the pH be a precise value at the operating temperature, you should ensure that the pH is checked with the buffer at that temperature.

Table 2 gives a list of buffers suitable for use in ion-exchange chromatography. An ionic strength of 0.01 (10 mM) has been chosen as the minimum concentration advised. The buffers suggested have pK_a values at or close to the pH indicated; those marked * are within 0.2 units of the pK_a at 20°C. In order that the ionic strength is exactly defined, basic buffers are made by titrating 10 mM HCl to the required value with the buffer base, and acidic buffers by titrating 10 mM KOH with the buffer acid.

2.2. Operation of an ion-exchange column
(a) The column must be equilibrated with the starting buffer, either by washing at least 10 volumes through, or by first applying about one volume of 5 times concentrated buffer, followed by two to three volumes of correct strength buffer. As a final check, the pH of the eluting buffer should be the same as that being applied.
(b) The sample to be applied should be in exactly the same buffer conditions, if the adsorption is weak. This may be achieved either by prior dialysis or by gel filtration. The sample must be

Table 2. Buffers for use in ion-exchange chromatography

For anion exchange

pH 9.0	*20 mM Diethanolamine	Adjust 10 mM HCl to pH 9.0 with diethanolamine
pH 8.5	40 mM Tris	Adjust 10 mM HCl to pH 8.5 with Tris
pH 8.0	*20 mM Tris	Adjust 10 mM HCl to pH 8.0 with Tris
pH 7.5	*15 mM Triethanolamine	Adjust 10 mM HCl to pH 7.5 with triethanolamine
pH 7.0	*20 mM Imidazole[a]	Adjust 10 mM HCl to pH 7.0 with imidazole
pH 6.5	*20 mM Bis-Tris[b]	Adjust 10 mM HCl to pH 6.5 with Bis-Tris
pH 6.0	*20 mM Histidine[a]	Adjust 10 mM HCl to pH 6.0 with histidine (base)
pH 5.5[c]	*10 mM Piperazine	Adjust 15 mM HCl to pH 5.5 with piperazine

For cation exchange

pH 4.0	*20 mM Lactate	Adjust 10 mM KOH to pH 4.0 with lactic acid
pH 4.5	*30 mM Acetate	Adjust 10 mM KOH to pH 4.5 with acetic acid
pH 5.0	15 mM Acetate	Adjust 10 mM KOH to pH 5.0 with acetic acid
pH 5.5[d]	60 mM Mes	Adjust 10 mM KOH to pH 5.5 with Mes
pH 6.0	*25 mM Mes	Adjust 10 mM KOH to pH 6.0 with Mes
pH 6.5	15 mM Mes	Adjust 10 mM KOH to pH 6.5 with Mes
pH 7.0	*25 mM Mops	Adjust 10 mM KOH to pH 7.0 with Mops
pH 7.5	15 mM Mops	Adjust 10 mM KOH to pH 7.5 with Mops
pH 8.0	*20 mM Tricine	Adjust 10 mM KOH to pH 8.0 with Tricine

[a] Imidazole and histidine are efficient complexing agents for certain metal ions, especially Zn^{2+}. Metalloenzymes could be inactivated by removal of metal by these buffers.

[b] Bis-Tris is bis-(2-hydroxyethyl)imino-tris(hydroxymethyl)methane.

[c] This is a difficult pH at which to obtain a satisfactory buffer. Pyridine has the ideal properties, but is volatile and poisonous. The acidic species of piperazine at pH 5.5 is doubly charged, so the buffering power is less for the given ionic strength. The composition suggested has 5 mM buffering power, and an ionic strength of 0.025.

[d] pH 5.5 is a difficult value: if this precise pH is required with a low ionic strength, then this Mes buffer can be used; it should be noted that for large-scale operation, this concentration of Mes may be costly. If a starting ionic strength of over 0.02 can be tolerated, then an acetate buffer (e.g. >20 mM KOH to pH 5.5 with acetic acid) provides enough buffering power.

free of particulate material, and so should be pre-centrifuged, or filtered through a 0.4 μm membrane. The protein concentration in the sample should not be more than 10 mg ml^{-1}, and preferably not less than 1 mg ml^{-1}. It is best to concentrate large volumes of dilute sample by ultrafiltration before application to the column.

(c) After applying the sample, a starting buffer wash of several column volumes should be carried out to remove all unbound and weakly bound proteins. The gradient can then be applied.

(d) The gradient should not be too steep; aim to pass at least 10 column volumes through by the end of the process. Using high-performance systems with fast flow rates, as many as 100 column volumes may be used. Gradients are most commonly created using NaCl or KCl, but there can be circumstances when other salts give better results (e.g. $MgCl_2$, Na_2SO_4, K-phosphate). Gradients in pH are not commonly employed because of irregular behavior at low buffering power, but they can be useful if the total buffer concentration is high.

(e) After seeing how the protein component that you want has performed, it may be possible to take a few short cuts to shorten the process next time. If the protein elutes late, consider either running at a different pH (lower for anion exchange, higher for cation exchange), or using a higher ionic strength buffer to start with. These options will reduce the amount of protein adsorbing to the column in the first place, and may allow the use of a smaller column or a higher loading. Anion exchangers operate very successfully at pH levels below 7.0, but 'standard practice' has been to use a Tris buffer at pH 8.0, when most proteins bind. Alternatively, for a strongly binding protein, it may be possible to eliminate the step that gets the buffer conditions right; provided that the pH is correct, the sample could be applied in a quite different buffer to that being used on the column. If your protein does not bind at all, then there is the option of using the opposite charged exchanger.

Ion-exchange chromatography is a resolution-based technique; it does not aim to adsorb the proteins selectively (apart from the crude selection on the basis of charge), but achieves the main separation during the elution process. Much effort has gone into developing highly sophisticated ion exchange materials which give optimum resolution, and on a small scale these should be used whenever possible. Large-scale processing of proteins is generally a cost-sensitive exercise, and some compromises need to be made between optimum performance and capital outlay, as the more sophisticated adsorbents are also the most expensive. No general advice can be given, as each protein separation task is unique, and many trials may be needed before a satisfactory method is discovered.

2.3. Chromatofocusing—general principles

Chromatofocusing (4,5) is an extension of ion-exchange chromatography. The adsorbent relies on electrostatic interactions between the positively charged matrix and the negatively charged proteins. It is carried out at very low ionic strength, using polymeric zwitterionic buffer species. These buffers, although they have a high buffering power, do not contribute much to the overall ionic strength. Proteins are bound even at pH levels very close to their isoelectric points, as there is little to disrupt the electrostatic interactions. Because of the high buffering power, and a comparable buffering of the adsorbent, it has been possible to generate a pH gradient that is smooth; as the pH falls proteins become less negative, and eventually reach zero charge, at which point there is theoretically no interaction with the column. Because there is quite a steep pH gradient in the column, the displaced protein moves down to a higher pH point, where it acquires negative charges again and so is retarded. As the pH catches up with it again, it moves on. This results in a concentration of specific protein components in narrow bands on the column at or about their isoelectric points. The whole pH gradient moves down the column, and these bands are eluted as sharp, discrete protein components (*Figure 4*).

The requirements for chromatofocusing are:

(i) a chromatofocusing adsorbent;
(ii) polymeric buffers (ampholytes);
(iii) the protein you want must be soluble at its isoelectric point even at low ionic strength;
(iv) the protein you want must be stable at its isoelectric point.

Not all these things are necessarily known at the start. For optimum operation, all proteins in the sample should be soluble and stable at the pH levels that they are going to be exposed to; if not, precipitation may clog the column. Consequently, the technique is best suited to a late stage in purification, when few proteins are still present.

2.4. Operation of chromatofocusing

In general, the important property of your protein, namely its isoelectric point, is likely to be known at least roughly, since it is probable that you have already used ion exchange. A variety of ampholyte buffer combinations are available to cover different pH ranges; one should be used that starts at a pH at least one unit above the suspected isoelectric point, and goes to at least one unit below. The commercial buffers available from Pharmacia, who introduced chromatofocus-

Figure 4. Progress of a chromatofocusing column. The sample is applied at pH 8.0, and all protein components adsorb (left). As a pH 6.0 buffer is applied at the top of the column, a pH gradient forms, and gradually spreads down the column; the top quickly becomes pH 6.0. Individual proteins move down the column at a position corresponding to each one's isoelectric point, in the illustration about pH 7.5, 7.0, 6.5 and 6.3, respectively. One component with an isoelectric point below 6.0 remains on the top.

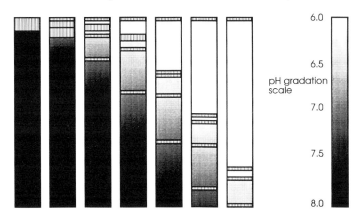

ing, cover the pH ranges 11.0–8.0, 9.0–6.0, and 7.0–4.0. The column is equilibrated with a high pH buffer. This may be the polymeric ampholyte at the top of its range, but it has been found that a simple buffer (such as listed above) can be used; this cuts down on the use of the expensive ampholytes. The protein sample, appropriately dialyzed into the same starting buffer, is then applied. Ampholyte buffer is adjusted with HCl to a suitable value below the expected isoelectric point of the desired protein, and then this is applied directly to the top of the column. No gradient device is required, because the gradient is self-establishing in the column. The more acidic form of the buffer gradually titrates the column (and the proteins adsorbed to it) starting from the top. The eluting buffer pH gradually decreases and individual protein components come out at close to their isoelectric point. It is claimed that similar proteins differing by as little as 0.2 pH units in isoelectric point can be totally resolved; this will depend partly on loading, higher resolution being achieved with lower loading.

The big advantage of chromatofocusing is in the focusing aspect. Protein bands are kept compact because a molecule in the rear of the band experiences a lower pH, interacts with the adsorbent more weakly, and so moves faster. Conversely at the front of the band the pH is a little higher, and these protein molecules bind more strongly. Similar arguments can be made for salt gradient elution in conventional ion-exchange chromatography, but the effect is not so sharp. The disadvantages are several; apart from possible problems of solubility and stability as indicated above, the removal of the polymeric ampholytes from the purified protein is not always simple. Despite the potential of high resolution and ease of operation, the technique has surprisingly not been used very extensively.

3. HYDROPHOBIC CHROMATOGRAPHY

Of the several properties of proteins that allow separation, hydrophobicity of the surface is also one that can be exploited using chromatography. Whereas in ion exchange opposite charges attract, in hydrophobic interaction chromatography (HIC), like surfaces interact, in particular the patches of hydrophobic residues that are exposed to the solvent. Amino acid side-chains such as phenylalanine, leucine, isoleucine, valine, and so on, are hydrophobic and 'prefer' to interact with other hydrophobic substances rather than water (*Figure 5*). These interactions are in fact weak in pure water, but strengthen as salt concentration increases, due to clustering of water molecules around salt ions, exposing the hydrophobic patches and allowing them to get together. This is the principle behind salting out precipitation in which proteins stick together; with HIC the interaction is between a protein and a hydrophobic ligand attached to the adsorbent matrix.

3.1. Hydrophobic interaction chromatography—practical aspects

The main hydrophobic adsorbents consist of aliphatic chains linked to the matrix by an uncharged oxygen atom; however, hydrophobics linked through nitrogen, with a positive charge, have also been used widely, as have sulfur-linked ligands (6). In addition, one of the most popular adsorbents has a benzene ring (ether-linked phenyl: Pharmacia Phenyl-Sepharose). The strength of interaction depends on:

(i) the hydrophobicity of the protein;
(ii) the hydrophobicity of the ligand;
(iii) the concentration and nature of the salt in the buffer;
(iv) pH and temperature.

It is variations in factor (i) that are being exploited. Factor (ii) can be selected, from weak butyl, through hexyl, to octyl and phenyl, with variations in between. In addition, different behavior has been found according to the chemistry of how the ligand is attached to the matrix (O-, N-, S-linked). Factor (iii) determines the strength of the interactions. Normally a salt such as Na_2SO_4 or $(NH_4)_2SO_4$ is used at 0.5 to 1.0 M concentration, in a suitable neutral buffer. However, the more hydrophobic proteins may adsorb without any additional salt. Binding is generally stronger at lower pH and at *higher* temperature. Elution is sometimes carried out in the cold after application at room temperature. For very strongly binding proteins, solutes that disrupt hydrophobic forces can be used, such as water-miscible alcohols (2-propanol is one of the best) and ethylene glycol (up to 50% v/v). However, it might be better to use a less hydrophobic adsorbent for such proteins. The usual procedure for elution of proteins that have been bound in the presence of salt is to apply a decreasing salt gradient, down to zero. This elutes in increasing order of hydrophobicity. It should be noted that in general, hydrophobic chromatography does not

Figure 5. Sketch of the surface of a typical protein, with positive and negative charges, and hydrophobic patches exposed to the solvent.

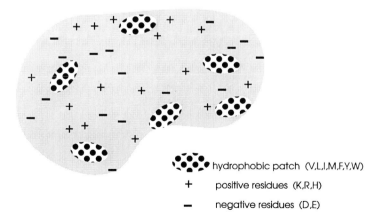

hydrophobic patch (V,L,I,M,F,Y,W)

+   positive residues (K,R,H)

−   negative residues (D,E)

produce the same sharp resolution as does ion exchange. This can largely be attributed to the fact that adjacent hydrophobic proteins have an affinity for each other, as well as the adsorbent. This is in contrast with ion exchange, in which adjacent proteins are of the same charge and so do not attract one another.

3.2. Operation of a hydrophobic interaction column

The conditions will vary according to the situation, but here it is assumed that the proteins are adsorbed in the presence of 1 M Na_2SO_4.

(a) Equilibrate the column with the chosen buffer, containing 1 M Na_2SO_4, using several column volumes.
(b) Prepare the sample. If it does already not have salt in, then sufficient should be added. If the sample is a re-dissolved fraction from an ammonium sulfate precipitation, then it will have salt present; it is a good plan to try to calculate the amount. For instance, suppose that the ammonium sulfate precipitate was formed at 75% saturation, that is 3 M. You use 20 ml of buffer to re-dissolve the precipitate, and find that the volume is now 25 ml. This means that there is the equivalent of 5 ml of 3 M ammonium sulfate present in the 25 ml, which works out at 0.6 M final concentration. Since the effects of Na_2SO_4 and $(NH_4)_2SO_4$ are about the same in HIC, you now have to add a further 0.4 M Na_2SO_4 (or $(NH_4)_2SO_4$) to bring the total salt concentration to 1 M. After centrifuging to ensure clarity, this can be applied directly to the column.

Note that not all salts are equivalent. Chlorides are much less effective than sulfate or phosphate; thus, 1 M NaCl will not cause as much hydrophobic interaction as even 0.5 M Na_2SO_4.
(c) After washing in the sample with about one column volume of starting buffer (1 M salt), a gradient to zero salt is applied. This should aim to reach zero in about five column volumes, and be followed with a further two column volumes of salt-free buffer. If your protein is not eluted by this stage, a second gradient, using ethylene glycol can be applied. However, it is usually better to do this stepwise, since little separation of individual proteins can be expected at this stage. So a direct application of 30% (v/v) ethylene glycol in the buffer (one column volume) should be made, followed, if necessary, by a further application of 50% (v/v) ethylene glycol. When resolution is poor, and the peaks are spread out in a gradient, stepwise elution is often a better option. This keeps the active fraction in a small volume (*Figure 6*). Alternatively, a very steep gradient (e.g. in less than two column volumes) can achieve the same result.

Hydrophobic chromatography is an excellent partner to ion exchange, since each exploits a completely different property of proteins. There are more variables to try with HIC, since there are many different adsorbent types, and experimentation with different salts in elution schemes can be productive (7). Resolution is not so good, but selectivity can be better than ion exchange.

3.3. Reverse-phase chromatography

Reverse-phase (RP) chromatography is an alternative method of hydrophobic chromatography that involves the use of organic solvents in the buffers. It has been developed exclusively with HPLC equipment, and to many people, HPLC *is* RP chromatography. Because organic solvents (and the acids that are often used) may cause denaturation of proteins, it is most useful for small, sturdy proteins that can resist these solvents, and in particular for peptide separation, for which more details are given in Section 9. It is also useful if recovery of bioactivity is not an objective.

The adsorbents used are, as in HIC, aliphatic chains attached to a matrix, usually made of silica. They are mostly longer chains than in HIC, typically either 8 or 18 carbons long. A typical

Figure 6. Typical appearance of a chromatogram of a hydrophobic column. Some proteins are not bound, others are eluted by the decreasing salt, and a final stepwise addition of ethylene glycol removes the most hydrophobic components.

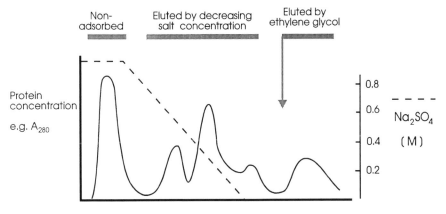

Time; volume

operation is to prepare the sample in 0.1% (v/v) trifluoracetic acid; this causes opening up of the protein structures, allowing the internal hydrophobic residues to be exposed to the adsorbent. Practically all proteins bind to the adsorbent under these conditions, and separation depends on the elution scheme. This uses an increasing gradient in organic solvent, generally acetonitrile. As the proteins elute, they regain their native structure, after being effectively denatured on the adsorbent. But some proteins, especially larger ones, do not regain bioactivity. Resolution is extremely sharp for small proteins, but larger ones (greater than about 30 kDa) are retained until late in the gradient, at which point the organic solvent concentration may prevent refolding into a native state, and cause spreading of the peak.

RP-HPLC is principally used for isolation of small amounts of proteins and peptides for amino acid analysis, sequencing and mass spectrometry (see Chapters 16 and 23). It is not a method widely used for preparation of significant amounts of bioactive proteins. Some attempts have been made to scale up, using organic solvent elution from hydrophobic columns, but the denaturing influence of the solvent is a major disadvantage. For larger scale work HIC on more weakly hydrophobic adsorbents is preferable.

4. AFFINITY CHROMATOGRAPHY

Affinity chromatography is the term used for chromatography that is based on highly specific interactions between a protein (or other biomolecule) and the immobilized ligand. The ligands range from antibodies and other proteins that naturally combine together, through enzyme substrate molecules, to generalized ligands which are not biological molecules themselves, but which have a strong interaction with certain protein types. The basis of affinity chromatography is selectivity *at the adsorption stage*, in contrast with ion-exchange and reverse-phase techniques in which selectivity is achieved through resolution during the elution step. However, affinity principles can be applied during elution, as described below.

4.1. Affinity chromatography theory
The steps in an affinity chromatographic procedure are illustrated in *Figure 7*. Usually a large

Figure 7. Stages in affinity chromatography. Most workers use pre-synthesized adsorbents, made as illustrated at top. The immobilized ligand recognizes only the target proteins with binding sites for that ligand.

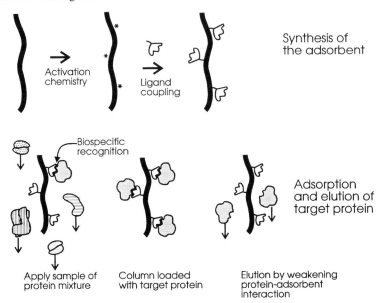

amount of sample is applied, since with high selectivity only the target protein should bind to the adsorbent (in practice there may be nonspecific interactions making some contaminating proteins bind also). After washing the adsorbent, the bound component must be released by a change in the buffer. This change will depend on the nature of the protein–ligand interaction, and some general examples are given in *Table 3*.

The theory behind affinity adsorbents is not as simple as it might appear. There are numerous problems and reasons why a carefully designed adsorbent does not work as expected. These include steric problems, and lack of sufficiently tight binding between the ligand and target protein. The introduction of 'spacer arms' which ensure that the ligand is well away from the base matrix has overcome many such problems. In many cases the reason for the success has been not so much the steric location of the ligand, but instead the introduction of nonspecific hydrophobic interactions with the aliphatic chains used, thereby increasing the overall strength of binding of the target protein (8).

4.2. Activation of the matrix

There are numerous 'group' affinity adsorbents available commercially, which interact with certain groups or classes of proteins. These include immunoglobulins (Proteins A and G), glycoproteins (lectins), dehydrogenases (5'-AMP, Cibacron Blue), and so on (see *Table 3*). Otherwise, it might be necessary to synthesize one's own affinity adsorbent. 'Activated' matrices, which react readily with simple ligands and with proteins, can be purchased, to prepare an affinity adsorbent. Alternatively, it is relatively simple (with a few skills in organic chemistry) to activate the base matrix oneself, using one of a variety of chemistries (*Table 4*). However, it may take considerably more effort to synthesize a ligand that is not commercially available.

Table 3. A selection of some commercially available group affinity adsorbents

Ligand	Protein group and comments
Protein A	Immunoglobulins. Not all immunoglobulin types bind
Protein G	Immunoglobulins. Better for nonhuman immunoglobulins
Lectins (various)	Glycoproteins. Bind to carbohydrate moiety
Avidin	Biotin-containing proteins
Glutathione	Glutathione-S-transferase fusion recombinant proteins
Amylose	Maltose-binding protein fusion recombinant proteins
Ni-IDA	Polyhistidine-fusion recombinant proteins (IMAC)
5′-AMP	ATP and NAD enzymes
2′,5′-ADP	NADP enzymes
3′,5′-ADP	CoA enzymes
5′-ATP	ATPases, chaperone proteins
Benzamidine	Serine proteases
Heparin	Nucleic acid-binding proteins and enzymes
Cibacron Blue	ATP and NAD enzymes (and other proteins)

Table 4. Some activation chemistries for affinity chromatography and for protein immobilization

Chemical	Spacer atoms[a]	Linkage[b]	Stability
Cyanogen bromide (9)	1	Iso urea	Moderate; poor at high pH
Epichlorhydrin (10)	3	Secondary amine	Very good
Bisoxirane (11)	12	Secondary amine	Very good
Carbonyldiimidazole (12)	1	Amide	Good, except at high pH
N-Hydroxysuccinimide (13)	1–8	Amide	Good, except at high pH
Divinyl sulfone (14)	5	Secondary amine	Good

[a] Number of atoms between the oxygen on the matrix and the nitrogen of the ligand. With N-hydroxysuccinimide, the length of spacer depends on the method of synthesis. All of these numbers can be increased by inclusion of an additional spacer arm.
[b] Assuming that ligand is a primary amine. All except the amide linkages have a positive charge at neutral pH.

4.3. Operation of affinity chromatography

This is not a section in which a general operating procedure can be given, as the behavior of different affinity adsorbents varies considerably. The position in a purification scheme can also vary. If the adsorbent is highly selective, it can be used at the first step, passing crude, clarified extract through the column. On the other hand, expensive adsorbents which become fouled by crude extracts are best used at a later stage, perhaps the last step, when there is only a little protein left. The main principles to observe are as follows.

(i) Adsorption should occur under buffer conditions which maximize the affinity interaction; for example an enzyme may not recognize its ligand at a pH outside its normal operating range. The exact composition of the buffers used may not matter much provided that the interaction is specific and strong.

(ii) If the interaction is weak, then the target protein may elute with the starting buffer wash. In

Chromatography

this case one should aim to apply the sample in as small a volume as possible, keep the flow rate down, and monitor the wash fractions carefully. This advice would apply to any column chromatography with weak interactions.

(iii) To elute strongly binding proteins, the interaction with the adsorbent must be weakened, and some thought should be put into how to do this. In many cases, an increase in salt concentration or a small change in pH is sufficient to loosen binding between protein and small ligand, such as a nucleotide. But with protein–protein interactions such as those occurring between antibody and antigen or Protein A/G, more vigorous conditions are needed. Generally antibody interactions are weakened using a low pH buffer that partially denatures the antibody, destroying the specific interactions. Fortunately this partial denaturation is reversible. Buffers used include glycine or citrate at pH values between 2.5 and 3.0. Sometimes even more vigorous methods are used including chaotropic agents such as guanidine hydrochloride. It is best to use the buffer recommended by the manufacturer. Of course, if your process is purifying a protein *antigen* on an antibody column, your protein must be able to survive these denaturing conditions as well as the antibody.

If you have designed your own adsorbent, or have a novel application, then it will be necessary to do some trials for optimum elution. Unless the affinity adsorbent has exquisitely high selectivity, there will be other unwanted proteins present. So if you can get the target protein off without these contaminants, it is worth the effort. Stepwise elution, rather than a gradient, is generally more satisfactory. Better still, an 'affinity elution' procedure can be used. In this case, the target protein is displaced by a specific compound, usually related either to the natural ligand immobilized on the column, or the interacting portion of the protein; this procedure does not apply to protein–antibody interactions. Thus glycoproteins can be eluted from lectin columns with the carbohydrate that the lectin recognizes; dehydrogenases can be eluted from 5'-AMP or from dyes using NAD^+, NADH, and so on. In the latter case, and similar examples, in order to minimize the amount of expensive substrate used, some pre-conditioning of the column should take place. A buffer is used that is just unable to elute the target protein by itself, but it weakens the binding. Then, only a low concentration of substrate (say 10 times its dissociation constant) is needed to elute the protein specifically.

4.4. Recombinant proteins

Affinity chromatography is a major method of choice for the purification of recombinant proteins. Many eukaryotic proteins are either poorly expressed, or only expressed in insoluble form in *Escherichia coli*. By constructing a fusion with a natural *E. coli* protein, much higher expression levels are obtained, and often in a soluble form (see Chapter 11). The other major benefit is that the purification from host proteins is very easy using affinity chromatography. Some of the more successful systems include fusion to the glutathione-*S*-transferase protein, which binds to a glutathione adsorbent; to maltose-binding protein, which binds to an amylose adsorbent; Protein A fusion which binds to immobilized immunoglobulin, or an antibody to Protein A; and a fusion which includes a biotinylation recognition site, so that the recombinant product is biotinylated in the *E. coli*, and will bind to an avidin column. In the latter case the problem is to weaken the biotin–avidin interaction, but if bioactivity is not required, the elution can be accomplished using sodium dodecyl sulfate. A list of some commonly used fusion protein systems is given in *Table 5*; however, it should be noted that new products are being introduced fairly frequently.

4.5. Affinity elution techniques

As indicated above, elution from an affinity adsorbent can be achieved using a specific method involving a substance related to one of the interacting components. By adding a ligand that interacts specifically with the adsorbed target protein, elution can be achieved even if the

Table 5. Some recombinant-fusion proteins designed for affinity purification

Fusion with	Affinity adsorbent	Elution with
Poly Arg (15)	Cation exchanger	Salt
Glutathione-S-transferase (16)	Glutathione	Glutathione
Protein A (17)	Immunoglobulin G	Cleave fusion
Maltose-binding protein (18)	Amylose	Maltose
Poly His (19)	Ni-IMAC	Histidine buffer

For further details see Chapter 11.

adsorption is not in itself specific. This can occur if the binding of the ligand to the protein causes some change which reduces the interaction between the protein and the adsorbent. The change may be conformational, or a blocking of access between the protein's surface and the adsorbent. Some of the best examples of elution with a specific ligand are when using a negatively charged enzyme substrate, which on binding to the enzyme, decreases the overall positive charge which was causing adsorption to a cation exchanger (*Figure 8*) (20). Other important uses of this technique, called variously 'affinity elution', or 'biospecific elution', are when using biomimetic, or pseudo-affinity adsorbents such as dyes (see Section 5).

Affinity elution, whatever the adsorbent, depends on the extent of interaction between target protein and the added ligand in the elution buffer. If the dissociation constant between these two is known, then at that concentration, half of the protein *that is not adsorbed* will have the ligand bound. Except in the case of nonspecific adsorption, the protein cannot interact with the adsorbent and the free ligand at the same time. Thus it is important, when attempting an affinity elution, that the buffer conditions are such that a proportion of the target protein is not adsorbed at any instant, so that there can be some interaction with the ligand (*Figure 9*). Then, by adding ligand at a concentration significantly higher than the dissociation constant, the protein is 'pulled off' the adsorbent by mass-action equilibrium.

So for affinity elution to succeed, when the ligand is added to the elution buffer, the target protein should already be partially desorbed, and slowly moving down the column. After applying the sample, it will often be necessary to find these conditions by trial-and-error. For instance, assume

Figure 8. Principle of affinity elution from a cation exchanger. On binding of a negatively charged ligand, the protein becomes less positive overall, and so its interaction with the cation exchanger (represented by the minus signs) is weakened.

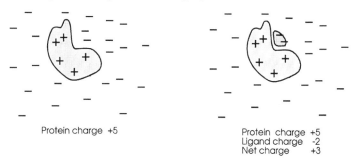

Protein charge +5

Protein charge +5
Ligand charge −2
Net charge +3

Figure 9. Principle of affinity elution from an affinity adsorbent. For the free ligand to bind to the target protein, there must be some free in solution: K_{ads} (bound/free) should be in the range 10–30 for optimum operation, and the concentration of free ligand should be at least $10 \times K_d$.

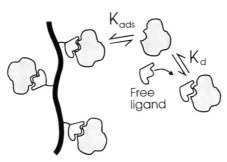

that application to a cation exchanger at pH 6.0 adsorbs all of the target protein strongly. Adding ligand to the buffer does not make any difference, because there is no protein free to bind with it. Raising the pH to 7.0 decreases the overall positive charge on the protein (*Figure 8*), so that it is now only weakly adsorbed, and a small proportion, say 10%, is free in solution. In this case, the protein will be moving down the column at 10% of the rate of buffer flow. Addition of negative ligand further decreases the overall charge, and the protein–ligand complex has little interaction with the exchanger, and so runs down the column with the buffer. Similarly with a dye adsorbent, raising either or both of pH and ionic strength weakens the interaction, allowing ligand to bind to the protein. Elution from affinity adsorbents is generally an 'on-off' process, using stepwise changes, rather than gradients.

5. DYE–LIGAND CHROMATOGRAPHY

Dye adsorbents result from the serendipitous discovery that many textile dyes bind strongly to proteins, and especially enzymes that bind nucleotides (21). Most dyes have in fact been designed to stick to biological materials such as proteins and cellulose. The interaction may be specific at the ligand binding site, or relatively nonspecific at various spots on the protein surface. By far the most extensively used dye is known as Cibacron Blue 3GA, Reactive Blue 2, and other pseudonyms. This was the dye that was originally found to interact with enzymes that had ATP- or NAD-binding sites, and was shown to bind competitively with these nucleotides. Since then, many other dyes have been used, since a great deal of specificity is shown between the dyes. In fact, some bind exclusively one protein out of a crude mixture.

5.1. Selection of dye adsorbent

Commercially available dye adsorbents start with Cibacron Blue and in many cases end there. Ranges of 5 to 10 dye adsorbents are available from Amicon, Sigma and Affinity Chromatography Ltd. The adsorbents in their simplest form are not difficult to synthesize, since it requires only a base matrix, usually agarose, and stirring this with the textile dye for a period of hours to days, depending on dye type (22). The dyes can be purchased from local distributors. They are in the category of 'reactive dyes', which form a covalent bond with the textile, or in this case with the agarose. Consequently one can accumulate a personal range of a large number of dye adsorbents. However, the questionable purity, batch variation and aging of stock dye powders can lead to variable results, and a more standardized set is advisable. Those available from Affinity Chromatography Ltd. (MIMETIC™ AFFINITY LIGANDS) are prepared in carefully controlled conditions from purified dyes, and represent the top of the range in quality. Sets of 40 dyes at two levels of substitution are also available from the Centre for Protein and Enzyme Technology, La Trobe University, Australia.

To choose the ideal dye, a screening process must be carried out (23). This can be done in one of several ways. Pre-packed columns with 1–2 ml of adsorbent can be used; in this case at least 5 mg, and preferably up to 50 mg, of protein in each sample is needed for each column. After equilibrating the columns with starting buffer, the sample is applied, and washed in with at least three column volumes of buffer. A trial stepwise elution is then made with 1 M NaCl. The nonadsorbed and the eluted fractions are then tested for the target protein. When suitable adsorbents have been found, the protein content of the eluted fractions may be compared, to select the adsorbent that gave the best specific activity (with acceptable recovery).

This is satisfactory when using a crude extract of readily available material, but is not so good at a later stage of purification when the amount of protein available is small. An alternative is to try a batchwise procedure. A small amount of drained adsorbent, about 20–40 mg, is weighed into the tubes, and 1 ml of the starting buffer mixed in; after a brief centrifugation this is decanted (*Figure 10*). Up to 1 ml of sample containing between 0.5 and 5.0 mg of protein is then added, shaken for 10–20 min, then the supernatant collected. The adsorbent can then be eluted with 0.5 ml of 1 M NaCl, and both supernatant and NaCl eluate analyzed for the target protein.

Other screening procedures using 96-well plates have been described, in which case a very large number of adsorbents are needed (24). In addition, it is possible to make very small columns containing only 100–200 μl out of pipette tips for small-scale work.

The aim of the screening process is to select an adsorbent that binds as little of the total protein as possible, while binding all of the target protein. It should also be eluted with high specific activity, and high % recovery.

A shortlist of some frequently used dyes is given in *Table 6*.

Figure 10. Batchwise screening of dye–ligand adsorbents. A small amount (20–40 mg damp weight) of each adsorbent is weighed into microfuge tubes, a sample containing 0.2–2.0 mg of protein added, shaken for 20–30 min, and the supernatant assayed for the target protein and total protein. A suitable adsorbent is chosen on the basis of these results.

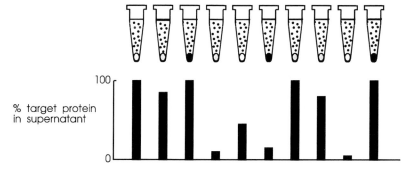

Table 6. Some dyes commonly used in dye–ligand chromatography[a]

Cibacron Blue 3GA [2]	Remazol Yellow GGL
Procion Blue MX-3G	Procion Yellow H-5G
Procion Blue MX-R [4]	Procion Yellow H-A [3]
Procion Blue H-EG(N) [187]	Procion Yellow MX-4R [14]
Remazol Brilliant Blue R	Procion Yellow MX-6G [1]
	Cibacron Yellow F-3R
Cibacron Red 3BA [4]	
Procion Red H-3B	Procion Brown H-2G
Procion Red H-E3B [120]	Procion Turquoise H-A
Procion Red MX-5B [2]	Procion Green H-4G [5]
Procion Scarlet H-3G	Procion Green H-E4BD [19]

[a] By far the most frequently used dye is Cibacron Blue 3GA (Ciba-Geigy), also known as Reactive Blue 2, and several other trade names. Unless otherwise stated, all commercially available 'Blue' adsorbents use this dye. The most commonly used 'Red' dye is Procion H-E3B. Since most of the work using multiple dyes has been with the ICI products (trade name Procion), these are over-represented in this list; many dyes by other manufacturers can be used, and in several cases the functional chromophore is identical with a Procion dye. Remazol dyes are manufactured by Hoechst. Where known, the alternative classification of dyes as 'Reactive <color><number>' is indicated in square brackets (e.g. Procion Yellow H-A [3] is also known as Reactive Yellow 3).

5.2. Two-adsorbent procedures—differential chromatography

Although binding can be very specific, in many cases several proteins can bind to a particular dye. An investigation of a range of 50 dyes using a bacterial extract in a pH 6.0 buffer designed for optimum adsorption (see below) found that the amount of protein bound varied from less than 10% to over 90%, with an average of around 40%. There was clearly little specificity in binding to those dyes which bound more than about 25% of the applied protein from a complex mixture. However, the nature of those proteins which bound tended to be consistent, with the 'most sticky' proteins binding to all the dyes, and the least sticky to none. If the target protein was in the middle range of 'stickiness', it was possible to find dyes which bound similar amounts of protein in general, but one of these would bind the target protein totally, and the other one not bind it at all. This led to the development of differential column chromatography (or tandem columns), in which the first adsorbent does not bind the target protein, and the second one does (25). However, the first column does bind many proteins that otherwise would have bound to the second column (*Figure 11*). Thus there is a high degree of selectivity achieved at the adsorption stage, which is a prime consideration in affinity chromatography. Dye–ligand can be called pseudo-affinity chromatography as a result of its selectivity, even if it does not involve interaction through a natural ligand-binding site. When the interaction is through the ligand-binding site, the dye is mimicking the natural ligand, and the term 'biomimetic' has been coined to describe such materials.

5.3. Buffers for use in dye–ligand chromatography

The buffer composition is important, but not as important as in ion-exchange chromatography. A number of factors can lead to increased binding to a dye. These are:

(i)   low ionic strength;
(ii) low pH;
(iii) presence of divalent metal ions;

Figure 11. Principle of differential tandem column chromatography. Sample passes through two or more columns. The first is called the 'negative' adsorbent; it removes unwanted proteins, but the target protein passes through. The second column contains the 'positive' adsorbent, which binds the target protein. With the correct choice of adsorbents, few proteins bind to the positive column, thereby achieving a high degree of purification at the adsorption stage.

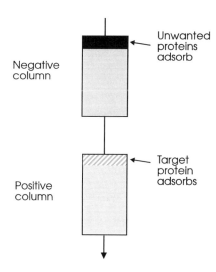

(iv) lack of divalent anions;
(v) low protein loading.

Factors which do not make much difference include temperature and the presence of neutral solutes and miscible solvents (e.g. urea at 1 M; 20% v/v ethanol; 20% v/v glycerol). *Table 7* gives some suggested buffers, but most others can be used.

Elution can be carried out by a variety of processes, especially using biospecific affinity elution. Mostly, dye adsorbents are eluted by stepwise rather than gradient procedures, but gradients can be particularly useful when both salt concentration and pH are increased simultaneously. Apart from biospecific methods, the commonest procedure is to use salt (NaCl up to 1.5 M), increased pH up to 9.0, or switching to high pH phosphate buffer. In some cases where the dye–protein interaction has a hydrophobic character, chaotropic salts such as NaBr or NaSCN have been found to be efficient at eluting proteins.

The principal problems with dye–ligand chromatography are the limited availability of a range of adsorbents, variable quality of the dyes used, and sometimes leakage of the dye. The latter has probably been exaggerated as a problem specific to dyes, because the leakage is so readily detected! The better quality materials have reduced this problem to negligible amounts. Another operational point that should be mentioned is the dependence of reproducible results on the loading of the columns. The amounts and types of protein binding are much more dependent on the ratio of protein applied to adsorbent amount than with other adsorbents. Thus underloading can result in more difficulty in eluting; overloading of course may result in not all of the target protein binding. So this point should not be overlooked; if you have less protein than you expected, make the column smaller.

CHROMATOGRAPHIC PROCEDURES

Table 7. Buffers for use in dye–ligand chromatography

Maximize binding
20 mM K-Mes pH 6.0 + 5 mM $MnCl_2$
20 mM K-Mes pH 6.0 + 5 mM $MgCl_2$
20 mM K-Mes pH 6.5
20 mM K-phosphate pH 6.5
50 mM K-phosphate pH 7.0
50 mM Tris–Cl pH 8.0

Minimize binding
100 mM K-phosphate pH 8.0

Elution schemes
(a) Increase pH, e.g. in 0.5 pH unit steps, up to pH 9.0
(b) Increase I up to 1.5 M NaCl
(c) Increase phosphate, up to 0.2 M
(d) Use chaotropic salts such as NaSCN, NaBr, LiBr in place of NaCl

Or combinations of the above

A gradient suggestion: apply in low I buffer, pH 6.0; gradient with 0.2 M phosphate + 0.5 M NaCl, pH 8.0.

6. IMMOBILIZED METAL AFFINITY CHROMATOGRAPHY

Immobilized metal affinity chromatography (IMAC), involves utilizing the strong interaction between certain amino acid residues in proteins, notably histidines, and metal ions that are immobilized on an adsorbent matrix (26).

6.1. IMAC principles and practice

The adsorbent consists of a metal ion-complexing agent attached to the matrix in such a way that when metal ions are introduced, they are complexed leaving part of the coordination sphere free, or rather, occupied by water molecules. These metals are then searching for alternative ligands to complete their coordination, and residues in proteins may be suitable. In particular it has been found that surface histidine residues, and to a lesser extent cysteines and tryptophans, are mainly responsible for the interactions. This is not a true affinity technique, since most proteins have histidines, and there is no expectation that the interaction will be at a natural ligand binding site. In particular it should be noted that the method is not especially useful for metalloproteins, since such proteins either will already have their metal bound, or if in an apo-form, may require to coordinate the metal totally. Indeed, apo-metalloproteins have been known to extract the metal from an IMAC column rather than be adsorbed. The interaction between the protein and the metal is not upset by salts, and in fact it is advisable to operate in a high salt mode (e.g. 0.5 to 1 M NaCl) to prevent any ion-exchange effects.

There are many different metal ions that can be used; the most widely used are Zn^{2+}, Ni^{2+}, Cu^{2+}; more recently Fe^{3+} has found uses for phosphoproteins. To start the process, the adsorbent contains free ligand, and must be charged with metal ions. A solution of 50 mM metal salt is passed through until the whole column is charged – in many cases this can be observed by color. The column is washed with starting buffer, the sample applied, washed in and an elution scheme

Table 8. Buffer systems for IMAC

Starting buffer	Eluting buffer [a]
20 mM K-phosphate + 0.5 M NaCl, pH 7.0	50 mM Histidine-Cl + 0.5 M NaCl, pH 5.5
50 mM K-phosphate + 0.5 M NaCl, pH 8.0	50 mM KH_2PO_4 + 0.5 M NaCl, pH 4.5
20 mM K-phosphate + 1.0 M NaCl, pH 7.0	20 mM K-phosphate + 1 M NH_4Cl, pH 7.0
20 mM Tris–Cl + 0.5 M NaCl, pH 8.0	20 mM Tris–Cl + 0.5 M NaCl + 0.2 M glycine, pH 8.0
20 mM K-phosphate + 1 M NaCl, pH 7.0	20 mM K-phosphate + 1 M NaCl + 20 mM imidazole, pH 7.0

[a] Typically a gradient of 10 column volumes or more is used, mixing the starting buffer with the eluting buffer. In some cases a simple stepwise change to the eluting buffer is appropriate.

commenced. Because some eluting buffers also cause some leakage of metal ions, it can be a good idea to have some adsorbent that has not been charged with the metal on the bottom of the column to trap this leakage.

The buffers used are all-important. Only uncharged histidines will coordinate with the metal ion, so the pH for adsorption should be above 7.0. Elution can be with a gradient of decreasing pH, or with a gradient of imidazole or histidine buffer. Also useful is a gradient of a weaker metal coordinating buffer such as one containing ammonium or glycinate ions. Some examples are given in *Table 8*.

7. GEL FILTRATION

All of the methods of chromatography in Sections 2–6 involve adsorption and desorption. Gel filtration chromatography differs in that there is no adsorption; the proteins remain dissolved in solution at all times. The principle is that the column material consists of bead particles which are only partially penetrable by proteins; large proteins not at all. The pore sizes cover a range which results in very small proteins and other low molecular mass materials penetrating throughout the beads, middle-sized proteins can find their way only to parts of the beads, and very large proteins cannot get into the beads at all (*Figure 12*). The latter, excluded from the beads, pass down the column with the flowing buffer, and emerge in a fraction at about 25–35% of the column volume. This is called the void volume, the amount that is outside the beads. Small molecules penetrate throughout, and so spend a proportion of their time not moving down with the buffer (the buffer does not move inside the beads except by diffusion). Consequently, they emerge at a late stage, at 90–100% of the total volume of the column. Once a complete column volume has passed through, the chromatography is over. Medium-sized molecules emerge at between 35 and 90% of the column volume. The separation is strictly on the basis of size, or more correctly Stokes radius (the radius of an equivalent spherical molecule). The flow rate should allow near equilibrium to be reached as the proteins pass in and out of the beads.

7.1. Gel filtration: operating conditions for separating proteins

Each component will emerge in a volume considerably greater than the applied volume, because of diffusion and chromatographic spreading. So it is important that the applied sample is small,

Figure 12. Diagram to represent the separation of proteins on the basis of size by a gel filtration bead. The larger proteins can only access the few large pores in the bead, whereas the smallest proteins can penetrate into most areas. Black dots represent small buffer molecules which spread evenly throughout the beads.

Outside   Inside

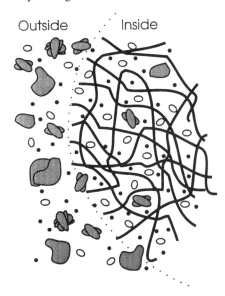

preferably between 1 and 5% of the column volume. Emerging fractions will be spread into a volume of about 3–15% of the column, depending on extent of retardation and the applied volume.

Details of the buffers used in the column are not important. Buffers are chosen so that the proteins remain stable and soluble, and protein–protein interactions should be minimized. Gel filtration is commonly used both to separate proteins and to change the buffer that they are in, because the buffer in which the protein is eluted is that which the column has been pre-equilibrated with. Similarly it is not important what the applied sample is dissolved in. It is common to precipitate the sample with ammonium sulfate, and redissolve in a small volume of buffer; this is a very effective way of concentrating the sample (see Section 8). A couple of warnings, however; gels that swell and contract may do so on adding a high concentration of protein, through osmotic effects, and if the sample is a redissolved ammonium sulfate precipitate, make sure it is well redissolved – add a little extra buffer to be sure, because precipitation effects on contact with the gel medium can occur at very high salt levels. Flow rate should be restricted to the lower value that the manufacturer recommends, to ensure sufficient equilibrium is achieved to optimize separation.

7.2. Gel filtration to separate proteins from low molecular mass compounds

A frequent requirement in a protein purification protocol is to remove low molecular mass salts and other compounds. This can be achieved rapidly using gel filtration on a medium that does not allow even small proteins to enter the beads. The most commonly used is Sephadex G-25 (Pharmacia), which excludes proteins totally (molecules above about 5000 Da; polypeptides smaller than this cannot really be considered as proteins). The proteins elute around the 25–35% column volume mark, but small molecules and ions penetrate the beads and elute around the 90% column volume mark. By applying a relatively large sample, the elution spreads, but does not mix, up to an applied sample of about 20% of the column volume. Protein then elutes between 25 and 60%, and the salts start at about 70%, giving a sufficiently complete separation. Thus a 1 l column can completely desalt 200 ml of sample, and it should not take more than about 30 min; the column should have a height about five times its diameter.

Gel filtration is a gentle technique, and quite rapid even on a large scale (using modern materials). For protein separation it is limited in terms of the amount of sample it can handle (compared with similar-sized adsorption columns), and is usually best applied near or at the end of a purification procedure. As a final step, the shape of the eluted fraction on a chart is a fair indication of its purity.

7.3. Media for gel filtration

There are many different media available for gel filtration. *Table 9* lists some of the most frequently used media, and their main uses. At the time of writing, the most *selective* in terms of obtaining or separating reasonable amounts of proteins of similar size are the materials designed for low-pressure operation. HPLC-style gel filtration achieves rapid results with very small samples, but the materials used cover a very wide size range, and so do not discriminate well between proteins of similar sizes. They are more suited for analytical-scale separations and for molecular mass determination, unless the separation required is of widely differing sized proteins.

Table 9. Gel filtration media[a]

Manufacturer	Medium	Protein separation range (kDa)	Comments
Pharmacia	Sephadex G-25	Proteins excluded	Mainly used for desalting
Pharmacia	Sephadex G-50	5–30	
Pharmacia	Sephadex G-75	10–100	
Pharmacia	Superdex 75	5–80	Pre-packed column
Pharmacia	Superdex 200	10–600	Pre-packed column
Pharmacia	Sephacryl S-100	1–100	
Pharmacia	Sephacryl S-200	2–300	
Pharmacia	Sephacryl S-300	10–1500	
Pharmacia	Superose 6	5–5000	HPLC, pre-packed
Pharmacia	Superose 12	1–30	HPLC, pre-packed
Bio-Rad	Bio-Gel P-2, P-4	Proteins excluded	Mainly used for desalting
Bio-Rad	Bio-Gel P-30	2–40	
Bio-Rad	Bio-Gel P-100	5–100	
Bio-Rad	Bio-Gel A-0.5m	10–500	
Bio-Rad	Bio-Gel A-1.5m	20–1500	
Bio-Rad	Bio-Gel SEC 30XL	1–800	HPLC, pre-packed
Bio-Rad	Bio-Gel SEC 40XL	10–1500	HPLC, pre-packed
Merck/	Fractogel TSK HW-50	1–80	
Toyo Soda	TSK HW-55	2–800	
	TSK HW-65	30–6000	

[a] This list is not comprehensive, and excludes many materials that are outdated or rarely used. It also excludes very recent products which have not so far proven themselves. Many manufacturers of HPLC equipment (e.g. Waters, Beckman) provide pre-packed HPLC gel filtration columns covering protein separation ranges from 10–1000 kDa. Some products are available as free beads; others only in pre-packed columns, as indicated.

8. METHODS FOR CONCENTRATING PROTEINS

At various stages in handling proteins, a reduction of the volume that the sample is dissolved in is desirable. Many proteins can become inactivated at low concentration. This can be due to several factors, which include:

(i) adsorption to the walls of the container – if 10 μg adsorbs out of 10 mg, it would not be noticed, but if 10 μg adsorbs out of only 20 μg to start with, there is a problem;

(ii) subunit dissociation – many oligomeric proteins are stabilized by forming oligomers; at high dilution these can dissociate, and the monomer may inactivate (denature) readily;

(iii) oxygen and contaminants in the buffers – trace amounts of undesirable contaminants may be able to inactivate proteins and enzymes; the proportion of contaminant to protein is obviously higher at low protein concentration. This also applies to oxidative damage. Consequently it is best to keep protein concentrations as high as possible during handling procedures. It is easy to dilute a solution, but not simple to concentrate it.

Using the methods described in Sections 2–6, it is possible to concentrate a protein solution without necessarily going through all the steps involved in a separation procedure. But the composition of the solution must first be considered.

8.1. Concentration by precipitation

Precipitation using ammonium sulfate is commonly used as a concentration step. The solution should preferably not have large amounts of other solutes such as glycerol or detergents, and the protein concentration should not be too low to start with. As a rule, the starting protein concentration should not be less than 0.5 mg ml^{-1}, and in some cases should be even higher. Problems arise at high dilution due to denaturation by surface tension at bubbles that form as the salt is being dissolved, and due to the fact that even at high salt concentration, proteins do have a definite solubility, and some will remain unprecipitated.

Add 60 g solid $(NH_4)_2SO_4$ to every 100 ml of solution, stir gently to dissolve, but ensure that all the salt has dissolved. Keep on ice for at least 30 min (much longer times will normally do no harm; this is an excellent stage to leave it overnight or over the weekend). Centrifuge, at least 10 000 g for 15 min, decant the liquid and remove as much liquid as possible from the surface of the precipitate with a Pasteur pipette. Redissolve the precipitate with a volume of suitable buffer that is at least twice the volume of the precipitate. If the protein you want is still not precipitated, up to 10 g further $(NH_4)_2SO_4$ can be added to every 100 ml.

8.2. Concentration by adsorption

By running a sample through a column under conditions that result in complete adsorption, a dilute solution can be concentrated, since the protein can then be eluted stepwise in a relatively small volume. Possible adsorbents include ion exchangers, in which case the salt concentration of the sample must be low, and hydrophobic adsorbents, for which the salt concentration must be high. The latter is more flexible, since it is easy to add additional salt. The column should have a large cross-section compared with its height, so that the sample will not take too long to pass through: the column size will be dictated by the amount of protein being adsorbed, and generally this means about 1 ml for every 10 to 20 mg protein.

Ion-exchange concentration
For 100 ml of a 1 mg ml^{-1} sample, on a DEAE- or Q-adsorbent. Adjust the pH to 8.0 with 1 M Tris. Set up a column of dimensions c. 25 mm diameter and 25 mm height, equilibrated with 20

mM Tris–Cl buffer, pH 8.0. Run the sample through at not more than 2 ml min^{-1}. When all has gone on, elute using 10 ml of 1 M NaCl in the Tris buffer, and collect 2 ml fractions from the column. Measure the protein in the fractions and pool the main ones.

Hydrophobic concentration
For 100 ml of a 1 mg ml^{-1} sample, using a C8 (octyl) adsorbent. Add Na_2SO_4 (or $(NH_4)_2SO_4$) to 1 M, and apply to the column (for dimensions, see above), equilibrated with a suitable buffer containing 1 M sulfate salt. For elution, use a very dilute buffer with no salt, and monitor fractions as above.

8.3. Concentration by removal of water
The most satisfactory way of concentrating a protein solution is to remove excess water and buffer salts, without directly affecting the proteins themselves. There are numerous ways of doing this, and each has uses under different circumstances.

Direct drying by evaporation rarely finds use in concentrating proteins, except by freeze drying. Apart from the instability of proteins at high temperatures, boiling solutions under reduced pressure is impracticable because of bubbles and frothing. Freeze drying is also fairly impractical except on a small scale because it is very slow. If a product needs to be freeze dried, it is usual to concentrate it by another technique first.

Removal of water occurs by ultrafiltration, in which a membrane or other porous structure is used, where the pore sizes are too small to allow the proteins through, but the water and other salts can pass. The technique of desalting gel filtration (Section 7) is comparable, except that no concentration takes place. Some force must be applied to drive the water through the pores against the osmotic pressure caused by the proteins (this is a very low pressure at low protein concentration). This force can be centrifugal, gaseous pressure, or competing osmotic force by polymers on the other side of the membrane.

8.4. Concentration by centrifugal force
The technique of concentration by centrifugal force is very popular for small, dilute samples. Generally it is used to take 1–3 ml of sample down to a few 100 μl. Several manufacturers make tubes with appropriate membranes, that fit into bench centrifuges; the liquid is forced through the membrane and the remaining solution contains all the protein (*Figure 13a*). Typically this process will take from 30 to 60 min, depending on the starting volume.

8.5. Concentration by gas pressure
Concentration by gas pressure is the most widely used method for small to mid-scale (10 to 500 ml) concentration. The sample is placed in a concentration cell (*Figure 13b*), which has a membrane at its base. To minimize slowing due to accumulation of protein molecules on the membrane, a stirrer close to the membrane surface is used. Gas pressure, air or nitrogen at 3–5 atm. (300–500 kPa), is applied, and the excess water is forced through the membrane. Rapid reduction in volume, for instance 100 ml to 10 ml, can be obtained in 30–60 min, though this will depend on the starting concentration. The rate of water loss decreases as the protein concentration increases. Warnings:

(i)   watch the last stages carefully; it is easy to concentrate the sample down to dryness;
(ii) be very careful when getting the sample out; the membranes are easily damaged. Use a Pasteur pipette with soft silicone tubing attached to the end. After taking out the concentrated protein, rinse with a small volume of the diffusate to maximize overall recovery;

Figure 13. Ultrafiltration to concentrate proteins. (a) Centrifugal method; (b) gas pressure method; (c) powdered polymer (osmotic pressure) method. In each case water, and buffer molecules, pass through a membrane that is impermeable to proteins, leaving a concentrated protein solution in a buffer of the same composition as at the start.

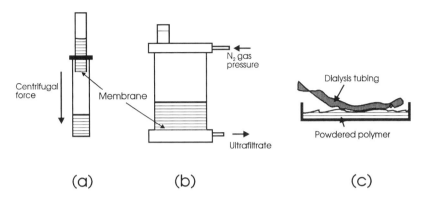

(a)   (b)   (c)

(iii) do not throw away the diffusate until you have checked that no protein has leaked through, due to holes in the membrane.

8.6. Concentration by water pressure

For large-scale work (liters of solution) use concentration by water pressure. Sheets of membranes or tubular structures are used in equipment that generates a pressure difference between the outside and inside, forcing the water through out of the protein solution. There are several commercial systems available.

8.7. Concentration by osmotic pressure

Osmotic pressure is a useful way of concentrating small samples quickly. The protein solution is placed in dialysis tubing, and this is placed in a solution, or a powder which draws the water out by osmotic force. Solutions of high molecular weight polyethylene glycol (PEG) have been used (e.g. 20% w/v of PEG 20 000), but simply placing the dialysis tubing into powdered PEG, or high-substitution CM-cellulose, is often better and quicker (*Figure 13c*). These materials are known as 'Aquacide', available from Calbiochem. Warning: take care when removing the concentrated protein that the polymer on the outside does not contaminate your sample; remove as much as possible of the outside solution/solid/gel.

9. PURIFICATION OF SMALL PEPTIDES FROM BIOLOGICAL SOURCES

Contributed by R.J. Simpson and R.L. Moritz

9.1. Introduction

The dramatic advances in the isolation and characterization of small biologically active peptides made over the past 20 years owes much to the improvements in separation technologies, especially RP-HPLC and characterization technologies such as N-terminal sequence analysis by Edman degradation and mass spectrometry (Chapter 23). Indeed, the rate-limiting step in identifying a small biological peptide or protein is not the sensitivity of the structural characterization procedures available (sub-picomole for N-terminal sequence instruments and low femtomole for the current generation mass spectrometers), but the ability to purify and handle low-abundance bioactive materials of interest. Bioactive peptides, unlike DNA, more often than not have physical properties that make their isolation and characterization elusive. Invariably, they require a multiplicity of techniques for their purification; thereby, presenting the researcher with a formidable technical challenge.

Since the first demonstration of the efficacy of RP-HPLC for the isolation of both synthetic and native peptides (27,28), the number of instruments available, as well as the diversity of chromatographic supports and mobile phases that are now commercially available, has increased dramatically. In fact, RP-HPLC has found universal acceptance as the technique of choice for peptide isolation and purification. Most commercially available columns today provide rapid, high-resolution separations together with generally excellent recoveries.

9.2. Practical considerations

The most important feature of a successful purification is that it is a convergent process. Each chromatographic step should ideally be based upon principles entirely or radically different from the preceding step. The use of size-exclusion chromatography (SEC), ion-exchange chromatography and RP-HPLC is a useful combination which exploits in a predictable manner the size, charge and hydrophobic character of the peptide in question (for general reviews, see refs 29 and 30, and references therein).

Most isolation schemes are generally not driven by their elegance, but by very pragmatic criteria such as the required degree of purity of the desired biological peptide and the nature of the impurities. These criteria are often dictated by the technologies utilized for determining the amino acid sequence of the purified native peptide. For instance, many protein-sequencing instruments such as the gas-phase or pulsed-liquid instruments can tolerate only small volumes (typically, 30–100 μl) free from buffer salts, whereas the recent generation of sequencing instruments such as the bi-phasic reaction column sequencers (Hewlett-Packard) can tolerate large volumes and a wide variety of buffer salts and detergents due to the advantage of a reversed-phase sample applicator. In the case of high-sensitivity sequence analysis by tandem mass spectrometry (e.g. electrospray ionization triple quadrupole (ESIMS) instruments), it is essential that peptides be introduced into the ESI source by RP capillary column liquid chromatography (31–33). For this reason samples must be in relatively small volumes, (<1 ml) and lack detergents which, otherwise, make ESIMS spectra difficult to interpret. It is now becoming clear for ultra-high sensitivity studies (~25 fmol) that peptide samples be subjected to multiple chromatographic steps prior to ESIMS (34).

Many biologically active peptides are present in bulk source material (e.g. tissue extract, cell conditioned medium, etc.) in only trace amounts (<50 μg). This necessitates the inclusion of a rapid preliminary low-resolution purification (or general protein concentration step) in order both to:

(i) reduce the overall protein content and
(ii) concentrate the bioactive peptide to levels compatible with subsequent high-resolution purification methods.

This can be achieved by judicious alteration of some property of the initial extraction reagent (e.g. salting-out using high concentrations of salts such as ammonium sulfate; precipitation with organic compounds such as PEG, trichloroacetic acid, rivanol, etc.); precipitation with organic solvents such as methanol, ethanol or acetone; precipitation by selective denaturation (e.g.

temperature, pH); or selective adsorption onto an interactive matrix (e.g. ion-exchange or RP supports). Many of these classical separation techniques are discussed earlier in this chapter and elsewhere (35).

In most cases, where the structure of a biologically active peptide has been determined, the peptide had to be essentially pure in order to exclude the possibility that the biological activity was not associated with some minor contaminant. Typically, total chemical synthesis and exhaustive comparison of both synthetic and native compounds is essential for proof of structure.

9.3. Parameters affecting RP-HPLC separation of peptides

To maximize the effect of RP-HPLC on a peptide purification scheme, each chromatographic step should ideally be based upon principles which differ significantly from the preceding step. RP-HPLC separations can be manipulated in two main ways:

(i) the type of column employed (i.e. stationary-phase);
(ii) altering the nature of the solvent system (i.e. mobile-phase).

The mobile phase however can be subdivided also into two further classes where:

(i) manipulating the ionic component of the solvent system may give rise to large selectivity differences of the analytes;
(ii) the organic modifier used may also contribute to differences in the chromatographic selectivity.

First, with respect to the stationary-phase, there is a wide range of bonded-phase silica, polymer and several other exotic materials such as porous carbon available (for annual reviews on recently introduced stationary-phase material, see ref. 36). The way in which peptides interact with these supports is often subtle, and in most cases, unpredictable. However, moderately successful attempts have been made to rationalize the retention times of peptides of interest (37). Apart from the nature of the chromatographic support, there are several other parameters that need to be taken into consideration in any column selection process. These include: particle size (μm), pore size (Å), surface area of the stationary-phase particles ($m^2\ g^{-1}$), the amount of reversed-phase coating per gram of stationary-phase particle (μmol m^{-2}), the stability of the support to aqueous conditions and pH, and last, but not least, the column efficiency (N) (for a review see ref. 38). Similarly, altering the nature of the organic component of the solvent system can alter the behavior of a column, but often in an unpredictable way. In general, the UV-transparent solvents, methanol, acetonitrile, and 1- or 2-propanol are the most widely used solvents; the retention of peptides generally follows an inverse relationship to the elutropic value of these solvents. The most successful way in which RP-HPLC peptide separation behavior is altered is by manipulating the ionic component of the solvent system. By altering both the nature and the strength of the ion-pairing reagent in the mobile phase and the pH of the solvent system, it is possible to exploit both the acidic and basic character of peptides which, in turn, will affect their chromatographic behavior.

The most common RP-HPLC solvent system in use today is aqueous acetonitrile containing 0.1% trifluoroacetic acid (TFA), pH $\sim$2.1, a weak hydrophobic ion-pairing reagent which exploits the basic character of peptides. For instance, at this pH all carboxyl groups are protonated while the basic amino acids and the terminal amino group are fully ionized. Under these conditions a hydrophobic anion like trifluoroacetate associates or 'ion pairs' with the basic charges on the peptide thereby leading to an increased retention time. TFA, at low concentrations, is UV transparent at 214 nm, a wavelength that is particularly useful for proteins and peptides since it corresponds to an isobestic point at which random and helical peptide bond absorptions are equal (39); at this wavelength the UV absorption is independent of conformation and, ignoring absorption tails of aromatic residues, the molar absorptivity is about 10^3 M^{-1} cm^{-1} per residue. Useful alternatives to TFA are two other perfluorinated carboxylic acids, pentafluoropropionic acid (PFPA) and heptafluorobutyric acid (HFBA). Like TFA, they are fully dissociated acids and volatile. In progressing from TFA>PFPA>HFBA, the hydrophobicity of these perfluorinated carboxylic acids increases, which, in turn, leads to an increase in retention time of peptides (40). More recently, the use of long chain perfluorinated acids have been shown to be beneficial for the retention of very small hydrophilic peptides (41).

It is also possible to exploit the acidic character of peptides by using a volatile weak ion-pairing reagent at a pH elevated above that of the pK value of carboxylic acids (e.g. 0.01 M triethylamine acetate (TEA), pH 5.5). Under these conditions, peptides usually exhibit increased hydrophilic character, and retention times are significantly decreased relative to the TFA system. Alternative buffers that can be used at high pH (pH ~ 7.0) are:

(i) 0.01 M tetrabutylammonium phosphate (TBAP); or
(ii) 0.025 M triethylamine phosphate (TEAP).

Since both TBAP and TEAP are nonvolatile, a final desalting step using a volatile buffer system (e.g. TFA/acetonitrile) may be required. In some cases, TEAP has an additional advantage in that it can transiently end cap any remaining underivatized free silanol groups on silica-based supports, thereby resulting in sharper peaks.

Many of the above-mentioned ion-pairing reagents can be obtained commercially in ultra-pure form suitable for high-sensitivity peptide separation protocols. For a detailed list of buffer systems employed in various bioactive peptide purification schemes see refs 29, 42 and 43.

9.4. Isolation of guinea-pig adrenocorticotrophin (ACTH)

As an example of a purification protocol for a small biologically active peptide by RP-HPLC, we describe the isolation and purification of guinea-pig ACTH, previously reported by Smith *et al.* (44). Prior to chromatography, guinea-pig anterior pituitary lobes (previously stored frozen at $-80°C$) were thawed, homogenized in 2 ml of 2 M acetic acid, centrifuged (30 000 g, 30 min) and the supernatant applied to a Sep-Pak C18 reversed-phase cartridge (Waters Assoc.). Immunoradioactive (Ir) ACTH was then eluted from the cartridge with 80% acetonitrile/0.1% TFA, the eluate dried, and then reconstituted in 500 μl of 0.1% TFA for sequential RP-HPLC as follows (*Figure 14*).

The Sep-Pak eluate of pituitary extract was applied to an Aquapore RP-300 reverse-phase column (Applied Biosystems). The column was developed, at 1 ml min^{-1}, with a linear 30 min gradient from 0–100% B, where solvent A was 0.08% TFA and solvent B was 70% acetonitrile/30% water containing 0.08% TFA (*Figure 14a*). Four major ACTH immunoreactive peaks (solid bars in *Figure 14a*) were evident; these Ir-ACTH peaks, consistent with glycosylated and/or phosphorylated and/or C-terminally truncated forms of ACTH, are also found in other species (44). The major Ir-ACTH-containing fractions, shown by the horizontal bar in *Figure 14a*, were pooled, dried, reconstituted in 500 μl of 0.1% TFA and applied to a Nova-Pak C18 reverse-phase column (Waters Assoc.). This column was developed with a linear 30 min gradient from 30–50% B (*Figure 14b*), using the same solvents described in *Figure 14a*. Major Ir-ACTH-containing fractions from this column were pooled, dried, reconstituted in 500 μl aqueous 13 mM HFBA and applied to the same column which was developed with a linear 30 min gradient from 40–60% B, where solvent A was aqueous 13 mM HFBA and solvent B was 70% acetonitrile/30% water containing 13 mM HFBA (*Figure 14c*). Ir-ACTH-containing fractions from *Figure 14c* were pooled, dried, reconstituted in 500 μl of 0.08% TFA and applied to the same Nova-Pak column and eluted with a linear 30

Figure 14. Purification of guinea-pig ACTH from anterior pituitary extracts by sequential HPLC steps. Adapted from ref. 44 by permission of the Journal of Endocrinology Ltd.

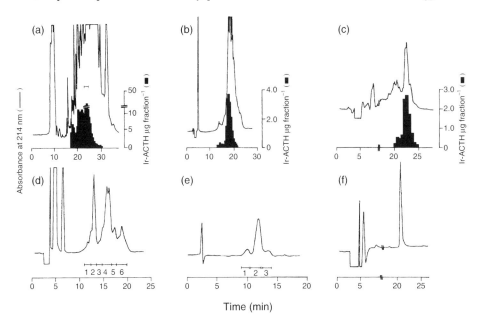

min gradient from 40–60% B, using the same solvents as for *Figure 14a*. The Ir-ACTH material recovered from this column (fractions 1 and 2, *Figure 14e*) was dried, reconstituted in 0.5 mM phosphate buffer, pH 3.8, and applied to a TSK SP5PW cation-exchange column (Toya Soda). The column was developed with a linear 15 min gradient from 0–100% B, where solvent A was 0.5 mM phosphate buffer, pH 3.8 and solvent B was 0.75 mM phosphate buffer, pH 3.8 (*Figure 14e*). The Ir-ACTH-containing peak from this column (peak fraction 2 in *Figure 14e*) was dried, reconstituted in 0.08% TFA and applied to the Nova-Pak column which was developed with a linear 30 min gradient from 0–100% B with the same solvents as for *Figure 14a*. The single peak obtained (*Figure 14f*), corresponding with homogeneous guinea-pig ACTH, which was fully bioactive as assessed by stimulation of aldosterone synthesis in rat adrenal glomerulosa cells, was employed for structural analysis (44).

9.5. Conclusion

With the advent of bonded-phase RP-HPLC, purification of biologically important peptides from fresh tissue sources are now seen as simple short-step procedures. By the judicious choice of both stationary and mobile phases, it is possible to obtain single compounds of high purity from complex mixtures containing thousands of related and unrelated compounds. Pure peptides obtained in this manner can readily be identified by chemical Edman degradation or mass spectrometric techniques.

10. REFERENCES

1. Nachman, M., Azad, A.R.M. and Bailon, P. (1992) *Biotechnol. Bioeng.*, **40**, 564.

2. Afeyan, N.B., Gordon, N.F., Mazaroff, I., Varady, L., Fulton, S.P., Yang, Y.B. and Regnier, F.E. (1990) *J. Chromatogr.*, **519**, 1.

3. Martin, A.J.P. and Synge, R.L.M. (1941) *Biochem. J.*, **35**, 1358.

4. Sluyterman, L.A.A. and Elgersma, O. (1978) *J. Chromatogr.*, **150**, 17.

5. Wagner, G. and Regnier, F.E. (1982) *Anal. Biochem.*, **126**, 37.

6. Maisano, F., Belew, M. and Porath, J. (1985) *J. Chromatogr.*, **321**, 305.

7. El Rassi, Z., De Ocampo, L.F. and Bacolod, M.D. (1990) *J. Chromatogr.*, **499**, 141.

8. Lowe, C.R. (1977) *Eur. J. Biochem.*, **73**, 265.

9. March, S.C., Parikh, I. and Cuatrecases, P. (1974) *Anal. Biochem.*, **60**, 149.

10. Ersson, B. (1977) *Biochim. Biophys. Acta*, **494**, 51.

11. Sunderberg, L. and Porath, J. (1974) *J. Chromatogr.*, **90**, 87.

12. Bethel, G.S., Ayers, J.S., Hancock, W.S. and Hearn, M.T.W. (1979) *J. Biol. Chem.*, **254**, 2572.

13. Weetall, H.H. (1976) *Methods Enzymol.*, **44**, 134.

14. Pepper, D.S. (1992) *Methods Mol. Biol.*, **11**, 173.

15. Brewer, S.J. and Sassenfeld, H.M. (1985) *Trends Biotechnol.*, **3**, 119.

16. Smith, D.B. and Johnson, K.S. (1988) *Gene*, **67**, 31.

17. Nilsson, B., Holmgren, E., Josephson, S., Gatenbeck, S., Philipson, L. and Uhlen, M. (1985) *Nucleic Acids Res.*, **13**, 1151.

18. Moks, T., Abrahamsen, L., Holmgren, E., Bilich, M., Olsson, A., Uhlen, M., Pohl, G., Sterky, C., Hultberg, H., Josephson, S., Holmgren, A., Jornvall, H. and Nilsson, B. (1987) *Biochemistry*, **26**, 5239.

19. Smith, M.C., Furman, T.C., Ingolia, T.D. and Pidgeon, C. (1988) *J. Biol. Chem.*, **263**, 7211.

20. Scopes, R.K. (1977) *Biochem. J.*, **161**, 253.

21. Haeckel, R., Hess, B., Lauterborn, W. and Wurster, K.-H. (1968) *Hoppe-Seyler's Z. Physiol. Chem.*, **349**, 699.

22. Atkinson, T., Hammond, P.M., Hartwell, R.D., Hughes, P., Scawen, M.D., Sherwood, R.F., Small, D.A.P., Bruton, C.J., Harvey, M.J. and Lowe, C.R. (1981) *Biochem. Soc. Trans.*, **9**, 290.

23. Scopes, R.K. (1986) *J. Chromatogr.*, **376**, 131.

24. Kroviarski, Y., Cochet, S., Vadon, C., Truskolaski, A., Bovin, P. and Bertrand, O. (1988) *J. Chromatogr.*, **449**, 403.

25. Scopes, R.K. (1984) *Anal. Biochem.*, **136**, 525.

26. Porath, J. (1992) *Protein Express. Purif.*, **3**, 263.

27. Burgus, R. and Rivier, J. (1976) in *Peptides 1976* (A. Loffet, ed.) Editions de l'Universite de Bruxelles, Brussels, p. 427.

28. Hancock, W.A. and Bishop, C.A. (1976) *FEBS Lett.*, **72**, 139.

29. Rivier, J. and McClintock, R. (1989) in *The Use of HPLC in Receptor Biochemistry* (A.R. Kerlavage, ed.). Alan R. Liss, Inc., New York, p. 77.

30. Simpson, R.J. and Nice, E.C. (1989) in *The Use of HPLC in Receptor Biochemistry* (A.R. Kerlavage, ed.). Alan R. Liss, Inc., New York, p. 201.

31. Moritz, R.L. and Simpson, R.J. (1992) *J.Chromatogr.*, **599**, 119.

32. Moritz, R.L. and Simpson, R.J. (1993) in *Methods in Protein Sequence Analysis* (F. Sakiyama and K. Imahori, eds). Plenum Publishing Corporation, New York, p. 3.

33. Moritz, R.L., Reid, G.E., Ward, L.D. and Simpson, R.J. (1994) in *Methods, A Companion to Methods in Enzymology*, **6**, p. 213.

34. Cox, A.L., Skipper, J., Chen, Y., Henderson, R.A., Darrow, T.L., Shabanowitz, J., Engelhard, V.H., Hunt, D.F., and Slingluff, C.L., Jr. (1994) *Science*, **264**, 716.

35. Scopes, R. (1993) *Protein Purification, Principles and Practice, 3rd edn.* Springer-Verlag, New York.

36. Majors, R.E. (1995) *LC-GC*, **13(3)**, 202.

Chromatography

37. Meek, J.L. (1980) *Proc. Natl Acad. Sci. USA*, **77**, 1632.

38. Hearn, M.T.W. (1983) in *High Performance Liquid Chromatography*, Vol. 3 (C. Horvath, ed.). Academic Press, New York, p. 87.

39. Gratzer, W.B. (1967) in *Poly-α-amino Acids, Biological Macromolecules Series* (G.D. Fasman, ed.). Dekker, New York, p. 177.

40. Bennett, H.P.J. (1991) in *High- Performance Liquid Chromatography of Peptides and Proteins: Separation, Analysis, and Conformation* (C.T. Mant and R.S. Hodges, eds). CRC Press, Boca Raton, FL, p. 319.

41. Pearson, J.D. and McCroskey, M.C. (1994) Poster 216. *19th Lorne Protein Conference*, Lorne, Australia.

42. Hancock, W.S. and Sparrow, J.T. (1983) in *High Performance Liquid Chromatography*, Vol. 3 (C. Horvath, ed.). Academic Press, New York, p. 49.

43. Hancock, W.S, (1984) in *CRC Handbook of HPLC for the Separation of Amino Acids, Peptides, and Proteins*, Vol. I. CRC Press, Boca Raton, FL, p. 127.

44. Smith, A.I., Wallace, C.A., Moritz, R.L., Simpson, R.J., Schmauk-White, L.B., Woodcock, E.A. and Funder, J.W. (1987) *J. Endocrinol.*, **115**, R5.

11. FURTHER READING

Deutscher, M.P. (ed.) (1990) Guide to Protein Purification. *Methods Enzymol.*, **182**.

Scopes, R.K. (1993) *Protein Purification, Principles and Practice.* Springer-Verlag, New York.

Wheelwright, S.M. (1991) *Protein Purification; Design and Scale up of Downstream Processing.* Carl Hanser Verlag, Munich.

Kenney, A. and Fowell, S. (eds) (1992) *Practical Protein Chromatography.* Humana Press, NJ.

Harris, E.L.V. and Angal. S. (eds) (1990) *Protein Purification Applications: A Practical Approach.* IRL Press, Oxford.

CHAPTER 7
COVALENT CHROMATOGRAPHY BY THIOL–DISULFIDE INTERCHANGE USING SOLID-PHASE ALKYL 2-PYRIDYL DISULFIDES

K. Brocklehurst

1. INTRODUCTION

Covalent chromatography (reviewed in ref. 1) involves the specific chemical reaction of the chromatographic material with one or more components of a mixture. In one version, specificity is achieved in the bonding step and, after removal of the other components by washing, the covalently bonded component is released by reaction with a suitable reagent. The elution reaction should leave the chromatographic material in a reactivatable form.

This type of chromatography differs from conventional affinity chromatography by its reliance on scission and formation of covalent bonds during the separation process and by the lack of a requirement for specific adsorptive binding areas on the activating ligand designed from knowledge of the structure of the binding site of the component to be isolated.

In another version, sequential elution covalent chromatography, several components of the mixture react with the chromatographic material and specificity is achieved in the elution step.

2. COVALENT CHROMATOGRAPHY BY THIOL–DISULFIDE INTERCHANGE (TDCC)

Thiol groups exist in many important proteins and peptides and the gels for TDCC have general applicability for all of these systems in:

(i) fractionation and specific isolation of thiol-containing proteins and peptides;
(ii) reversible immobilization of enzymes;
(iii) peptide sequence analysis;
(iv) removal of terminated from nonterminated peptides during solid-phase peptide synthesis;
(v) synthesis of adsorbents for conventional affinity chromatography.

TDCC was introduced in 1973 (2), discussed in *Methods in Enzymology* in 1974 (3) and further discussed in 1985 in a more general review of covalent chromatography (1).

Insoluble mixed disulfides containing the pyridine-2-mercaptide leaving group (Gel-spacer-S-S-2-Py) react rapidly and specifically with thiol-containing molecules (RSH) to produce Gel-spacer-S-S-R with the release of the chromophoric pyridine-2-thione, which permits the attachment to be monitored by measurement of A_{343} ($\epsilon_{343} = 8080$ M^{-1} cm^{-1}). After washing to remove pyridine-2-thione and components of the initial mixture other than RSH (e.g. RSO$_2$H), RSH is released by elution with cysteine or some other mercaptan and separated from low M_r material by gel filtration. The 'spent' gel (Gel-spacer-SH) may be reactivated by reaction with 2,2' dipyridyl disulfide (2-Py-S-S-2-Py).

An alternative version of TDCC involves derivatization of the thiol-containing molecules to be isolated (RSH) by reaction with 2-Py-S-S-2-Py to produce R-S-S-2-Py followed by attachment of the electrophilic products to thiol-containing gels (Gel-spacer-SH).

Both thiolated gels and disulfide gels (Gel-spacer-S-S-2-Py) are available commercially, for example Thiol Sepharose, Activated Thiol Sepharose (a glutathione agarose derivative with the thiol group derivatized as a 2-pyridyl disulfide (2)) and Thiopropyl Sepharose (a 2-pyridyl disulfide-containing beaded cross-linked agarose derivative (4)) from Pharmacia (see (1) for other commercially available products).

TDCC may be carried out in weakly alkaline media (e.g. pH 8.0) such that most thiol-containing molecules react rapidly, or in acidic media (e.g. pH 4.0) where additional selectivity of attachment of thiol groups with abnormally low pK_a values is achieved. Low pK_a values can occur in macromolecules as a result of intramolecular interaction as is the case, for example, in the catalytic sites of many cysteine proteinases (5–7).

The selectivity at low pH arises from the co-existence of significant concentrations of the nucleophilic (RS⁻) form of a low pK_a thiol group and the protonated form of the gel (Gel-space-S-S-Py⁺H) which possesses substantially enhanced reactivity. This type of TDCC is referred to as proton(or hydron)-activated covalent chromatography.

3. OTHER APPLICATIONS OF 2-PYRIDYL DISULFIDES AND THE ORIGINS OF TDCC

Applications of soluble reagents containing the pyridine-2-mercaptide group (R-S-S-2-Py) as enzyme active site titrants, reactivity probes, delivery vehicles for spectroscopic reporter groups, and heterobifunctional cross-linking reagents have been reviewed (8).

TDCC was devised (2) as a logical extension to the work on 2-Py-S-S-2-Py as a catalytic site titrant and reactivity probe for papain (9–11). This disulfide had been shown to permit titration of catalytic sites in papain even in the presence of low M_r mercaptan or denatured enzyme that still retained its thiol group. Kinetic analysis of reactions of cysteine proteinases with 2-Py-S-S-2-Py and with a series of disulfides of the type R-S-S-2-Py, particularly when R contains various selections of molecular recognition features, is a valuable approach to characterization of active center chemistry (e.g. see refs 5, 6, 12–16).

4. SEQUENTIAL ELUTION TDCC

This method of producing selectivity (an alternative to proton-activated TDCC, see Section 2) involves allowing all thiol-containing proteins in a mixture to react with the 2-pyridyl disulfide gel and selectively eluting each protein in turn by using either different mercaptans of increasing reducing strength or increasing concentrations of the same mercaptan as eluant. This technique was used to separate bovine liver protein disulfide isomerase and glutathione insulin transhydrogenase and thus to establish that they are distinct proteins (17,18) and, more recently, to isolate gingivain, the extracellular proteinase produced by *Porphyromonas gingivalis*, and to establish this enzyme, previously considered a trypsin-like enzyme, as a cysteine proteinase (19,20).

5. COVALENT AFFINITY CHROMATOGRAPHY

Specificity in covalent chromatography relies, initially at least, on the chemistry involved in producing covalent attachment of the protein or other molecule to be isolated to the chromatographic material. By contrast, traditional affinity chromatography is based on biospecific association in couples such as enzyme–inhibitor, hormone–receptor and antibody–antigen, where binding usually involves several functional groups in both molecules, usually in noncovalent interaction. Both techniques may be combined in covalent affinity chromatography by using a chromatographic ligand that can form an adsorptive complex with a particular binding area of a particular target molecule (e.g. a protein) and thus provide for juxtaposition of the electrophilic and nucleophilic sites necessary for covalent bond formation. An example is the use of Gly-Phe-Phe-cystamine immobilized on Affi-Gel 10 (BioRad) for the isolation of cathepsin B (21).

6. COVALENT CHROMATOGRAPHY AND SALT-PROMOTED THIOPHILIC ADSORPTION

Competition between covalent reaction with a gel containing disulfide sites and 'thiophilic' adsorption (22) may need to be considered in some cases, as has been found in connection with analysis of the proteins present in human serum (23). The ligands in thiophilic adsorbents consist of at least two functional groups, one of which contains sulfur and the other may be a π-electron-rich moiety. Covalent attachment of serum proteins to a 2-pyridyl disulfide gel is optimized by control of water-structuring salts.

7. SOME EXAMPLES OF THE APPLICATIONS OF TDCC

7.1. Isolation of thiol-containing proteins

Following the introduction of TDCC in 1973 for the purification of papain (2) many other thiol-containing enzymes and other thiol-containing proteins have been isolated by procedures involving TDCC, sometimes making use of the additional selectivity provided by proton (strictly hydron)-activated TDCC at low pH or by the sequential elution technique. The range of proteins that have been purified by covalent chromatography is demonstrated by the examples shown in *Table 1*. In many cases covalent chromatography was introduced at a relatively late stage in the purification procedure. The possibility of using TDCC with crude extracts in the very early stages of purification however, was demonstrated in the isolation of bovine mercaptalbumin (24) and perhaps this approach should be tried more often.

7.2. Isolation and sequencing of thiol-containing peptides

Purification of thiol-containing peptides from proteolytic digests is often difficult and TDCC is of considerable value. In one method the thiol group of the protein is allowed to react with the disulfide gel and this is followed by proteolysis. In the other, the protein is derivatized by reaction with 2-Py-S-S-2-Py followed by proteolysis in solution. The peptides containing the mixed disulfide groups are then isolated by reaction with the thiolated gel.

Table 1. Illustration of the range of thiol-containing proteins that have been isolated by covalent chromatography

Papain (2,3)	Thermostable cysteine proteinases from
Bovine mercaptalbumin (24)	*Ficus carica* callus cultures (36)
Ficin (25)	Cathepsins B and H (37,38)
Soft tissue collagens (26)	Chymopapain A (39)
Histone F3 (27)	Adipocyte lipid-binding proteins (40,41)
Prealbumin (28)	Chemoattractant from *Lumbricus*
Procallagen type II (29)	*terrestris* for *Thamnophis sirtalis* (42)
Components of the pyruvate	Fatty acid binding protein (43)
dehydrogenase complex (30)	Gingivain (19,44)
Human fibroblast interferons (31)	2-Oxoglutarate reductase from
1,25-Dihydroxy vitamin D3 receptor (32)	*Fusobacterium nucleatum* (45)
Copper thionein (33)	Thymidylate synthase (46)
Band 3 proteins (34)	Separation of chymopapain B from
α-Chains of bovine hemoglobin (35)	chymopapain M (7)

References to papers reporting the establishment of these techniques together with some examples are given in *Table 2*.

7.3. Removal of prematurely terminated peptides in solid-phase peptide synthesis

In solid-phase peptide synthesis byproducts arise in part as a result of premature chain termination of the peptide by blocking of the free terminal amino group. These prematurely terminated peptides, which can accumulate in unacceptable amounts, are readily separated from nonterminated peptides by using TDCC (54). The dipeptide Cys-Met is added to the free amino group of the nonterminated peptide before cleavage from the solid-phase matrix.

Table 2. Examples of the application of covalent chromatography in the isolation and sequencing of peptides

Demonstration of the general techniques for thiol peptide isolation using covalent chromatography (47–49)

Immobilization of proteins via a thiol group to Sepharose, followed by proteolysis by pepsin with and without urea or guanidine hydrochloride, and eventual release of the thiol-containing peptide by elution with a reducing agent exemplified by determination of the amino acid sequences around the thiol groups of the major parvalbumin from hake muscle, bovine serum mercaptalbumin and human serum ferroxidase (48)

Isolation of peptides prepared by solid-phase peptide synthesis (50)

Comparison of sequences of potato and rabbit muscle phosphorylase (51)

Isolation of a thiol-containing peptide from cerruloplasmin (49)

Replacement of thiol-agarose by thiol-silicates to permit a greater range of chemical fragmentation reagents to be used (52)

Identification of the cysteine residue contributing the reactive thiol group in human liver Mn superoxide dismutase (53)

The pyridyl disulfide gel reacts with the modified peptide and the terminated peptides, which do not contain the Cys-Met addition, are washed away. The peptides containing the Cys-Met additions are released from the gel in the usual way by elution with a reducing agent and the Cys-Met dipeptide is removed by treatment with cyanogen bromide. An example of the use of this technique is the purification of a synthetic hexapeptide with a sequence present in cod allergen M (50).

7.4. Enzyme immobilization with associated purification

Covalent immobilization by thiol–disulfide interchange provides a convenient method of reversible immobilization under mild conditions This technique of reversible immobilization contrasts with most methods commonly used (see ref. 1 for examples) where immobilization frequently involves irreversible attachment to a polymer matrix. TDCC has the additional advantage that when the enzyme is eventually released from the matrix by thiolysis, a purified sample of enzyme is usually produced. The technique may be used also for the immobilization of enzymes that (originally) lack thiol groups by thiolation, for example by using *N*-succinimidyl-3-(2-pyridyl disulfanyl) propanoate (a compound that can be used also as a heterobifunctional cross-linking reagent) (55).

8. THIOL-SPECIFIC COVALENT CHROMATOGRAPHY BY METHODS OTHER THAN THIOL-DISULFIDE INTERCHANGE

Chromatographic materials containing organomercurial groups provide an alternative approach to the isolation of thiol-containing molecules. A number of systems have been described (see ref. 1 for a review).

Although isothiocyanates react with both amines and mercaptans, arylisothiocyanates are considerably more reactive towards aliphatic thiol groups than towards aliphatic amines and this selectivity permits their use in thiol-selective covalent chromatography (56).

Use of an immobilized tervalent organoarsenical, 4-aminophenylarsenoxide-agarose, provides a covalent chromatography technique for the selective isolation of molecules possessing vicinal thiol groups, such as lipoic acid (57).

Immobilization involves cyclic dithioarsenite formation and elution is achieved by using 2,3-dimercaptopropane-1-sulfonic acid. This releases the dithiol compound (e.g. the reduced form of lipoic acid) and forms a more stable complex with the tervalent arsenical. The technique was shown to be effective with lipoic acid itself, the 2-oxoglutarate dehydrogenase multi-enzyme complex and whole cells of *Escherichia coli*.

9. SELECTIVITY IN COVALENT CHROMATOGRAPHY

Selectivity in TDCC for thiol groups with abnormally low pKa values achieved by carrying out the attachment to 2-pyridyl disulfide gels at low pH is discussed in Section 2.

It can be envisaged that selectivity might be achieved more generally by the matching or mismatching of relevant structural features of protein and immobilized ligand. Steric and electrostatic restrictions are obvious factors determining selectivity.

Recently, examples of the marked effects that both of these factors can exert have been reported (58). All but one of the cysteine proteinases of *Carica papaya* (i.e papain, chymopapains A and B and papaya proteinase Ω (caricain)), react with Sepharose-glutathione-2-pyridyl disulfide gel. In marked contrast, neither chymopapain M, nor actinidin, the cysteine proteinase from *Actinidia chinensis,* react with this gel, although they do react with Sepharose-2-hydroxypropyl-2'-dipyridyl disulfide gel. The latter is both electrically neutral and sterically less demanding than the 'glutathione' gel. Electrostatic potential calculations, minimization and molecular dynamics simulations provided explanations for the specificities exhibited by actinidin and chymopapain M. Electrostatic repulsion prevents reaction of actinidin with the 'glutathione' gel, whereas steric 'cap' resulting from a unique Arg 65–Glu 23 interaction in chymopapain M prevents reaction of chymopapain M and also accounts for its lack of inhibition by cystatin and its specificity in catalysis.

10. ACKNOWLEDGMENTS

I thank Hasu Patel for searching the literature, Joy Smith and Rita Dobrin for rapid production of the typescript and SERC, MRC, AFRC and BBSRC for project grants and Earmarked Studentships.

11. REFERENCES

1. Brocklehurst, K., Carlsson, J. and Kierstan, M.P.J. (1985) *Top. Enzyme Fermentation Biotechnol.,* **10**, 146.

2. Brocklehurst, K., Carlsson, J., Kierstan, M.P.J. and Crook, E.M. (1973) *Biochem. J.,* **133**, 573.

3. Brocklehurst, K., Carlsson, J, Kierstan, M.P.J. and Crook, E.M. (1974) *Methods Enzymol.,* **34**, 531.

4. Axén, R., Drevin, H. and Carlsson, J. (1975) *Acta Chem. Scand.,* **B29**, 471.

5. Mellor, G.W., Patel. M., Thomas, E.W. and Brocklehurst, K. (1993) *Biochem. J.,* **294**, 201.

6. Brocklehurst, K. (1994) *Protein Engng.,* **7**, 291.

7. Thomas, M.P., Topham, C.M., Kowlessur, D., Mellor, G.W., Thomas, E.W., Whitford, D. and Brocklehurst, K. (1994) *Biochem. J.,* **300**, 805.

8. Brocklehurst, K. (1982) *Methods Enzymol.,* **87**, 427.

9. Brocklehurst, K. and Little, G. (1970) *FEBS Lett.,* **9**, 113.

10. Brocklehurst, K. and Little, G. (1972) *Biochem. J.,* **128**, 471.

11. Brocklehurst, K. and Little, G. (1973) *Biochem. J.,* **133**, 67.

12. Brocklehurst, K., Brocklehurst, S.M., Kowlessur, D., O'Driscoll, M., Patel, G., Salih, E., Templeton, W., Thomas, E., Topham, C.M. and Willenbrock, F. (1988) *Biochem. J.,* **256**, 543.

13. Kowlessur, D., Topham, C.M., Thomas, E.W., O'Driscoll, M., Templeton, W. and Brocklehurst, K. (1989) *Biochem. J.,* **258**, 755.

14. Kowlessur, D., O'Driscoll, M., Topham, C.M., Templeton, W., Thomas, E.W. and Brocklehurst, K. (1989) *Biochem. J.,* **259**, 443.

15. Topham, C.M., Salih, E., Frazao, C., Kowlessur, D., Overington, J.P., Thomas, M., Brocklehurst, S.M., Patel, M., Thomas, E.W. and Brocklehurst, K. (1991) *Biochem. J.,* **280**, 79.

16. Patel, M., Kayani, S., Templeton, W., Mellor, W., Thomas, E. and Brocklehurst, K. (1992) *Biochem. J.,* **287**, 881.

17. Hillson, D.A. and Freedman, R.B. (1980) *Biochem. J.,* **191**, 373.

18. Hillson, D.A. (1981) *J. Biochem. Biophys. Methods,* **4**, 101.

19. Shah, H.N., Gharbia, S.E., Kowlessur, D., Wilkie, E. and Brocklehurst, K. (1991) *Microbiol. Ecol. Health Disease,* **4**, 319.

20. Sreedharan, S.K., Patel, H., Smith, S., Gharbia, S.E, Shah, H.N. and Brocklehurst, K. (1993) *Biochem. Soc. Trans.*, **21**, 218S.

21. Evans, B. and Shaw, E. (1983) *J. Biol. Chem.*, **258**, 10227.

22. Porath, J., Maisano, F. and Belew, M. (1985) *FEBS Lett.*, **185**, 306.

23. Oscarsson, S. and Porath, J. (1989) *Anal. Biochem.*, **176**, 330.

24. Carlsson, J. and Svenson, A. (1974) *FEBS Lett.*, **42**, 183.‘

25. Malthouse, J.P.G. and Brocklehurst, K. (1976) *Biochem. J.*, **159**, 221.

26. Sykes, B.C. (1976) *FEBS Lett.*, **61**, 180.

27. Oster, O. and Buchlow, G. (1977) *Z. Naturforsch.*, **32C**, 72.

28. Fex, G., Laurell, C.B. and Thulin, E. (1977) *Eur. J. Biochem.*, **75**, 181.

29. Angermann, K. and Barrach, H.J. (1979) *Anal. Biochem.*, **94**, 253.

30. de Graaf-Hess, A.C. and de Kok, A. (1982) *FEBS Lett.*, **143**, 261.

31. Senussi, O.A., Cartwright, T. and Thompson, P. (1979) *Arch. Virol.*, **62**, 323.

32. Wecksler, W.R., Ross, F.P., Okamura, W.H. and Norman, A.W. (1979) in *Vitamin D Basic Research and its Clinical Applications* (A.W. Norman, ed.). Walter de Gruyter and Co., p. 663.

33. Rydén, L. and Deutsch, H.F. (1978) *J. Biol. Chem.*, **253**, 519.

34. Kahlenberg, A. and Walker, C. (1976) *Anal. Biochem.*, **74**, 337.

35. de Bruin, S.H., Joordens, J.J. and Rollema, H.S. (1977) *Eur. J. Biochem.*, **75**, 211.

36. Cormier, F., Charest, C. and Dufresne, C. (1989) *Biotech. Lett.*, **11**, 797.

37. Willenbrock, F.W. and Brocklehurst, K. (1985) *Biochem. J.*, **223**, 511.

38. Popovic, T., Brzin, J., Kos, J., Lenarcic, B., Machleidt, W., Ritonja, A., Hanada, K. and Turk, V. (1988) *Biol. Chem. Hoppe Seyler*, **369** Suppl., 175.

39. Baines, B.S., Brocklehurst, K., Carey, P.R., Jarvis, M., Salih, E. and Storer, A.C. (1986) *Biochem. J.*, **233**, 119.

40. Matarese, V. and Bernlohr, D.A. (1988) *J. Biol. Chem.*, **263**, 14544.

41. Baxa, C.A., Sha, R.S., Buelt, M.K., Smith, A.J., Matarese, V., Chinander, L.L., Boundy, K.L. and Bernlohr, D.A. (1989) *Biochemistry*, **28**, 8683.

42. Wang, D., Chen, P., Xian, C.J. and Halpern, M. (1988) *Arch. Biochem. Biophys.*, **267**, 459.

43. Armstrong, M.K., Bernlohr, D.A., Storch, J. and Clarke, S.D. (1990) *Biochem. J.*, **267**, 373.

44. Shah, H.N., Gharbia, S.E. and Brocklehurst, K. (1993) in *Porphyromonas gingivalis* (H.N. Shah, D. Mayrand and R. Genco, eds). CRC Press, Boca Raton, FL, 245.

45. Gharbia, S.E. and Shah, H.N. (1991) *FEMS Microbiol. Lett.*, **64**, 282.

46. Bradshaw, T.P. and Dunlop, R.B. (1993) *Biochem. Biophys. Acta.*, **1163**, 165.

47. Svenson, A., Carlsson, J. and Eaker, D. (1977) *FEBS Lett.*, **73**, 171.

48. Egorov, T.A., Svenson, A., Rydén, L. and Carlsson, J. (1975) *Proc. Natl Acad. Sci. USA*, **72**, 3029.

49. Rydén, L. and Norder, A. (1981) *J. Chromat.*, **215**, 341.

50. Lindeberg, G., Tergborn, J., Bennich, H. and Ragnarsson, U. (1978) *J. Chromat.*, **156**, 366.

51. Nakano, K., Fukui, T. and Matsubara, H. (1980) *J. Biochem.*, **87**, 919.

52. Egorov, T.A. and Shakhparonov, M. (1980) in *Methods in Peptide and Protein Sequence Analysis* (C. Birr, ed.). Elsevier, Amsterdam, p. 395.

53. Matsuda, Y., Higashiyama, S., Kijima, Y., Suzuki, K., Kawano, K., Akiyama, M., Kawata, S., Tarui, S., Deutsch, F. and Taniguchi, N. (1990) *Eur. J. Biochem.*, **194**, 713.

54. Krieger, D.E., Erikson, B.W. and Merryfield, R.B. (1976) *Proc. Natl Acad. Sci. USA*, **73**, 3160.

55. Carlsson, J., Drevin, H. and Axén, R. (1978) *Biochem. J.*, **173**, 723.

56. Gemeiner, P., Drobnica, L. and Polakova, K. (1977) *J. Solid-Phase Biochemistry*, **2**, 289.

57. Pratt, K.J., Carles, C., Carne, T.J., Danson, M.J. and Stevenson, J. (1989) *Biochem. J.*, **258**, 749.

58. Thomas, M.P., Verma, C., Boyd, S.M. and Brocklehurst, K. (1995) *Biochem. J.*, **306**, 39.

Covalent Chromatography

CHAPTER 8
CRITERIA OF PROTEIN PURITY
A. Berry

1. INTRODUCTION

The purification of a protein is usually only the starting point for a more detailed study of its structural and functional characteristics. The nature of these further studies will determine the quantity of purified protein, whether the protein is required in an active form, the time and cost of the purification and the level of purity required. If the protein is to be used in research, the amounts needed may be small but the purity is of the utmost importance (usually $\geqslant 95\%$) and the removal of interfering contaminants is essential. However, if the protein is needed for industrial applications, large quantities are required and the purity of the sample (typically 80–90%) may have to be compromised. For therapeutic uses, all contaminants must be removed, and the criteria used to check this should include assays for self-aggregation, contaminating proteins, DNA, lipid, carbohydrate, endotoxins and any additive used during the preparation. Specific steps may be needed to remove such contaminants. In general, the researcher should attempt to obtain a sample of the protein of the highest possible purity and biological activity. This chapter deals with the methods available to assess these features (*Table 1*).

1.1. Detection of nonprotein contaminants

Since most procedures for isolating proteins are highly successful at removing noncovalent, non-protein contaminants, this chapter will concentrate on a discussion of the techniques which can be used to detect other contaminating proteins. However, the measurement of the amount and nature of nonprotein material in a sample may be critical in some cases. The determination of the amount of nucleic acid in a protein sample can be easily achieved by the spectrophotometric method of Warburg and Christian (1) (*Figure 1*) and this method may also be used to estimate the protein concentration of the sample. The absorbance of a suitably diluted aliquot of protein solution is measured at both 260 nm and 280 nm and the A_{280}/A_{260} ratio is calculated. *Figure 1* relates the values of A_{280}/A_{260} ratio to the % nucleic acid in the sample and a factor, F, for calculating the protein concentration.

The values of F in *Figure 1* were calculated from the extinction coefficients of solutions (each 1 mg ml^{-1}) of pure crystalline yeast enolase ($\varepsilon_{280} = 2.06$; $\varepsilon_{260} = 1.18$) and yeast nucleic acid ($\varepsilon_{280} = 24.8$; $\varepsilon_{260} = 50.8$) and the method is therefore liable to some small error, as other proteins and nucleic acids will have different extinction coefficients. Despite this it is a generally useful method for estimating the amount of nucleic acid in a protein sample.

Detailed discussion of the methods of detecting and analyzing carbohydrate and lipid components in a preparation is beyond the scope of this chapter and the reader is referred to refs 2–4 for descriptions of these methods.

1.2. Detection of protein contaminants

Many of the preparative methods used during enzyme purification may be used on an analytical

Table 1. Methods for assessing the purity of protein preparations

Method	Quantity required	Comments
Specific activity measurement	Varies	Essential measurement for enzymes. Purification to a constant specific activity should be attained. Useful as a comparison with previously reported values
Electrophoresis (non-denaturing)	ng–μg	Samples should be electrophoresed at several pH values as two proteins might run together at a single pH
Electrophoresis in the presence of SDS	ng–μg	Very useful for detecting impurities that differ in subunit molecular mass. Useful for detecting proteolysis of samples. Problems with enzymes composed of non-identical subunits
Isoelectric focusing/ chromatofocusing	ng–μg	Very sensitive method for detecting impurities. Very small differences in isoelectric point (about 0.01 pH units) can be detected. Some problems caused by presence of ampholytes after technique has been applied
Capillary electrophoresis (CE)	ng–pg	Rapidly becoming popular. Variety of modes appropriate for proteins. Very small amounts of sample required (pg by mass, nl by volume). Rapid separations (1–30 min)
Chromatography (HPLC/ FPLC/conventional)		Analytical scale separations possible based on:
Gel filtration	μg	Molecular size
Ion exchange	μg	Charge
Affinity chromatography	μg	Specific binding
Ultracentrifugation		Difficult to detect low levels of impurities (e.g. < 5%). Some problems with associating/dissociating systems
Velocity	μg	
Equilibrium	μg	
Mass spectrometry	ng–μg	Accurate measurement of mass possible (± 0.01%). Can resolve species of very similar mass (± 0.1%). May detect minor species such as proteolytic fragments and post-translational modifications. Small amounts of protein required. Rapid analyses (15–30 min)
Amino acid analysis/ N-terminal analysis/ N-terminal sequencing		Compare with amino acid composition inferred from DNA sequence or previously purified enzyme. N-terminal analysis or sequencing should show the expected number of polypeptide chains, with the expected sequences. Problems with blocked N-termini
Active site titrations	Varies	Useful to determine whether an apparently homogeneous preparation contains inactive forms of the protein

Figure 1. The quantification of nucleic acid contaminants in protein samples. The absorbance of a suitably diluted aliquot of protein solution is measured at both 260 nm and 280 nm and the A_{280}/A_{260} ratio is calculated (1). From the value of A_{280}/A_{260} ratio, the % nucleic acid in the sample can be found and the factor, F, can be used to calculate the protein concentration according to the equation:

Protein concentration (mg ml^{-1}) = $A_{280} \times F \times 1/d$, where d is the path length in cm.

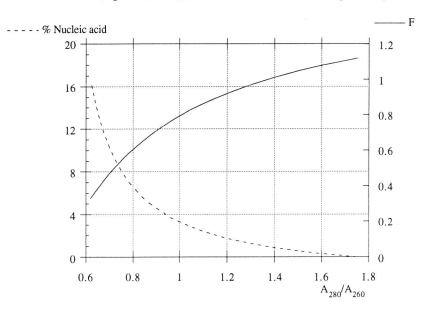

scale to detect contaminating proteins (*Table 1*). Each method is based on a different physical property of the protein. The choice between the methods will depend on the following parameters:

(i)   the quantity of protein available;
(ii) the nature of the suspected impurity;
(iii) the accuracy of the test needed;
(iv) the sensitivity of the test needed;
(v)   specific factors depending on the particular protein.

Each technique can only demonstrate the presence of an impurity rather than prove its absence, and a single test will not, therefore, be sufficient to prove that the preparation is homogeneous. However, since each method will establish a certain degree of probability that the preparation is pure (dependent on the sensitivity of the method), a combination of two or more methods should demonstrate whether the preparation is homogeneous to a very high degree and hence that the protein can be accepted as pure.

Many of the tests of purity are associated with fractionation of the sample with the criterion of purity being the presence of a single component. Similarly, the determination of some of the molecular properties of the purified protein and the comparison with those expected is important in ensuring that the protein is in the required, active form. Practical details of many of these methods are given elsewhere in this book and are not repeated here. Instead, this chapter will attempt to provide guidance in the selection of appropriate methods and will discuss some of the problems and pitfalls associated with each method.

2. DETERMINATION OF PURITY

2.1. Specific activity measurement

In addition to assessment of purity based on physical properties, other tests based on the activity of the protein are fundamentally important. The choice of the precise method of determining biological activity will vary for different proteins (see Chapter 1). However, no matter how this activity is measured, it may be combined with a determination of protein concentration (see Chapter 4) to yield a measure of the specific activity of the protein (activity per mg protein) and this may be used as a very sensitive assay of the efficiency of any purification procedure and the quality of the product. Given that the level of contamination may be very low ($< 1\%$), it may be useful to measure specifically the biological activity of suspected interfering proteins and such tests are often carried out and reported by commercial suppliers of proteins.

The purification of a protein should be followed at every step, if at all possible, by measurement of the activity of the preparation and of the amount of protein present. These measurements can then be combined to provide information of the specific activity of the protein of interest and of the degree of purity achieved at each step in the purification:

$$\text{Specific activity} = \frac{\text{Biological activity of protein of interest (units)}}{\text{Total protein (mg)}}.$$

Equation 1

$$\text{Degree of purification} = \frac{\text{Specific activity at step } (n+1)}{\text{Specific activity at step } (n)}.$$

Equation 2

The specific activity of the protein should increase during the purification procedure until a constant specific activity is obtained for the pure protein. The calculation of the degree of purification obtained at each step can be used to develop purification procedures to an optimum yield and degree of purification by attempting to modify or replace steps which yield only low degrees of purification. The specific activity measurement is an indirect measure of the purity of the enzyme since it must be compared with that *expected* or previously reported for the 'pure' enzyme.

2.2. Electrophoretic methods

Gel electrophoresis

Useful electrophoretic methods (5) for assessing the purity of enzymes include non-denaturing gel electrophoresis, denaturing electrophoresis in the presence of SDS, isoelectric focusing and combinations of these in two dimensions (for practical details of these methods see Chapter 9 of this volume and refs 6–8. Any of these methods can be used independently or in combination to detect contaminating proteins and one of these methods is likely to be first choice when assessing the purity of a protein because of the simplicity, cost and sensitivity. *Table 2* gives an outline guide to the choice of technique depending on the nature of the expected contaminants and the amount of protein available for analysis. The method of sample preparation, electrophoresis and visualization of the protein in each of these methods is described in Chapter 9. The protein of interest is electrophoresed according to the appropriate method and the protein is judged as pure if no other protein band is detected after electrophoresis. Of course, the higher the loading of

Table 2. The choice of electrophoretic method for the determination of protein purity

Difference between protein of interest and expected contaminants	Amount of protein needed	Method of choice
Difference in molecular mass	Low (ng)	SDS–PAGE plus silver stain, *or* SDS–protein molecular mass analysis by CE
	High (ng–μg)	SDS–gel electrophoresis
Different amino acid composition	Low (pg–ng)	Free solution capillary electrophoresis (FSCE) in coated or uncoated capillaries
	High (ng–μg)	Native gel electrophoresis
Differences in pI	Low (pg–ng)	CE isoelectric focusing
	High (ng–μg)	Isoelectric focusing

sample during electrophoresis, the more readily contaminating proteins will be visualized and this should be taken into consideration when judging protein purity by these methods. For this reason, the most sensitive staining techniques (see Chapter 9) should be utilized.

Denaturing gel electrophoresis is the most commonly used technique for assessing a protein's purity because of its ease. However, the choice of gel concentration to be used in the experiment is an important decision and the use of gradient gels may be appropriate (Chapter 9). Similarly, isoelectric focusing (5, 9, 10) may also be used to separate the required protein from any contaminant and hence assess purity. Care must always be taken, however, in the interpretation of the results of these experiments. *Table 3* lists some of the causes of possible mistakes and the steps which should be taken to ensure correct interpretation of the results (see Chapter 9 for a more detailed discussion).

Capillary electrophoresis
Capillary electrophoresis (CE) is a recently introduced technique which is gaining widespread use in the characterization of proteins and will soon be a commonly used method in the analysis of protein purity. It is interesting to note that the first electrophoretic separations were in a free solution mode (11) and the gels now commonly used were introduced to control the convective currents which were generated and which resulted in poor resolution. CE overcomes the problems of convection by the use of narrow bore (< 150 μm internal diameter) capillaries which dissipate heat very efficiently, obviating the need for a gel matrix. *Table 4* summarizes some of the possible modes of use of CE and each is described in *Figure 2*. CE is now finding a much increased role in purity assessment because of the ease, speed and low amounts of material needed for the analysis. In all cases, impurities are detected in the sample by the presence of extra peaks in the electrophoretogram. In free solution capillary electrophoresis (FSCE) in uncoated capillaries (*Figure 2a*) substances migrate in the electric field as a result of their charge, frictional coefficient, the strength of the electric field and the endosmotic flow (the flow of liquid through the capillary due to the electric field). In FSCE in coated capillaries (*Figure 2b*) the surface wall of the capillary is coated with positively charged molecules, which minimizes the adsorption of the positively charged proteins onto the wall over a wide range of pHs.

Table 3. Some common problems encountered in electrophoresis and their remedy

Erroneous conclusion	Cause	Remedy
False positive (i.e. pure sample appears contaminated)	Nonuniform gel	Clean gel plates and repour gel. Ensure good mixing of gel components and stable temperature for polymerization. Ensure persulfate does not exceed recommended quantity
	Residual oxidant in gel	Allow gel to set for 24 h. Pre-run gel. Include thioglycollate (2 mM) in upper reservoir
	Thermal effects; gel runs too fast	Run at lower current
False negative (i.e. contaminated sample appears pure)	Contaminant comigrates with protein of interest	Try different pH gel
	Contaminant too big to enter gel	Stain and examine whole gel, including stacking gel for staining material. Use lower % gel
	Contaminant protein does not stain	Stain with more than one stain, e.g. Coomassie brilliant blue followed by silver stain
	Net charge on contaminant is zero or opposite in charge to that of interest	Important in native gels. Run at more than one pH

For further details of electrophoretic methods see Chapter 9.

Further refinements of CE techniques have involved the introduction of CE isoelectric focusing, described in *Figure 2c*, and the introduction of matrices of linear molecules that form an entangled network into the capillary in order to carry out sodium dodecyl sulfate (SDS)–protein molecular weight analysis by CE. In this latter method, the sieving matrix separates proteins using the same principle as SDS–polyacrylamide gel electrophoresis (PAGE). The accuracy and resolution are comparable with SDS–PAGE but the amount of sample required and the time of a single separation are both improved.

2.3. Chromatographic techniques

Many chromatographic techniques can be used to detect contaminating proteins. These methods include gel filtration, ion-exchange, reverse-phase and affinity chromatography (see Chapter 6) and all may be carried out either in a conventional manner, by high performance liquid chromatography (HPLC) or by fast protein liquid chromatography (FPLC). These methods form some of the simplest methods for assessing the levels of impurities and are often used. In general they are less sensitive than the electrophoretic methods described above, however they have the advantage that they are usually nondestructive and the sample can be recovered. The amount of

Table 4. The uses of capillary electrophoresis

Method	Direction of flow	Basis of separation	Sample types
FSCE in uncoated capillaries or micellar CE	Towards cathode	Net charge Field strength Frictional coefficient Endosmotic flow	Inorganic ions Small charged molecules Proteins Peptides Carbohydrates
FSCE in coated capillaries	Generally towards anode (depends on coating)	Net charge Field strength Frictional coefficient Endosmotic flow	Inorganic ions Small charged molecules Proteins Peptides Carbohydrates
CE isoelectric focusing	Towards cathode	pI	Proteins Peptides
SDS–protein molecular mass by CE	Towards anode	Size	Large peptides Proteins

material required varies on the type of separation being effected and the type of detection being employed, but since the sample will be diluted during passage through the column, the starting concentration must be well above the minimum detectable.

Gel filtration
Molecules are separated on the basis of molecular size and shape (12, 13). The reader is referred to Chapter 6 for a complete description of the method, the preparation of sample for gel filtration and of the choice of column. Detection methods, such as continual measurement of the absorbance of eluate (e.g. at 280 or 220 nm) are suitable for detecting eluting proteins. A further refinement can be incorporated by specific assay of the fractions from the separation. The elution profile of a pure protein should be Gaussian, but often a slight skew towards the trailing edge is observed. Impurities are detected either as separate peaks in the chromatogram, as a shoulder on the peak of interest, as a strongly skewed peak or by the observation that the profiles monitored by absorbance and specific assay do not overlap. One of the major drawbacks of gel filtration is that proteins may associate with each other or with the column matrix and this will produce results similar to sample heterogeneity. Rechromatography or chromatography under different conditions of ionic strength or pH should allow one to distinguish between these two conditions.

Ion-exchange and reverse-phase chromatography
These forms of chromatography (see Chapter 6) depend on interactions between the protein and the gel matrix of the ion-exchange or reverse-phase resin (5, 14, 15). Ion-exchange resins may serve on an analytical scale to assess the purity of a protein. In this respect HPLC or FPLC ion-exchange chromatography have the best resolution and speed of analysis. The presence of a single peak of protein in the eluate suggests the sample is pure *as long as* it can be shown that all of the applied protein has been eluted from the column.

Figure 2. CE can be used in a variety of modes. The principles underlying these modes are illustrated here. (a) FSCE in uncoated capillaries. Charged samples are attracted to the oppositely charged electrode. Endosmotic flow is greater than the migration towards either electrode and all samples generally move towards the cathode. (b) FSCE in coated capillaries. The wall of the capillary is coated with material to alter the charge properties of the wall (e.g. to reverse the charge). This minimizes the adsorption of positively charged protein samples on to the wall and allows the use of a wide range of buffer pH values. (c) CE isoelectric focusing. For details of the uses of the various modes see *Table 4*.

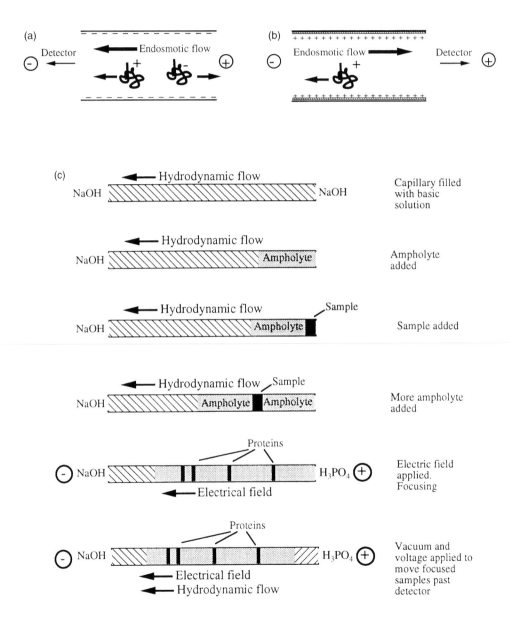

Affinity chromatography

Affinity chromatography (16–19), taking advantage of the specific interaction of a protein with its target molecule, is a very powerful method for purifying proteins (see Chapter 6). In principle the method could be used to evaluate the purity of a protein, although it is not often used for this purpose. As with ion-exchange chromatography, it is fundamentally important when assessing protein purity by this method to ensure that all the protein elutes from the column under the conditions chosen.

Chromatofocusing

In chromatofocusing (20) proteins are absorbed onto a poly(ethyleneimine) agarose ion-exchange resin by electrostatic interactions. A pH gradient is then generated by the addition to the column of the acid form of a mixture ampholyte-type buffers of high buffering capacity. This results in titration of the groups on the ion-exchange resin and this generates a decreasing pH gradient. As the pH falls, the proteins are eluted in the order of their isoelectric points (see Chapter 6). The method has high resolving power, but some of the materials are costly. Chromatofocusing has found use therefore in the later stages of protein purifications. Because of this it is rarely used to assess protein purity, although in special cases its application may be crucial in deciding the purity of a sample.

2.4. Centrifugation methods

Sedimentation velocity and sedimentation equilibrium centrifugation (21–23) may be used to detect the presence of contaminating components in a protein sample (24). These methods are described in detail in Chapter 16. Neither method is ideally suited to detecting small amounts of impurities ($< 5\%$) and the use of the ultracentrifuge has, to some extent, become less common in assessment of purity of proteins, being surpassed by other techniques which are less demanding in equipment, amount of material required and expertise in physical chemistry. However, in some cases the method can provide very important information on the purity of a protein sample.

Sedimentation velocity

Sedimentation velocity is a relatively simple, nondestructive method for measuring the molecular mass of a protein, and as a consequence of this, of assessing the purity of protein preparations. Impurities in the protein preparation will be detected either as a distinct sedimenting species or by a discrepancy between the expected molecular mass and that determined experimentally. A major limitation of the method is its insensitivity to small differences between molecules.

Sedimentation equilibrium

Another nondestructive method for detecting impurities based on their molecular mass, sedimentation equilibrium may be used to detect impurities in a sample and to detect heterogeneity in a sample which undergoes self-association. If the sample is pure, the concentration dependence of the apparent molecular mass should be independent of rotor speed. If this does not occur, sample heterogeneity or complications in the experiment are indicated (see Chapter 16). The acquisition and analysis of ultracentrifugation data has been automated and nonlinear fitting techniques (25) are also available to speed up and simplify the analysis of sedimentation equilibrium data. With the advent of modern machines this is a very powerful technique and will find ever-increasing use in modern biochemistry.

2.5. Mass spectrometry

The recent development of methods for producing intact molecular ions such as electrospray ionization mass spectrometry (ESIMS) (26) and matrix-assisted laser desorption ionization-time of flight mass spectrometry (MALDITOF-MS) (27) have revolutionized the use of MS in biochemistry to the point at which it has become a popular technique. The MS techniques are

21. Schachman, H.K. (1959) *Ultracentrifugation in Biochemistry*. Academic Press, New York.

22. Williams, J.W. (1972) *Ultracentrifugation of Macromolecules*. Academic Press, New York.

23. Birnie, J.D. and Rickwood, D. (1978) *Centrifugal Separations in Molecular and Cell Biology*. Butterworth, London.

24. Chervenka, C.H. (1970) *A Manual of Methods for the Analytical Ultracentrifuge*. Spinco Div., Beckman Instruments, Palo Alto, California.

25. Johnson, M.L., Correia, J.J., Yphantis, D.A. and Halvorson, H.R. (1981) *Biophys. J.*, **36**, 575.

26. Meng, C.K., Mann, M. and Fenn, J.B. (1988) *Z. Phys. D.*, **10**, 361.

27. Karas, M. and Hillenkamp, F. (1988) *Anal. Chem.*, **60**, 2299.

28. Loo, J.A., Udseth, H.R. and Smith, R.D. (1989) *Anal. Biochem.*, **179**, 404.

29. Karas, M., Bahr, U., Ingendoh, A. and Hillenkamp, F. (1989) *Angew. Chem. Int. Ed. Engl.*, **28**, 760.

30. Smith, R.D., Loo, J.A., Barinaga, C.J., Edmonds, C.G. and Udseth, H.R. (1990) *J. Am. Soc. Mass Spectrom.*, **1**, 53.

31. Hirayama, K., Akashi, S., Furuya, M. and Fukuhara, K. (1990) *Biochem. Biophys. Res. Commun.*, **173**, 639.

32. Poulter, B.N., Green, B.N., Kaur, S. and Burlingame, A.L. (1990) in *Biological Mass Spectrometry* (A.L. Burlingame and McCluskey, eds). Elsevier, Amsterdam, p. 477.

33. Page, M.J., Aitken, A., Cooper, D.J., Magee, A.I. and Lowe, P.N. (1990) *Methods: A Companion to Methods in Enzymology*, **1**, 221.

34. Harbour, G.C., Garlick, R.L., Lyle, S.B., Crow, F.W., Robins, R.H. and Hoogerheide, J.G. (1992) in *Techniques in Protein Chemistry III* (R.H. Angeletti, ed.). The Protein Society, Academic Press, London, p. 487.

35. Reinhold, B.B., Reinherz, E.L. and Reinhold, V.N. (1992) in *Techniques in Protein Chemistry III* (R.H. Angeletti, ed.). The Protein Society, Academic Press, London, p. 287.

36. Wang, Y.K., Liao, P.C., Allison, J., Gage, D.A., Andrews, P.C., Lubman, D.M., Hanash, S.M. and Strahler, J.R. (1993) *J. Biol. Chem.*, **268**, 14269.

37. Papac, D.I., Oatis, J.E., Crouch, R.K. and Knapp, D.R. (1993) *Biochemistry*, **32**, 5930.

38. Fersht, A.R. (1985) *Enzyme Structure and Mechanism*. W.H. Freeman & Co., New York.

CHAPTER 9
ELECTROPHORESIS METHODS
D. Patel and D. Rickwood

1. INTRODUCTION

This chapter does not seek to duplicate the detailed methods for fractionating and analyzing proteins offered in many manuals; the reader should consult these manuals for such information. Rather, it aims to provide key information required to analyze proteins by gel electrophoresis.

Many chemicals commonly used for gel electrophoresis are toxic, whilst the status of others is unknown. Researchers must acquaint themselves with the precautions required for handling the chemicals mentioned in this chapter. Acrylamide and bisacrylamide are both toxic. Acrylamide is a known potent neurotoxin. Great care must be taken when handling such reagents. All reagents must be of the highest purity available. It is recommended that the reagents are purchased from suppliers who have quality-tested the products for use in electrophoretic techniques.

It is essential to avoid any protease contamination and therefore, where possible, all solutions should be autoclaved. Solutions which cannot be autoclaved should be prepared in glass double-distilled, autoclaved water and then filtered through a Millipore filter (refer to Chapter 2 for inhibitors of proteases).

The following tables are not intended to be exhaustive and there are recipes and referenced techniques other than those listed. Manufacturers and suppliers are given in abbreviated form; Section 6 should be consulted for the complete names and addresses.

2. RECIPES FOR GELS AND BUFFERS

2.1. One-dimensional polyacrylamide gel electrophoresis
The analysis of complex mixtures of proteins and the determination of the purity of protein fractions has been possible using polyacrylamide gel electrophoresis (PAGE). The many methods can be categorized broadly into three main types:

(i)   SDS-denaturing gels;
(ii) nondenaturing gels;
(iii) isoelectric focusing gels.

SDS-denaturing gels
In sodium dodecyl sulfate (SDS) separations, migration is determined not by the intrinsic electrical charge of polypeptides but by the molecular mass. SDS is an anionic detergent which is bound by the protein, and in doing so, SDS confers a net negative charge to the protein. The proteins then have a mobility which is inversely proportional to their size. Caution must be exercised, however, as some proteins bind less SDS and so migrate anomalously; the accuracy of size estimates by SDS–PAGE is generally taken to be about 10%. The Laemmli system (1), is a discontinuous SDS system and is probably the most widely used electrophoretic system today.

Table 1. Recipe for gel preparation using the SDS–PAGE discontinuous buffer system (1)

Stock solution	Stacking gel[a]	%Acrylamide in resolving gel[a]			
		20.0	**15.0**	**12.5**	**10.0**
Acrylamide-bisacrylamide (30:0.8)	2.5	20.0	15.0	12.5	10.0
0.5 M Tris–HCl pH 6.8	5.0	–	–	–	–
3.0 M Tris–HCl pH 8.8	–	3.75	3.75	3.75	3.75
10% (w/v) SDS	0.2	0.3	0.3	0.3	0.3
Water	11.3	4.45	9.45	11.95	14.45
1.5% (w/v) APS	1.0	1.5	1.5	1.5	1.5
TEMED	0.015	0.015	0.015	0.015	0.015

Electrophoresis buffer: 0.025 M Tris, 0.192 M glycine, pH 8.3, 0.1% (w/v) SDS.
Sample loading buffer: use stacking gel buffer stock diluted 1/4–1/8, containing 10% (w/v) sucrose (or glycerol) and 0.002% tracking dye. Final concn of buffers:
    stacking gel: 0.125 M Tris–HCl pH 6.8;
    resolving gel: 0.375 M Tris–HCl pH 8.8.
[a]The columns represent volumes (ml) of stock solutions required to prepare the gel mixtures.
%Acrylamide = polyacrylamide gel concentration expressed in terms of *total* monomer (i.e. acrylamide and cross-linker); bisacrylamide = N,N'-methylene bisacrylamide; Tris = tris(hydroxymethyl)amino methane; TEMED = N,N,N',N'-tetramethylenediamine; APS=ammonium persulfate.

The discontinuous system is used for maximal resolution of protein bands because the proteins are stacked in a stacking gel before entering the smaller-pore resolving or separating gel. *Table 1* gives recipes for discontinuous systems that are likely to be useful for most types of separations. Recipes for the use of linear gradients as opposed to step gradients (in which gels of different concentration are layered one upon the other; *Table 2*) are also given.

Nondenaturing gels
Electrophoresis under native conditions is used in circumstances where the enzymatic activity or structure of a protein or protein complex under study is to be maintained. Native electrophoresis

Table 2. Gel mixtures for a 5-20% SDS–PAGE gradient gel

Stock solution	% Acrylamide[a]	
	5.0	**20.0[b]**
Acrylamide-bisacrylamide (30:0.8)	5.0	20.0
3.0 M Tris–HCl pH 8.8	3.75	3.75
10% (w/v) SDS	0.3	0.3
Water	20.25	2.75
1.5% (w/v) APS	0.7	0.7
TEMED	0.010	0.010

Electrophoresis buffer: 0.025 M Tris, 0.192 M glycine, pH 8.3, 0.1% (w/v) SDS.
[a] The columns represent volumes (ml) of stock solutions required to prepare 30 ml gel mixture.
[b] The 20.0% acrylamide mixture also requires the addition of 4.5 g sucrose (equivalent to 2.5 ml volume).

PROTEINS LABFAX

Table 3. Electrolytes used in IEF

Solution	Concn (M)	Use
CH$_3$COOH	0.5	Anolyte for alkaline pH ranges (pH > 7.0)[a]
H$_2$SO$_4$	0.1	Anolyte for very acidic pH ranges (pH < 4.0)[b]
H$_3$PO$_4$	1.0	Anolyte for all pH ranges
Histidine	0.2	Catholyte for acidic pH ranges (pH < 5.0)[b]
NaOH	1.0[c]	Catholyte for all pH ranges
Tris	0.5	Catholyte for acidic pH ranges (pH < 5.0)[b]

[a] Represents the lower limits of the pH range.
[b] Represents the higher limits of the pH range.
[c] Use air-tight plastic bottles for storage.

techniques can only be applied to protein samples which are soluble and which will not precipitate or aggregate during electrophoresis. As nondenaturing gels are not used on a day to day basis, recipes for such gels are not described here and the reader is advised to consult other texts (2, 3).

Isoelectric focusing
Isoelectric focusing (IEF) is a method in which enzymes are separated in a pH gradient according to their isoelectric points, that is, the pH at which the net charge of the enzyme molecule is zero. At this pH, the protein molecules have no electrophoretic mobility and will be concentrated or focused into narrow zones. The most popular method for generating pH gradients for IEF is the incorporation of low molecular weight amphoteric compounds, synthetic carrier ampholytes, into a polyacrylamide gel matrix. When an electric field is applied the ampholyte molecules 'migrate' to one or other of the electrodes depending on their net charge. *Table 3* lists solutions for IEF electrodes and *Table 4* gives recipes for IEF gels.

Table 4. Recipes for preparation of IEF gels (4–7%*T* and 4–8 M urea)

Gel vol. (ml)	30%*T* monomer solution[a] (ml)				2% Carrier ampholytes[b] (ml)		Urea (g)		TEMED (μl)	40% APS[c] (μl)
	4%*T*	5%*T*	6%*T*	7%*T*	A	B	4 M	8 M		
30	4.00	5.00	6.00	7.00	1.50	1.88	7.20	14.40	9.0	30
25	3.34	4.17	5.00	5.83	1.25	1.56	6.00	12.00	7.5	25
20	2.66	3.34	4.00	4.67	1.00	1.25	4.80	9.60	6.0	20
15	2.00	2.50	3.00	3.50	0.75	0.94	3.60	7.20	4.5	15
10	1.33	1.66	2.00	2.33	0.50	0.63	2.40	4.80	3.0	10
5	0.66	0.83	1.00	1.16	0.25	0.31	1.20	2.40	1.5	5

[a] Monomer solution (where %*T*=g monomers 100 ml^{-1}, and %*C*=g cross-linker 100 g^{-1} monomer):
30%*T*, 2.5%*C*; 29.25 g acrylamide and 0.75 g bisacrylamide mixed. Water added to final 100 ml volume.
30%*T*, 3.0%*C*; 29.10 g acrylamide and 0.90 g bisacrylamide mixed. Water added to final 100 ml volume.
30%*T*, 4.0%*C*; 28.80 g acrylamide and 1.20 g bisacrylamide mixed. Water added to final 100 ml volume.
[b] A, for 40% solution (Ampholine, Resolyte, Servalyte); B, for Pharmalyte.
[c] To be added after degassing the solution and just before pouring it into the gel mold.

Table 5. Recipes for first-dimensional IEF gel and second-dimensional SDS–PAGE gel (6)

(a) First dimension

Gel mixture

1.33 ml 28.38% acrylamide, 1.62% bisacrylamide; 5.5 g ultrapure urea; 2 ml 10% Nonidet P-40 (NP-40); 0.4 ml 40% Ampholines (pH 5.0–7.0); 0.1 ml 40% Ampholines (pH 3.5–10.0); 1.95 ml water; 10 μl 10% (w/v) APS; and 5 μl TEMED

Sample loading buffer: 9.5 M urea, 5% 2-mercaptoethanol, 2% NP-40, 1.6% Ampholines (pH 5.0–7.0), and 0.4% Ampholines (pH 3.5–10.0).
Electrolytes: anolyte, 10 mM H_3PO_4; catholyte, 20 mM NaOH.
Equilibration buffer: 2.5% (w/v) SDS, 5 mM dithiothreitol (DTT), 125 mM Tris–HCl pH 6.8, 10% (w/v) glycerol, 0.05% bromophenol blue.

(b) Second dimension

Stock solution	Stacking gel[a]	Resolving gel	
		Light solution[a]	Dense solution[a]
29.2% acrylamide, 0.8% bisacrylamide	0.75	5.3	4.3
0.5 M Tris–HCl pH 6.8, 0.4% (w/v) SDS	1.25	–	–
1.5 M Tris–HCl pH 8.8, 0.4% (w/v) SDS	–	4.0	2.0
75% glycerol	–	–	1.7
Water	3.0	6.7	–
10% (w/v) APS	0.015	0.025	0.010
TEMED	0.005	0.008	0.004

Electrophoresis buffer: 25 mM Tris, 0.192 M glycine, 0.1% (w/v) SDS.
Sealing gel: 0.1% (w/v) agarose in 0.125 M Tris–HCl pH 6.8. Melt the agarose in Tris buffer and allow to cool to 55°C for normal agarose, or 45°C for low gelling temperature (LGT) agarose, before adding a 10th volume 20% (w/v) SDS.

[a] Columns represent volumes (ml) of stock solutions required to prepare the gel mixtures.

Table 6. Reagents for immunoelectrophoresis

Gel mixture
1% (w/v) agarose[a] in Tris-barbital buffer pH 8.6
Tris-barbital buffer pH 8.6
2.4 g 5,5'-diethylbarbituric acid; 44.3 g Tris; distilled water to final 1 l volume. Dilute fivefold before use

Electrophoresis buffer: Tris-barbital buffer pH 8.6.
Agarose is dissolved by boiling for 5 min. Kept molten in a water bath at 50–60°C. Once cool and set, gel is stored at 4°C, and is ready for use once more after a short period of boiling.

[a] Use agarose with electroendosmosis $M_r = 0.13$ for normal procedures.

Table 7. Standard marker proteins used in either denaturing or IEF gels (2, 7-10). Several proteases, such as trypsin, chymotrypsin, and papain, have been used as molecular mass standards but these may sometimes cause proteolysis of other polypeptide standards and thus are omitted here

Polypeptide	Species	Tissue	Isoelectric point (pI)	No. of subunits	Subunit mass (kDa)
Acetylcholinesterase	*Electrophorus*	–	4.5	4	70.0
Adenine phosphoribosyltransferase	Human	Erythrocyte	4.8	3	11.0
Adenylate kinase	Rat	Liver (cytosol)	7.5	3	23.0
Agglutinin	Wheat	Germ	–	2	17.0
Alcohol dehydrogenase	Yeast	–	5.4	4	35.0
Aldolase	Yeast	–	5.2	2	40.0
Alkaline phosphatase	Calf	Intestine	4.4	2	69.0
Arginase	Human	Liver	9.2	4	30.0
Bovine serum albumin	Bovine	Serum	4.7	1	68.0
Carbonic anhydrase	Parsley	Leaf	–	6	28.7
Catalase	Cow	Liver	5.4	4	57.5
Ceramide trihexosidase	Human	Plasma	3.0	4	22.0
Chymotrypsinogen A	Bovine	Pancreas	9.2	1	25.7
Deoxyribonuclease I	Cow	Pancreas	4.8	1	31.0
Deoxyribonuclease II	Pig	Spleen	10.2	1	38.0
Enolase	Rabbit	Muscle	8.8	2	42.0
Fumarase	Pig	Heart	–	4	48.5
Galactokinase	Human	Erythrocyte	5.7	2	27.0
β-Galactosidase	*Escherichia coli*	–	–	4	130.0
β-Glucuronidase	Rat	Liver	6.0	4	75.0
Glyceraldehyde-3-phosphate dehydrogenase	Rabbit	Muscle	8.5	2	72.0
Glycerol-3-phosphate dehydrogenase	Rabbit	Kidney	6.4	2	34.0
Glycogen synthase	Pig	Kidney	4.8	4	92.0
Hemoglobin	Rabbit	Erythrocyte	7.0	4	16.0
Hexokinase	Yeast	–	5.3	2	51.0
Lactate dehydrogenase	Pig	Heart	–	4	36.0
β-Lactoglobulin	Bovine	Serum	5.2	2	17.5
Lipoxidase	Soybean	–	5.7	2	54.0
Lysine decarboxylase	*Escherichia coli*	–	4.6	10	80.0
Lysozyme	Chicken	Egg white	10.7	1	14.3
Malate dehydrogenase	Pig	Heart	5.1	2	35.0
Micrococcal nuclease	*Staphylococcus aureus*	–	9.6	1	16.8
Myoglobin	Horse	Heart	6.8	1	16.9
Myosin heavy chain	Rabbit	Muscle	–	2	212.0
Myrosinase	Rapeseed	–	5.0	2	65.0
Nerve growth factor	Mouse	Salivary gland	9.3	2	13.3
Pepsinogen	Pig	Stomach	3.7	1	41.0
Phosphoenolpyruvate carboxylase	*Escherichia coli*	–	5.0	4	99.6
Phosphoenolpyruvate carboxylase	Spinach	Leaf	4.9	2	130.0

Table 7. Continued

Polypeptide	Species	Tissue	Isoelectric point (pI)	No. of subunits	Subunit mass (kDa)
Phosphoglycerate mutase	Pig	Muscle	9.3	2	33.0
Phosphorylase *a*	Rabbit	Muscle	5.8	4	92.5
Pyruvate kinase	Rabbit	Muscle	6.6	4	57.2
Ribonuclease	Bovine	Pancreas	7.8	1	13.7
Thyroglobulin	Pig	Thyroid	4.5	–	330.0
Triosephosphate isomerase	Rabbit	Muscle	6.8	2	26.5
Trypsinogen	Cow	Pancreas	9.3	1	24.5
Tubulin	Pig	Brain	5.5	1	56.0
Urease	Jack bean	–	4.9	2	240.0
Uricase	Pig	Liver	6.3	4	32.0

Table 8. *In situ* direct polypeptide detection methods in SDS gels without staining

Detection method	Comments	Ref.
Chilling	SDS gels only. Low temperature crystallizes SDS	11
Precipitation with K^+ ions	SDS gels only. K-SDS is insoluble	12
Reaction with cationic surfactant	SDS gels only. Insoluble complex formed between SDS and surfactant	13
Sodium acetate	SDS gels only. Precipitates SDS not bound to protein	14
Via protein phosphorescence	Uses intrinsic phosphorescence of proteins	15

IEF using immobilized pH gradients (IPG), an alternative to IEF using carrier ampholytes, is often preferred. IEF using IPG is achieved by the incorporation of Immobiline reagents (Pharmacia Biosystems). The Immobiline reagents are a series of seven acrylamide derivatives, forming a series of buffers with different pK values distributed throughout the pH 3.0–10.0 range. For details, refs 2, 4 and 5 should be consulted.

2.2. Two-dimensional gel electrophoresis

Two-dimensional gel electrophoresis is widely used to separate complex mixtures of proteins into many more components than is possible in conventional one-dimensional electrophoresis. The commonest two-dimensional electrophoresis method for analyzing polypeptides is to separate the proteins in the first dimension on the basis of charge by IEF and then to separate the polypeptides in the second dimension in the presence of SDS, primarily on the basis of molecular mass of the polypeptides (*Table 5*). Recipes for electrophoresis buffers, equilibration buffers and sealing gels are also given.

2.3. Immunoelectrophoresis

Immunoelectrophoresis is a procedure in which proteins and other antigenic substances are characterized both by their electrophoretic migration in a gel and their immunological properties. There are many variations of the technique but they are all based on the electrophoretic migration

Table 9. Protein staining and detection methods used in IEF

Method	Application	Ref.
Alcian blue	Glycoproteins	16
Autoradiography	Radioactive proteins	17
Blotting	Antigens	18
Coomassie blue G-250	General use	19
Coomassie blue G-250/urea/perchloric acid	In presence of detergents	20
Coomassie blue R-250/CuSO$_4$	General use	21
	In presence of detergents	21
Coomassie blue R-250/sulfosalicylic acid	General use	22
Copper stain	General use	23
Fast green FCF	General use	24
Fluorography	Radioactive proteins	25
Immunoprecipation *in situ*	Antigens	26
Periodic acid - Schiff (PAS)	Glycoproteins	27
Print-immunofixation	Antigens	28
Silver stain	General use	29
Sudan black	Lipoproteins	30
Zymograms[a]	Enzymes	31

[a] The concentration of buffer in the assay medium usually needs to be increased to counteract the buffering action by carrier ampholytes.

of antigens in an antibody-containing gel and specific immunoprecipitation of the antigens by means of corresponding precipitating antibodies. *Table 6* gives the reagents necessary for immunoelectrophoresis.

3. SAMPLE PREPARATION

The preparation of protein samples and the use of protease inhibitors has been covered in Chapters 2, 6 and 7. Sample loading buffers (gel loading buffers) accompany recipes for gels (see Section 2). In order to calibrate gels, whether they are SDS–PAGE or IEF gels, standard proteins are used as markers (*Table 7*). It is usual to select three or four proteins that migrate similarly to the protein of interest.

4. ANALYSIS OF GELS

Proteins separated on gels can be visualized using a number of different procedures (*Tables 8–13*). Generally, staining using Coomassie blue is the most popular method. However, if a more sensitive stain is required, silver staining is recommended.

Table 14 lists the many matrices available for blotting gels. *Table 15* gives blocking solutions to prevent nonspecific binding of probe to matrix. Transfer buffers for the transfer of protein to the blotting matrix are given in *Table 16*. There are many detection procedures for proteins on blots, some of which are exemplified in *Table 17*. If the proteins are radioactive, they can be located by using autoradiography (*Tables 18* and *19*).

Electrophoresis Methods

Table 10. Staining procedures for two-dimensional polyacrylamide gels

Staining technique	Destaining technique	Comments
(a) 3–4 h in 0.1% Coomassie blue in methanol:water:acetic acid (5:5:1)	Overnight by diffusion against methanol:water:acetic acid (5:5:1)	–
(b) 20 min in 0.1% Coomassie blue in 50% trichloroacetic acid	Several changes of 7% (v/v) acetic acid	Removal of commercial ampholytes
(c) 3 h in 0.25% Coomassie blue in methanol:water:acetic acid (5:5:1)	Several changes of 5% (v/v) methanol, 10% (v/v) acetic acid	–
(d) 1–4 h in 0.1% Coomassie blue R-250 in 7.5% (v/v) acetic acid, 50% (v/v) methanol in water	Overnight in 7.5% (v/v) acetic acid, 50% (v/v) methanol in water	–
(e) Overnight in 25% (v/v) isopropyl alcohol, 10% (v/v) acetic acid, 0.025–0.05% Coomassie blue, followed by 6–9 h in 10% (v/v) isopropyl alcohol, 10% (v/v) acetic acid, 0.0025–0.005% Coomassie blue	Several changes of 10% (v/v) acetic acid	An additional optional staining overnight in 10% (v/v) acetic acid containing 0.0025% Coomassie blue helps intensify the gel pattern
(f) 15 min in 0.55% Amido black in 50% (v/v) acetic acid	40 h in 1% (v/v) acetic acid	–
(g) 3 h at 80°C, or overnight at room temperature in 0.1% Amido black in 0.7% (v/v) acetic acid, 30% (v/v) ethanol in water	Several changes of 7% (v/v) acetic acid, 20% (v/v) ethanol in water	–
(h) 1 h in 50% (v/v) methanol, stain with fresh alkaline 0.8% $AgNO_3$ solution. Wash with water for 5 min. To develop soak in fresh 0.02% HCHO, 0.005% citric acid for 10 min.	Wash gel in water, transfer to 50% (v/v) methanol	Much more sensitive than Coomassie blue methods

Table 11. Staining procedures used after immunoelectrophoresis

Stain	Staining procedure	Destaining procedure
Coomassie blue R-250	Stain gel for 5 min in 0.5% Coomassie blue R-250 in 96% ethanol: glacial acetic acid: water (4.5:1:4.5)	Destain 2–3 times, each for 5 min in 96% ethanol: glacial acetic acid: water (4.5:1:4.5)
Nigrosin	Stain gel until precipitates are stained sufficiently in 0.14% Nigrosin in glacial acetic acid:0.1 M sodium acetate: methanol: glycerol (2.15:5.75:1.4:0.7)	Destain in 20% (v/v) methanol in 5% (v/v) acetic acid followed by 5% (v/v) acetic acid and water

Table 12. *In situ* polypeptide detection methods in gels using fluorophore labeling

Fluorophore	Ref.
Labeling with fluorophore prior to electrophoresis	
Dansyl chloride	32–34
Fluorescamine	35–38
DACM (*N*-(dimethylamino-4-methylcoumarinyl) maleimide)	39
MDPF (2-methoxy-2, 4-diphenyl-3(2H)-furanone)	37, 40, 41
o-Phthaldialdehyde	42
Labeling with fluorophore after electrophoresis	
Anilinonaphthalene sulfonate (ANS)	43
Bis-ANS	44
Fluorescamine	45
p-Hydrazinoacridine	46
o-Phthaldialdehyde	47, 48

Table 13. *In situ* detection of specific classes of proteins in gels

Method	Ref.
Glycoproteins	
Crossed lectin electrophoresis	49
Fluorescent lectins	50
p-Hydrazinoacridine	46
Cell surface glycoproteins using galactose oxidase	51
Glycoproteins containing terminal *N*-acetylglucosamine using galactosyl transferase	52
Glycoproteins labeled *in vivo* using radiolabeled sugars	53–55
Lectins with covalently bound enzymes	56, 57
PAS	50, 58–60
Periodic acid-silver stain	61
Radiolabeled lectins	62-64
Stains-all	65
Thymol-sulfuric acid	50
Lipoproteins	
Silver stain	66, 67
Staining before electrophoresis	68
Staining after electrophoresis	69
Phosphoproteins	
Entrapment of liberated phosphate (ELP)	70, 71
Silver stain	72
Stains-all	73
Trivalent metal chelation for acidic phosphoproteins (phosvitins)	71

Table 14. Common blotting membranes

Blotting matrix	Additional information	Supplier (see Section 6)
Glass fiber	Recommended for amino acid sequence analysis of separated proteins. Protein-binding capacity between 10–20 $\mu g\ cm^{-2}$. The proteins can be blotted from SDS–PAGE gels and then acid-hydrolyzed directly whilst still immobilized on the filter	BML, JNP
Nitrocellulose membrane	Pure nitrocellulose membranes have good protein binding capacity (~80–100 $\mu g\ cm^{-2}$). Nitrocellulose membranes with 0.45 μm pore size are typically used but 0.22–0.1 μm has been recommended for lower molecular mass proteins. Recommended for immunodetection analyses due to reduced nonspecific binding of antibodies, and for the analysis of basic proteins. Mixed ester membranes which contain cellulose acetate have reduced capacity	AIP, BML, BRL, EKL, GBL, MPC, PMB, RSL, S&S, SCC, STG
Nylon membrane	Far stronger and more robust than conventional pure nitrocellulose sheets. Substantially higher protein binding capacity, e.g. Zeta-Probe (Bio-Rad) has approximately six-fold higher protein binding capacity (~480 $\mu g\ cm^{-2}$) than nitrocellulose. Due to its highly cationic nature, recommended for electroblotting of SDS–PAGE gels to maximize binding of highly anionic SDS–polypeptide complexes. Charged nylon filters such as Zeta-Probe recommended for electroelution of SDS–PAGE gels. Binding also much stronger than in the case of nitrocellulose. Uncharged nylon membranes should give higher binding of basic proteins. Major disadvantage of all nylon membranes is the lack of simple general staining procedures	AIP, BHM, BML, BRL, EKL, GBL, ICN, PMB, RSL, SCC, STG
Polyvinyldifluoride (PVDF)	Hydrophobic in nature. Compatible with commonly used protein stains as well as standard immunodetection methods	BRL, ICN, MPC, S&S, SCC

Table 15. Blocking solutions to prevent nonspecific binding of the probe to the matrix in protein blotting (74)

Blocker	Concn (% w/v)	Blocker	Concn (% w/v)
Bovine serum albumin (BSA)	0.5–10	Milk	5
Casein	1–2	Newborn calf serum (NCS)	5
Ethanolamine	10	Ovalbumin	1–5
Fetal calf serum (FCS)	10	Polyvinylpyrrolidone	2
Gelatin	0.25–3	Tween 20	0.05–0.5

Table 16. Transfer buffers used in blotting

Transfer buffer	Additional information	Ref.
25 mM Tris, 192 mM glycine, 20% (v/v) methanol, pH 8.3	0.05–0.1% (w/v) SDS can be included	18
48 mM Tris, 39 mM glycine, 20% (v/v) methanol, pH 9.2	0.0375% (w/v) SDS can be included	75
10 mM NaHCO$_3$, 3 mM Na$_2$CO$_3$, 20% (v/v) methanol, pH 9.9	–	76
0.7% (v/v) acetic acid	–	77

Table 17. General protein stains for blot transfers

Protein stain	Blotting matrix	Sensitivity
Amido black 10B	Nitrocellulose, PVDF	1.5 μg
Colloidal gold	Nitrocellulose, PVDF	4 ng
Colloidal iron	Nitrocellulose, nylon, PVDF	30 ng
Coomassie brilliant blue R-250[a]	Nitrocellulose, PVDF	1.5 μg
Fast green FC	Nitrocellulose, PVDF	–
India ink	Nitrocellulose, PVDF	100 ng
In situ biotinylation + HPR-avidin	Nitrocellulose, nylon, PVDF	30 ng
Ponceau S	Nitrocellulose, PVDF	–
Silver-enhanced copper	Nitrocellulose, nylon	–

[a] Results obtained in high background.

Table 18. *In situ* detection of radioactive proteins in gels

Detection method	Ref.
Double-label detection using X-ray film	78, 79
Electronic data capture	80, 81
Fluorography using 2,5-diphenyloxazole (PPO) in dimethylsulfoxide (DMSO)	17, 82
Fluorography using PPO in glacial acetic acid	83
Fluorography using sodium salicylate	82, 84, 85
Fluorography using commercial reagents	86, 87
Image intensification	88
Indirect autoradiography using an X-ray intensifying screen	89, 90
Quenching of radiolabeled proteins by gel conditions	91
Radiolabeling proteins *in vivo* prior to electrophoresis	92, 93
Radiolabeling proteins *in vitro* prior to electrophoresis	92
Radiolabeling proteins after gel electrophoresis	94, 95

Electrophoresis Methods

19. Blakesly, R.W. and Boezi, J.A. (1977) *Anal. Biochem.*, **82**, 580.

20. Vesterberg, O. (1972) *Biochim. Biophys. Acta*, **257**, 11.

21. Righetti, P.G. and Drysdale, J.W. (1974) *J. Chromatogr.*, **98**, 271.

22. Neuhoff, V., Stamm, R. and Eibl, H. (1985) *Electrophoresis*, **6**, 427.

23. Lee, C., Levin, A. and Branton, D. (1987) *Anal. Biochem.*, **166**, 308.

24. Allen, R.E., Masak, K.C. and McAllister, P.K. (1980) *Anal. Biochem.*, **104**, 494.

25. Laskey, R.A. (1980) *Methods Enzymol.*, **65**, 363.

26. Richtie, R.F. and Smith, R. (1976) *Clin. Chem.*, **22**, 497.

27. Hebert, J.P. and Strobbel, B. (1974) *LKB Application Note # 151*.

28. Arnaud, P., Wilson, G.B., Koistinen, J. and Fudenberg, H.H. (1977) *J. Immunol. Methods*, **16**, 221.

29. Merril, C.R., Goldman, D., Sedman, S.A. and Ebert, M.H. (1981) *Science*, **211**, 1438.

30. Godolphin, W.J. and Stinson, R.A. (1974) *Clin. Chim. Acta*, **56**, 97.

31. Harris, H. and Hopkinson, D.A. (1976) *Handbook of Enzyme Electrophoresis in Human Genetics*. Elsevier, Amsterdam.

32. Stephens, R.E. (1975) *Anal. Biochem.*, **65**, 369.

33. Schetters, H. and McLeod, B. (1979) *Anal. Biochem.*, **98**, 329.

34. Tijssen, P. and Kurstak, E. (1979) *Anal. Biochem.*, **99**, 97.

35. Eng, P.R. and Parker, C.O. (1974) *Anal. Biochem.*, **59**, 323.

36. Ragland, W.L., Pace, J.L. and Kemper, D.L. (1974) *Anal. Biochem.*, **59**, 24.

37. Douglas, S.A., La Marca, M.E. and Mets, L.J. (1978) in *Electrophoresis '78* (N. Catsimpoolas, ed.). Elsevier, Amsterdam, Vol. 2, p. 155.

38. Ragland, W.L., Benton, T.L., Pace, J.L., Beach, F.G. and Wade, A.E. (1978) in *Electrophoresis '78* (N. Catsimpoolas, ed.). Elsevier, Amsterdam, Vol. 2, p. 217.

39. Yamamoto, K., Okamoto, Y. and Sekine, T. (1978) *Anal. Biochem.*, **84**, 313.

40. Burger, B.O., White, F.C., Pace, J.L., Kemper, D.L. and Ragland, W.L. (1976) *Anal. Biochem.*, **70**, 327.

41. Urwin, V.E. and Jackson, P. (1991) *Anal. Biochem.*, **195**, 30.

42. Weidekamm, E., Wallach, D.F.H. and Flückiger, R. (1973) *Anal. Biochem.*, **54**, 102.

43. Hartman, B.K. and Udenfriend, S. (1969) *Anal. Biochem.*, **30**, 391.

44. Harowitz, P.M. and Bowman, S. (1987) *Anal. Biochem.*, **165**, 430.

45. Jackowski, G. and Liew, C.C. (1980) *Anal. Biochem.*, **102**, 34.

46. Carson, S.D. (1977) *Anal. Biochem.*, **78**, 428.

47. Liebowitz, M.J. and Wang, R.W. (1984) *Anal. Biochem.*, **137**, 161.

48. Andrews, A.T. (1986) *Electrophoresis: Theory, Techniques and Biochemical and Clinical Applications*. Clarendon Press, Oxford, p. 29.

49. West, C.M. and McMahon, D. (1977) *J. Cell Biol.*, **74**, 264.

50. Gander, J.E. (1984) *Methods Enzymol.*, **104**, 447.

51. Gahmberg, C.G. (1978) *Methods Enzymol.*, **50**, 204.

52. Wallenfels, B. (1979) *Proc. Natl Acad. Sci. USA*, **76**, 3223.

53. Gregg, J.H. and Karp, G.C. (1978) *Exp. Cell Res.*, **112**, 31.

54. Bradshaw, J.P. and White, P.A. (1985) *Biosci. Rep.*, **5**, 229.

55. Taylor, T. and Weintraub, B.D. (1985) *Endocrinology*, **116**, 1968.

56. Avigad, G. (1978) *Anal. Biochem.*, **86**, 443.

57. Moroi, M. and Jung, S.M. (1984) *Biochim. Biophys. Acta*, **798**, 295.

58. Fairbanks, G., Steck, T.L. and Wallach, D.L.H. (1971) *Biochemistry*, **10**, 2026.

59. Wardi, A.H. and Michos, G.A. (1972) *Anal. Biochem.*, **49**, 607.

60. Furlan, M., Perret, B.A. and Beck, E.A. (1979) *Anal. Biochem.*, **96**, 208.

61. Dubray, G. and Bezard, G. (1982) *Anal. Biochem.*, **119**, 325.

62. Burridge, K. (1978) *Methods Enzymol.*, **50**, 54.

63. Dupuis, G. and Doucet, J.P. (1981) *Biochim. Biophys. Acta*, **669**, 171.

64. Koch, G.L.E. and Smith, M.J. (1982) *Eur. J. Biochem.*, **128**, 107.

65. King, L.E. and Morrison, M. (1976) *Anal. Biochem.*, **71**, 223.

66. Goldman, D., Merril, C.R. and Ebert, M.H. (1980) *Clin. Chem.*, **26**, 1317.

67. Tsai, C.M. and Frasch, C.E. (1982) *Anal. Biochem.*, **119**, 115.

68. Ressler, N., Springate, R. and Kaufman, J. (1961) *J. Chromatogr.*, **6**, 409.

69. Prat, J.P., Lamy, J.N. and Weill, J.D. (1969) *Bull. Soc. Chim. Biol.*, **51**, 1367.

70. Debruyne, I. (1983) *Anal. Biochem.*, **133**, 110.

71. Cutting, J.A. (1984) *Methods Enzymol.*, **104**, 451.

72. Satoh, K. and Busch, H. (1981) *Cell Biol. Int. Rep.*, **5**, 857.

73. Green, M.R., Pastewka, J.V. and Peacock, A.C. (1973) *Anal. Biochem.*, **56**, 43.

74. Gershoni, J.M. (1987) in *Advances in Electrophoresis* (A. Chrambach, M.J. Dunn and B.J. Radola, eds). VCH, Weinheim, Vol. 1, p. 141.

75. Bjerrum, O.J. and Schafer-Nielsen, C. (1986) in *Electrophoresis '86* (M. Dunn, ed.). VCH, Weinheim, p. 315.

76. Dunn, S.D. (1986) *Anal. Biochem.*, **157**, 144.

77. In *Protein Blotting, a guide to transfer and detection.* (1993) Bio-Rad Laboratories, Bulletin 1721. Bio-Rad Laboratories Ltd, Hemel Hempstead, UK.

78. Walton, K.E., Styer, D. and Gruenstein, E. (1979) *J. Biol. Chem.*, **254**, 795.

79. Cooper, P.C. and Burgess, A.W. (1982) *Anal. Biochem.*, **126**, 301.

80. Davidson, J.B. and Case, A. (1982) *Science*, **215**, 1398.

81. Burbeck, S. (1983) *Electrophoresis*, **4**, 127.

82. Bonner, W.M. (1984) *Methods Enzymol.*, **104**, 461.

83. Skinner, K. and Griswold, M.D. (1983) *Biochem. J.*, **209**, 281.

84. Chamberlain, J.P. (1979) *Anal. Biochem.*, **98**, 132.

85. Heegard, N.H.H., Hebsgaard, K.P. and Bjerrum, O.J. (1984) *Electrophoresis*, **5**, 230.

86. Roberts, P.L. (1985) *Anal. Biochem.*, **147**, 521.

87. McConkey, E.H. and Anderson, C. (1984) *Electrophoresis*, **5**, 230.

88. Laskey, R.A. (1981) *Amersham Research News* No.23.

89. Laskey, R.A. and Mills, A.D. (1977) *FEBS Lett.*, **82**, 314.

90. Bonner, W.M. (1983) *Methods Enzymol.*, **96**, 215.

91. Harding, C.R. and Scott, I.R. (1983) *Anal. Biochem.*, **129**, 371.

92. Dunbar, B.S. (1987) *Two-dimensional Gel Electrophoresis and Immunological Techniques.* Plenum Press, New York.

93. Latter, G.I., Burbeck, S., Fleming, S. and Leavitt, J. (1984) *Clin. Chem.*, **30**, 1925.

94. Christopher, A.R., Nagpal, M.L., Carrol, A.R. and Brown, J.C. (1978) *Anal. Biochem.*, **85**, 404.

95. Zapolski, E.J., Gersten, D.M. and Ledley, R.S. (1982) *Anal. Biochem.*, **123**, 325.

96. Rickwood, D., Patel, D. and Billington, D. (1993) in *Biochemistry Labfax* (J.A.A. Chambers and D. Rickwood, eds). BIOS Scientific Publishers Ltd, Oxford.

97. Johns, E.W. (1976) in *Subcellular Components: Preparation and Fractionation* (G.D. Birnie, ed.). Butterworth, London, p. 202.

98. Kruh, J., Schapira, G., Lareau, J. and Dreyfus, J.C. (1964) *Biochim. Biophys. Acta*, **87**, 669.

99. Adamietz, P. and Hiltz, H. (1976) *Hoppe-Seylers Z. Physiol. Chem.*, **357**, 527.

100. Sinclair, J.H. and Rickwood, D. (1985) *Biochem. J.*, **229**, 771.

CHAPTER 10
INTEGRAL MEMBRANE PROTEINS
R.J. Cogdell and J.G. Lindsay

1. INTRODUCTION

Membrane proteins, in general, are much more difficult to study than their soluble counterparts. Detailed analysis of proteins requires their purification in milligram amounts in the native state and a convenient means of assay to ensure their functionality. Clearly, before this can be achieved, integral proteins must be released from their host membrane in a soluble, active form. Typical membrane proteins possess both hydrophobic and hydrophilic surfaces so detergents must be employed to screen the exposed hydrophobic segments from the aqueous phase. Disruption of the lipid bilayer by detergents is not a gentle procedure so that the ability of the protein to retain its native structure is critically dependent on the choice of detergent. Additionally, many important membrane proteins, for example receptors, are present in small amounts and at present no routine methods are available for successful overexpression of hydrophobic membrane proteins in their native state. Finally, many membrane polypeptides, for example transporters, only have an assayable function when located within the anisotropic environment of a lipid bilayer. This short contribution provides an introductory guide to overcoming these problems which should help workers in the field to make informed choices on the best general approaches to adopt when attempting to purify and analyze their favorite membrane protein.

2. PROBLEMS OF SOLUBILIZATION, PURIFICATION AND ASSAY OF MEMBRANE PROTEINS

2.1. Choice of detergent
There are now many detergents on the market, although there is still no obvious rule-of-thumb when choosing the most appropriate detergent. In *Table 1*, information is collated on some of the more common detergents along with comments on their uses. It is worth testing a wide range of detergents initially, although many of the more esoteric ones are expensive. It is often possible to maintain proteins in a common detergent in the early stages of a purification when large volumes are needed, and exchange into a more suitable (expensive) detergent for structure–function studies at a later stage.

2.2. Methods of purification
There is no special formula for purifying membrane proteins. Once solubilized in their native state, conventional techniques of protein purification can be applied, although it must be appreciated that the protein is present in a detergent micelle. Thus the properties of the detergent and the micelle must be taken into account when devising a purification strategy. Two examples will serve to illustrate this point. Solubilization of proteins in anionic detergents, such as cholate or SDS, exclude the employment of anion exchange columns which will absorb the detergent. The use of nonionic detergents like LDAO is not compatible with $(NH_4)_2SO_4$ precipitation, as high salt concentrations promote a 'phase separation' usually seen as an oily layer owing to the

Table 1. General properties of a range of detergents used with membrane proteins

Detergent type	m^a	n	M_r	CMC[b](mM)	Comments
Ionic detergents					
Na$^+$/Li$^+$ dodecyl sulfate			304/288	3.5	This is a cheap detergent but is denaturing for most membrane proteins. Commonly used in PAGE. In the presence of Li$^+$ it is more soluble at low temperatures. The CMC quoted was determined in the presence of 10 mM NaCl
Na$^+$ cholate			430	13–15	This is a mild, rather cheap detergent. Has rather poor solubility below pH 8.0 and is easy to remove by dialysis
Na$^+$ deoxycholate			414	4–6	Very similar to Na$^+$ cholate. Lower CMC, but still will dialyze away well. Like many ionic detergents it is more soluble in the presence of high salt
Nonionic detergents					
(a) Polyethyleneoxide (PEO) head groups $(C_mH_{m+1}-E_n-OH)$ where $E = -OCH_2CH_2$					
Brij–series (C_mE_n)	8	5	350	6.0	Rather mild detergents, not very commonly used. Can be expensive, but are useful in crystallization studies, e.g. porin (1)
	10	6	422	0.90	
	12	8	538	0.071	
	14	8	566	0.009	
	16	8	594	0.00021	
Triton series (E_n)					
Triton X-45	4–5		404	0.11	Mild detergents and very cheap. However, difficult to dialyze away, they absorb at 280 nm and there are mixtures with different chain lengths. Not so suitable for crystallization. Will bind to Biobeads
X-114	7–8		537	0.20	
X-100	9–10		625	0.24	
X-405	40		1966	0.81	

Tween series (PEO polysorbates)				
Tween 20	12	1288	0.08[c]	Mild, cheap detergents. Often used in purification of membrane proteins
40	16	–	0.012[c]	
60	18	–	0.0027[c]	
80	9=9	1310	0.10[c]	
(b) Sugar derivatives as head groups				
Alkyl-β-D-glucopyranosides	6	264	250	Mild, usually very expensive detergents. Very useful for crystallization studies. Easy to remove by dialysis—most common one is $m=8$
	7	278	79	
	8	292	30.3	
	9	306	6.5	
	10	320	2.6	
Aklyl-β-D-thioglucopyranosides	7	294	30	Mild, expensive detergents. Useful for crystallization studies
	8	308	9	
Alkyl-β-D-maltosides	10	483	1.6	Mild, expensive detergents. The $m=12$ is often used to stabilize solubilized membrane proteins, especially those from chloroplasts
	12	511	0.15	
Alkanoyl-N-methyl-glucamide (MEGA-m)	8	322	58	Expensive detergents. Mega 8 has been used to obtain crystals of bacterial reaction centers. Easily removed by dialysis
	9	336	25	
	10	350	7	
Zwitterionic/N-oxide detergents				
N-Alkyl-N,N-dimethylamine N-oxide	8	173	162	Quite strong detergents. LDAO is very cheap. Used with many photosynthetic membrane proteins. Good for crystallization, e.g. with bacterial reaction centers. Can be destroyed by chemical reduction of the amine oxide bond
	9	187	50.8	
Decyldimethylamine-N-oxide (DDAO)	10	201	22	
Lauryldimethylamine-N-oxide (LDAO)	12	229	1.4	
Alkyl/methylammoniopropane-sulfonate	8	280	200	
	10	308	20	
	12	336	2.7	
	14	364	0.28	
	16	392	0.03	

Membrane Proteins

Table 1. Continued

Detergent type	m^a	n	M_r	CMC^b (mM)	Comments
Dodecyldimethylaminopropyl-sulfoxide			335	3.0	
Bile acid derivative detergents					
3-(3-Cholamidopropyl)-dimethylammonio-1-propanesulfonate (CHAPS)			615	2–10	
3-(3-Cholamidopropyl)-dimethylammonio-2-hydroxypropane-1-sulfonate (CHAPSO)			631	4	

[a] m = number of carbon atoms in the alkyl chain.
[b] CMC=critical micelle concentration.
[c] %w/w for the Tween series only.
N.B. CMCs can vary depending upon the temperature, detergent purity and solution conditions, i.e. ionic strength and pH.
These data were compiled from refs 2–4.

formation of a detergent-rich phase which is often sufficient to denature the protein. Gel filtration techniques are also not normally applicable to the purification of integral hydrophobic proteins as the presence of these proteins in large mixed-detergent micelles renders their separation based on size ineffective. Another difficulty with many detergents is their absorption profile in the UV region, for example the Triton series, which presents difficulties in monitoring the elution profile of proteins from column matrices at 280 nm or analysis of circular dichroism (CD) spectra. Several common detergents also interfere to some extent with standard protein assays.

2.3. Assay: use of reconstitution

Many membrane proteins perform enzymatic functions which can be measured by conventional means; however, the majority of membrane-bound proteins are involved in recognition, transmembrane signaling or transport activities which are difficult to monitor routinely during purification. In this case, it may be necessary to use an indirect assay, although this will not provide definitive information on the functionality of the protein; for example ^{3}H-cytochalasin B binding to glucose transporters (5), or, ^{35}S-carboxyatractyloside binding to the mitochondrial adenine nucleotide translocase (6). When the vectorial properties of the protein need to be analyzed, reconstitution into artifical phospholipid vesicles will be necessary (see Further reading: Liposomes and reconstitution methods).

Reconstitution of a phospholipid bilayer around a detergent-solubilized integral membrane protein is usually achieved by the addition of the appropriate phospholipids and removal of the detergent. This frequently relies on dialysis, which is suitable for removal of detergents such as cholate or CHAPS with a high CMC (see *Table 1*). Certain detergents can also be absorbed selectively by treatment with polystyrene beads (e.g. SM-Biobeads; 7,8). As the detergent concentration falls below a critical level, sealed phospholipid vesicles form spontaneously. Another approach which has been used for the successful reconstitution of transport proteins involves sonication and freeze/thaw cycles to disrupt preformed vesicles, permitting the insertion of the protein into the membrane. All these methods are fully described in the references in Further reading: Liposomes and reconstitution methods.

Clearly, while artificial liposomes can be valuable tools for studying the function of membrane proteins, their routine use for monitoring protein purification regimes is probably impractical in many cases. In addition, the extent of reconstituted activity may reflect many factors other than the integrity of the protein in question, such as the efficiency and uniformity of insertion of the polypeptide into the lipid bilayer and the phospholipid composition of the liposomes. Moreover, there is always the possibility that partially denatured or inactive proteins will be renatured/activated on incorporation into the membrane. In spite of these limitations, reconstitution is the only alternative in many cases and it will undoubtedly remain a useful technique for the foreseeable future.

3. STRUCTURAL ANALYSIS OF MEMBRANE PROTEINS

With the application of molecular cloning techniques to integral membrane proteins, a large database of primary sequences has accumulated in recent years. However, it is not possible to determine 3D structures from amino acid sequences at present, although predictive computer algorithms can be employed to search for putative hydrophobic transmembrane α-helices or β-sheets, providing information on secondary structure and the possible transmembrane organization of the protein. Thus, detailed structural studies require production of 2D or 3D crystals for electron or X-ray diffraction analysis. For smaller proteins or overexpressed individual domains, 2D-nuclear magnetic resonance (NMR) is a viable alternative, although this

Membrane Proteins

approach is limited to polypeptides of less than 20 kDa at present. It is beyond the scope of this chapter to review molecular modeling techniques; however, the reviews by Eisenberg and Chothia (see Further reading) provide a useful introduction to this subject.

Electron crystallography of ordered 2D arrays of membrane proteins is increasingly being recognized as a viable alternative to convential X-ray diffraction techniques, which require the growth of well-ordered 3D-crystals. This technique has been applied successfully to the analysis of bacteriorhodopsin and the major plant antenna complex (9,10). It has three potential advantages: first, it requires less protein than the 3D approach; second, there is some justification for believing that ordered 2D arrays in a lipid bilayer provide an environment which more closely resembles the natural state; and third, this approach does not require the production of isomorphous heavy atom derivatives—often a major limitation in X-ray crystallography. The review by Kuhlbrandt (see Further reading) is an excellent guide to 2D-analysis by electron imaging techniques and its advantages and limitations.

Protocols for performing systematic trials designed to ensure optimal production of 3D crystals of integral membrane proteins are clearly described in the CRC handbook by Michel (see Further reading). This is still the method of choice for high-resolution structural determination if it proves possible to induce the growth of well-ordered crystals. Briefly, crystallization occurs as a consequence of reducing the solubility of the protein in solution. At the critical point when the solution becomes supersaturated, the protein either precipitates out or crystal formation is induced. Crystal formation is encouraged if this critical stage is passed through slowly enough to allow formation of ordered arrays rather than amorphous aggregates. The reduction in solubility is achieved by the addition of precipitants such as salts, organic solvents or long-chain polymers. However, with membrane proteins there is a major drawback, as detergents have a tendency to undergo phase separation in the presence of high concentrations of precipitants. Supersaturation is not achieved and hydrophobic proteins tend to be denatured as they partition into the detergent-rich phase where they are exposed to high concentrations of detergents. This problem can be overcome by the addition of small molecules, termed amphiphiles, such as heptane-1,2,3-triol or benzamidine-HCl to the crystallization mixture. In these cases, the phase diagram of the detergent is altered so that supersaturation of the protein occurs before phase separation, allowing the possibility of crystal formation.

The major limitation of this approach is that it requires regular supplies of pure, fully active protein, with reasonable long-term stability, as it often takes 3–4 weeks to produce crystals of appropriate size and quality for X-ray analysis. However, now that many of the critical ground rules for promoting crystallization of membrane proteins have been established, there will undoubtedly be increased interest in this area and we can look forward confidently to viewing the structures of many more important membrane proteins in the near future.

4. REFERENCES

1. Weiss, M.S., Kreusch, A., Schitz, E., Nestrel, H., Welte, W., Weckesser, J. and Schultz, G.E. (1991) *FEBS Lett.*, **280**, 379.

2. Lever, T.M., Cogdell, R.J. and Lindsay, J.G. (1994) in *Membrane Protein Expression Systems—A Guide* (G.W. Gould, ed.). Portland Press, London, p. 1.

3. Garavito, R.M. (1991) in *Crystallisation of Membrane Proteins* (H. Michel, ed.). CRC Press, Boca Raton, FL, p. 89.

4. Zulauf, M. and Michel, H. (1991) in *Crystallisation of Membrane Proteins* (H. Michel, ed.). CRC Press, Boca Raton, FL, p. 209.

5. Shanahan, M.F. (1982) *J. Biol. Chem.*, **257**, 7290.

6. Riccio, P., Aquila, H. and Klingenberg, M. (1975) *FEBS Lett.*, **56**, 129.

7. Allen, T.M., Romans, A.Y., Kercret, H. and Segrest, J.P. (1980) *Biochim. Biophys. Acta*, **601**, 328.

8. Holloway, P.N. (1973) *Anal. Biochem*, **53**, 304.

9. Henderson, R., Baldwin, J.M., Ceska, T.A., Zemlin, F., Beckmann, E. and Downing, K.H. (1990) *J. Mol. Biol.*, **213**, 899.

10. Kuhlbrandt, W., Wang, D.N. and Fuijyoshi, Y. (1994) *Nature*, **367**, 614.

5. FURTHER READING

5.1. Liposomes and reconstitution methods

This group of references will provide the experimenter with an excellent practical guide on the production and use of liposomes.

Gregoriadis, G. and Allison, A.C. (eds) (1980) *Liposomes in Biological Systems*. John Wiley and Sons, New York.

Priessen, A.J.M. and Konings, W.N. (1993) *Methods Enzymol.*, **221**, 394.

Wohlrab, H., Kolbe, H.V.J. and Collins, A. (1986) *Methods Enzymol.*, **125**, 697.

New, R.R.C (ed.) (1990). *Liposomes: A Practical Approach*. IRL Press, Oxford.

5.2. General

The first two references provide a comprehensive introduction to the main modelling strategies.

Chothia, C. (1984) *Ann. Rev. Biochem.*, **53**, 537.
An extremely useful guide to the principles of protein structure—important reading for all modellers.

Eisenberg, D. (1984) *Ann. Rev. Biochem.*, **53**, 595.
A lucid survey of modelling approaches to the analysis of structures of membrane proteins.

Gould, G.W. (ed.) (1994) *Membrane Protein Overexpression Systems—A User's Guide*. Portland Press, London.
An excellent laboratory guide of all the best current methods for obtaining functional overexpression of membrane proteins.

Kuhlbrandt, W. (1992) *Quart. Rev. Biophys.*, **25**, 1.
A comprehensive review of methods of 2D crystallography designed specifically for membrane proteins.

Michel, H. (ed.) (1990) *Crystallisation of Membrane Proteins*, CRC Handbook. CRC Press, Boca Raton, FL.
Many substantial reviews on the use of detergents and reviews on approaches to crystallizing membrane proteins. An excellent starting point for all prospective crystallizers of membrane proteins.

Tanford, C. (1980) *The Hydrophobic Effect*. John Wiley and Sons, New York.
Essential reading—still a landmark reference work providing a valuable introduction to the properties and use of detergents.

Membrane Proteins

single affinity chromatography step. In addition, the presence of a fusion partner can facilitate detection, improve stability, oligomerize monomeric species and provide a common purification method for mutant species of a given recombinant protein. The disadvantages of gene fusions including potential interference with function of the expressed product and an increase in molecular mass are either outweighed by the advantages or, by a judicious choice of vector and fusion partner, can generally be overcome.

2. PROTEINS COMMONLY USED IN GENE FUSION TECHNOLOGY

The basic properties of the most commonly used fusion partners are listed in *Table 2*. These proteins range from high molecular mass oligomeric species such as β-galactosidase from *E. coli*, to short peptide sequences, such as a sequence of six consecutive histidines. The corresponding vectors are listed in *Table 3*. The choice of fusion partner is governed largely by the properties of the recombinant species to be expressed and the experimental use to which the product will be put; for example, if the fusion protein is to be used primarily as a source of antigen for the production of antisera, less importance is attached to the biological activity of the expressed protein. In this case, there is no absolute need to remove the fusion partner. Most fusion partners can be used for the generation of antigenic protein. If the protein product is required in a biologically active form, then it is important to establish whether the addition of the fusion partner modifies activity. If the fusion partner does influence the activity of a recombinant protein, then it is important to choose a fusion partner that can be readily removed by proteolysis. If the fusion partner has no observable effect upon the function of the recombinant protein, then it is often convenient to leave the fusion partner attached. In our experience, there appears to be no single fusion partner that satisfies all requirements, and indeed, there are many reports, often unpublished, of proteins that are not expressed at significant levels as fusions with any of those proteins listed in *Table 2*.

In order to choose the most appropriate fusion partner a few general rules apply:

(i)   small polypeptides are probably best fused to high molecular mass fusion partners to prevent intracellular degradation by proteolysis;

Table 2. Physical properties of the most common proteins used in gene fusion technology

Fusion protein	Mol. mass (kDa)[a]	Orientation[b]	Quaternary structure[c]
β-Galactosidase	116	N or C	Tetramer
Maltose binding protein	40	N	Monomer
Protein A	31	N	Monomer
Glutathione-S-transferase	26	N	Dimer
Thioredoxin	12	N or C	Monomer
S-tag	2	N	Monomer
His-tag	1	N, S or C	Monomer
Strep-tag	1	N, S or C	Monomer

[a] Polypeptide molecular mass is rounded to the nearest kDa whole number.
[b] Orientation refers to the point at which the fusion protein (F) is attached to the recombinant protein (R). N refers to a protein with the organization F-R where F is N-terminal to (F); C to one in which the sequence is R-F and S is a sandwich with the sequence F-R-F.
[c] The fused product will be produced as an oligomer with either β-galactosidase or glutathione-S-transferase. Small peptides may be expressed in a form less susceptible to proteolysis with the larger fusion partners. If a naturally occurring homo-oligomer is fused to either β-galactosidase or glutathione-S-transferase, the fusion protein may form insoluble aggregates.

Table 3. Gene fusion vectors

Fusion protein	Vector[a]	Inducer[b]	Marker[c]	Supplier[d]
β-Galactosidase	pMC181	None	Tet	Pharmacia
Maltose binding protein	pMAL-c2[e]	IPTG	Amp	NEB
Protein A	pRIT2T	Thermal	Amp	Pharmacia
Glutathione-S-transferase	pGEX-2T[e]	IPTG	Amp	Pharmacia
Thioredoxin	pTruxFus[e]	Trp	Amp	Invitrogen
S-tag	pET29a-c(+)	Phage/IPTG	Kan	Novagen
His-tag	pET14b[e]	Phage/IPTG	Amp	Novagen
Strep-tag	pASK60	IPTG	Amp	Biometra

[a] The vectors listed are all commercially available and represent examples of those encoding the corresponding fusion partners. There are frequent developments in vector design which improve upon the features described here. All vectors are described in detail by the manufacturers, who often provide the nucleotide sequences on disk.

[b] Induction with IPTG is based on the release of the *lac* repressor from the *lac* or *tac* promoter in the presence of the inducer. This may be employed directly by activating transcription of the gene fusion, or alternatively by activating expression of a *lac* promoter–T7 RNA polymerase gene fusion which drives the specific transcription of the phage T7 promoters incorporated in the fusion vector.

[c] Antibiotics should be used at standard concentrations in all cell cultures.

[d] The list of suppliers is by no means exhaustive and those vectors chosen exemplify the types of fusion vectors which are readily available.

[e] These vectors are available in multiple forms and represent a series, all of which have been designed to suit a wide variety of experimental needs.

(ii) oligomeric polypeptides are probably best fused to monomeric fusion partners;

(iii) The biological activity of the recombinant protein should not be related to that of the fusion partner;

(iv) if there is no simple assay for the recombinant protein, it should be fused to a fusion partner for which either antibodies or a simple *in vitro* assay are available;

(v) if the fusion protein is required for antibody production, the fusion protein should be amenable to purification under strongly denaturing conditions (e.g. the histidine tag).

The smaller peptides that are employed as fusion partners are quite diverse in their properties. Antigenic peptide sequences have been widely employed as so-called epitope tags to facilitate the purification of a recombinant protein, although the high affinities often observed between such epitope tags and immunoglobulins often compromise recovery of the purified fusion protein. The S-tag and the Strep-tag are short peptide sequences which interact with the large proteolytic fragment of ribonuclease A (RNase A) and with Streptavidin, respectively. These protein–peptide interactions are strong enough to facilitate affinity chromatography, the Strep-tag–Streptavidin interaction can be disrupted by the addition of biotin, whilst a chaotropic agent is required to disrupt the formation of reconstituted RNase A. Thus whilst epitope tags are a convenient means of detecting protein fusions in immunocytochemical studies, they can present problems for purification of biologically active material. The addition of a hexahistidine tag to either end of a recombinant protein could, in principle, influence the activity of a metalloprotein, but the ability of the hexahistidine tag to act as a chelating agent in the presence of strong denaturants makes this short peptide fusion partner extremely useful for protein purification purposes.

3. INDUCTION OF EXPRESSION

Having decided upon a choice of vector and associated fusion protein (see *Tables 2* and *3*), the experimental conditions for optimizing yield of the fusion protein must be determined

Expression Systems

Table 4. Conditions that influence induction of a recombinant fusion protein

pH of the medium	Most commercial rich media are supplied as buffered lyophilizates and therefore the adjustment of pH is not necessary. However, if no expression is obtained, then alteration and monitoring of the pH has been shown to modify levels of protein expression in some cases. This is especially important when expression is attempted in a minimal medium (e.g. for the production of isotopically labeled protein for NMR work)
Temperature	In our experience this is one of the most critical variables in expression studies. The temperature of the growth experiment can influence both yield and solubility of the recombinant species. For induction of cells in most vectors a temperature range of 20–37°C should be explored. The lower temperatures tend to favor the production of soluble protein (52, 53)
Time	This parameter can similarly influence both the yield and the solubility of the expressed protein. In general, 1–4 h is sufficient time to expect induction to have occurred, although sometimes an overnight induction may be fruitful (54)
Cell density	Typically, cells are induced in mid-log phase when the optical density (absorbance) at 595 nm is around 0.5–0.7, but earlier or later times may be preferable
Strain	This is again an important parameter; the choice of strain employed may be restricted depending on the protein to be expressed and the nature of the promoter system employed. However, by using a vector such as one of the pGEX series, any strain can be used for expression since these vectors encode their own *lac* repressor gene and are therefore highly portable

empirically. The factors that influence the quantity and quality of the expressed polypeptide are discussed in *Table 4*, whilst the relationship between the promoters and inducers commonly used in expression vectors are summarized in *Table 5*.

4. PROTEOLYTIC CLEAVAGE OF FUSION PROTEINS

If the recombinant protein is in a biologically active form, it may be necessary to remove the fusion partner by proteolysis. In principle, the inclusion of nucleotide sequences encoding one or more protease cleavage sites facilitates such proteolysis. However, in our experience access to the cleavage site is just as important as its mere presence. Thus, cryptic cleavage sites are often generated if the fusion partner and the recombinant protein are connected by a short polypeptide linker. Alternatively, there may be nonspecific intramolecular contacts between the fusion partner and the recombinant protein. In both cases cleavage at the consensus site may be inhibited significantly. It may be possible to expose the site by the addition of low concentrations of urea or guanidine hydrochloride, but care must be taken not to denature the recombinant protein. Only an empirical approach will determine whether suitable cleavage conditions can be found. The properties of the most commonly employed proteases are given in *Table 6*. An excellent brochure describing properties and applications of proteases and their inhibitors has been produced by Boehringer Mannheim.

5. PURIFICATION OF FUSION PROTEINS

Probably the most convincing reason for using fusion vectors, as opposed to conventional expression vectors, is that the expressed protein can be purified by a highly specific, affinity

Table 5. Promoter and strain relationships

Promoter	Vector	Strain
Lac or *Tac* (a hybrid of the *Trp* and *Lac* promoters)	e.g. (a) pMAL-c2 (b) pGEX-2T	(a) Strain must carry *lacI* (*lac* repressor gene), preferably the *lacI^q* mutant allele (b) Any strain may be used when the plasmid (like the pGEX series) harbors the *lacI^q* gene. Examples of suitable strains include: JM109, TG1, XL1-Blue. The *lac* promoter is de-repressed by the inclusion of the lactose analog IPTG in the growth medium at a concentration of between 0.1–2.0 mM
Trp	pTruxFus	The pTruxFus vector is indirectly regulated by the *Trp* promoter which is coupled to a chromosomal copy of the *cI* gene. The λP_R promoter which drives expression of thioredoxin is therefore de-repressed in the presence of tryptophan
$\lambda P_L/P_R$	pRIT2T	The strain must contain a chromosomal, phage or plasmid-borne copy of the cI857 allele, e.g. HB101 (pNF2690) or N4830-1
T7	pET14b	The strain must harbor an inducible T7 RNA polymerase gene, and often the phage T7 lysozyme gene (pLysS and pLysE). The former is often under the control of the *lac* promoter, e.g. BL21, BL21 (DE3), HMS174 (42). This system can alternatively be provided as a phage (CE6) for infection of any other strain (42)

chromatography step. Indeed, the advantage of this approach extends to the purification of mutant proteins, thereby circumventing the need to develop a new purification schedule for each mutant protein. It is clear from *Table 7* that the vectors that incorporate glutathione-*S*-transferase or the maltose binding protein have advantages over the other fusion proteins, since conventional bio-affinity methods can be applied to the purification of the fusion protein. In addition, the incorporation of the urea- (and guanidine-) stable His-tag, lends itself well to the purification of denatured material for antigenic purposes. Alternatively, unfolded fusion protein may be refolded by a range of methods including stepwise reduction of the high urea concentration by dialysis.

6. DETECTION OF FUSION PROTEINS

One of the most useful byproducts of fusion technology is the ability to assay the recombinant protein indirectly. Moreover, by using antibodies directed against the fusion partner, many experimental opportunities are provided without the need to produce specific antibodies against

Table 6. Properties of some of those proteases frequently used to liberate recombinant proteins from a fusion partner

Protease[a]	Cleavage specificity[b]	Molecular mass (kDa)[c]
Thrombin	Gly-Pro-Arg↓	34
FactorXa	Ile-Glu/Asp-Gly-Arg↓	44
Enteropeptidase	(Asp)$_3$ -Asp-Lys↓	150
Collagenase	Pro-Val↓Gly-Pro	70–109
Formic acid[d]	Asp↓Pro	–

[a] Most proteases can be obtained from the major biochemical companies (e.g. Sigma, Pierce and Boehringer Mannheim) and immobilized or biotinylated versions may be preferable if contamination of the recombinant protein is to be minimized.
[b] The presence of a cleavage site does not imply that the site will be cleaved. Accessibility is a major factor in cleavage efficiency: denatured fusion proteins are cleaved more efficiently and all of the above proteases are resistant to mild denaturants. pGEX-KG has been specifically designed to maximize protease access (25).
[c] Molecular masses may vary depending on the source of the enzyme.
[d] Formic acid cleavage is less specific than proteolytic cleavage and is carried out under denaturing conditions.

Table 7. Purification of fusion proteins

Fusion protein (vector)	Affinity ligand[a]	Elution[b]	Cleavage[c]
β-Galactosidase (pMC181)	APTG[d]	100 mM Sodium borate, pH 10.5	None
Maltose binding protein (pMAL-c2)	Amylose	10 mM Maltose	Factor Xa
Protein A (pRIT2T)	IgG	100 mM Glycine pH 3.0	70% Formic acid
Glutathione-S-transferase (pGEX-2T)	Glutathione[e]	1–10 mM Reduced glutathione	Thrombin
Thioredoxin (pTruxFus)	None	Osmotic and thermal shock[f]	Enterokinase
S-tag (pET29a-c[+])	RNase A (S-protein)	2 M Guanidine thiocyanate	Thrombin
His-tag (pET14b)	Iminodiacetic acid	1 M Imidazole	Thrombin
Strep-tag (pASK60)	Streptavidin	1 mM Biotin	None

[a] In most cases the affinity ligand is coupled to either agarose or a form of Sepharose.
[b] Elution of bound fusion is either effected by competition between the bound and free ligand, although particularly tightly bound proteins can only be released by denaturants.
[c] Proteolytic cleavage is often used to liberate the recombinant protein *in situ* following immobilization on the affinity resin. This is often carried out during batch chromatography.
[d] APTG = *p*-aminophenyl-β-D-thiogalactoside.
[e] Glutathione may be replaced by *S*-hexylglutathione in both the elution and immobilization stages. Moreover, glutathione-S-transferase from the host strain may be removed by a second chromatography step.
[f] The targeting of the fusion protein to the periplasm is a feature of pTruxFus and pMAL-p2. In the former case, the suppliers claim that a significant purification is achieved through such targeting alone.

the recombinant polypeptide. Monitoring the success of an overproduction experiment is generally easier using fusion vectors, since SDS–PAGE analysis of the induced cell extract can often be combined with Western blotting in order to identify the induced species. The assumption that the presence of an induced protein species (detected using Coomassie blue or silver staining) implies successful expression of a recombinant protein can be significantly strengthened when using a fusion vector:

(i)   induction conditions can be monitored using the fusion vector without inserted DNA;

(ii) translational problems are apparent if the fusion protein is only expressed when a stop codon is inserted immediately 3′ to the original start codon of the recombinant gene. This is simply achieved by cleaving the recombinant plasmid at the site of insertion into the cloning vector, filling in the overhanging ends via the reaction catalyzed by the Klenow fragment of DNA polymerase I, re-ligating and introducing the modified plasmid into *E.coli*;

(iii) the absence of induced protein in the latter nonsense mutant may well signify that the mRNA species encoding the fusion protein adopts an unusual structure and deletions from the 3′ end of the recombinant gene may alleviate the problem, but may be inappropriate for the production of a biologically active species.

Since the analysis of most induction experiments is carried out initially by SDS–PAGE, the choice of a fusion partner that has a molecular mass which is readily resolved on a typical 10% sodium dodecyl sulfate (SDS) polyacrylamide gel is advantageous. Thus the recent introduction of glutathione-*S*-transferase (26 kDa) and the maltose binding protein as fusion partners which are readily resolved on gels, has gained widespread use in molecular biology.

In addition to the detection of fusion proteins by electrophoretic methods, the quantitation of expressed polypeptide can be performed indirectly if the fusion partner can be assayed by a sensitive spectrophotometric assay. Those enzymes for which a facile assay is available are listed in *Table 8*. This additional level of detection offers a very sensitive and reproducible way of determining the specific activity of a fusion protein and gives a reasonably reliable estimate of the

Table 8. Spectrophotometric assays for fusion proteins

Enzyme[a]	Assay buffer[b]	Comments
β-Galactosidase	60 mM $Na_2HPO_4.7H_2O$, 40 mM $NaH_2PO_4.H_2O$, 10mM KCl, 1 mM $MgSO_4.7H_2O$, 50 mM 2-mercaptoethanol, adjust pH to 7.0	(a) Do not autoclave the assay buffer (b) Store buffer at 4°C (c) Light sensitive (d) Should be freshly prepared for each application (e) Absorbance readings are taken at 420, 550 and 600 nm
Glutathione-*S*-transferase	100 mM potassium phosphate buffer, pH 6.5, 1 mM CDNB[c] 1 mM reduced glutathione	(a) CDNB is very toxic and should be used with care (b) Absorbance readings are taken at 340 nm
S-tag	200 mM Tris-HCl pH 7.5, 1 M NaCl, 1 mg ml^{-1} poly(C)	(a) Store assay buffer at -20°C (b) Absorbance readings are taken at 280 nm

[a] It is important to employ a control cell extract as a background control since *E.coli* will have a significant level of all three enzyme activities. It is also advisable to demonstrate a concentration dependence of activity by the cell-free extract. All assays are performed with cell-free extracts which have been prepared by sonication and centrifugation (or similar methods).
[b] Assays are performed by addition of undiluted and several dilutions (e.g. 10–100-fold) of the cell extract at the wavelength indicated, either using a recording spectrophotometer or by stopping the development of a color change at several time points to be determined empirically. In this way linear initial rates can be obtained.
[c] CDNB=1-chloro-2,4-dinitrobenzene.

EXPRESSION SYSTEMS AND FUSION PROTEINS

amount of folded protein in a given preparation (assuming, perhaps somewhat naively, that the fusion partner and the recombinant protein are similarly susceptible to denaturation).

7. REFERENCES

1. Muller-Hill, B. and Kania, J. (1974) *Nature*, **249**, 561.

2. Brake, A.A., Fowler, I., Zabin, J., Kania, J. and Muller-Hill, B. (1978) *Proc. Natl Acad. Sci. USA*, **75**, 4824.

3. Das, A. (1990) *Methods Enzymol.*, **182**, 93.

4. Gold, L. (1990) *Methods Enzymol.*, **185**, 11.

5. Uhlen, M. and Moks, T. (1990) *Methods Enzymol.*, **185**, 129.

6. MacFerrin, K.D., Terranora, M.P., Schreiber, S.L. and Verdine, G.L. (1990) *Proc. Natl Acad. Sci. USA*, **87**, 1937.

7. Shapira, S.K. Chou, J., Richaud, F.V. and Casadaban, M.J. (1983) *Gene*, **25**, 71.

8. Scholtissek, S. and Groose, F. (1988) *Gene*, **62**, 55.

9. Casadaban, M.J., Martinez-Arias, A., Shapira, S.K. and Chou, J. (1983) *Methods Enzymol.*, **100**, 293.

10. Jacobsen, R.H., Zhang, X.J., Dubose, R.F. and Matthews, B.W. (1994) *Nature*, **369**, 761.

11. Ullmann, A. (1984) *Gene*, **29**, 27.

12. Germino, J. and Bastia, D. (1984) *Proc. Natl Acad. Sci. USA*, **81**, 4692.

13. Miller, J. (1992) in *A Short Course in Bacterial Genetics*. CSHL Press, New York, p. 72.

14. Neilson, D.A., Chou, J., MacKrell, A.J., Casadaban, M.J. and Steiner, D.F. (1983) *Proc. Natl Acad. Sci. USA*, **80**, 5198.

15. Guan, C., Li, P., Riggs, P.D. and Inouye, H. (1988) *Gene*, **67**, 21.

16. Maina, C.V., Riggs, P.D., Grandea III, A.G., Slatko, B.E., Moran, L.S., Tagliamonte, J.A. and Guan, C. (1988) *Gene*, **74**, 36.

17. Duplay, P., Bedouelle, H., Fowler, A., Zabin, I., Saurin, W. and Hofnung, M. (1984) *J. Biol. Chem.*, **259**, 10606.

18. Kellerman, O.K. and Ferenci, T. (1982) *Methods Enzymol.*, **90**, 459.

19. Riggs, P.D. (1994) in *Current Protocols in Molecular Biology* (F.M. Ausebel *et al.*, eds). Greene Associates/Wiley Interscience, New York, Vol. 2, p. 16.6.

20. Nilsson, B., Abrahmsen, L. and Uhlen, M. (1985) *EMBO J.*, **4**, 1075.

21. Deisenhofer, J. (1981) *Biochemistry*, **20**, 2361.

22. Uhlen, M., Nilsson, B., Guss, B., Lindberg, M., Gatenbeck, S. and Philipson, L. (1983) *Gene*, **23**, 369.

23. Nilsson, B., Holmgren, E., Josephson, S., Gatenbeck, S., Philipson, L. and Uhlen, M. (1985) *Nucleic Acids Res.*, **13**, 1151.

24. Smith, D.B. (1993) *Methods Mol. Cell. Biol.*, **4**, 220.

25. Guan, K. and Dixon, J.E. (1991) *Anal. Biochem.*, **192**, 262.

26. Smith, D.B. and Johnson, K.S. (1988) *Gene*, **67**, 31.

27. Smith, D.B. and Corcoran, L.M. (1994) in *Current Protocols in Molecular Biology* (F.M. Ausebel *et al.*, eds). Greene Associates/Wiley Interscience, New York, Vol. 2, p. 16.7.

28. Frangiani, J.V. and Neel, B.G. (1993) *Anal. Biochem.*, **210**, 179.

29. Habig, W.H., Pabst, M.J. and Jakoby, W.B. (1974) *J. Biol. Chem.*, **249**, 7130.

30. Taylor, C., Ford, K., Connolly, B.A. and Hornby, D.P. (1993) *Biochem. J.*, **291**, 493.

31. Kaelin, W.G., Pallas, D.C., DeCaprio, J.A., Kaye, F.J. and Livingston, D.M. (1991) *Cell*, **64**, 521.

32. Chittenden, T., Livingston, D.M. and Kaelin, W.G. (1991) *Cell*, **65**, 1073.

33. La Vallie, E.R., DiBlasio, E.A., Kovacic, S., Grant, K.L., Schendel, P.F. and McCoy, J.M. (1993) *Bio/Technology*, **11**, 187.

34. McCoy, J.M. and LaVallie, E.R. (1994) in *Current Protocols in Molecular Biology* (F.M. Ausebel *et al.*, eds). Greene Associates/Wiley Interscience, New York, Vol. 2, p. 16.8.

35. Holmgren, A. (1985) *Ann. Rev. Biochem.*, **54**, 237.

36. Katti, S.K., LeMaster, D.M. and Eklund, H. (1990) *J. Mol. Biol.*, **212**, 167.

37. Richards, F.M. (1955) *Comp. Rend. Trav. Lab Carlsberg Ser. Chim*, **29**, 322.

38. Richards, F.M. and Vithayathil, P.J. (1959) *J. Biol. Chem.*, **234**, 1459.

39. Richards, F.M. and Wyckoff, H.W. (1971) in *The Enzymes* (P.D. Boyer, ed.). Academic Press, New York, p. 647.

40. Kim, J.-S. and Raines, R.T. (1993) *Protein Science*, **2**, 348.

41. Zimmerman, S.B. and Sandeen, G. (1965) *Anal. Biochem.*, **10**, 444.

42. Studier, F.W. and Moffatt, B.A. (1986) *J. Mol. Biol.*, **189**, 113.

43. Rosenberg, A.H., Lade, B.N., Chui, D., Lin, S., Dunn, J.J. and Studier, F.W. (1987) *Gene*, **72**, 179.

44. Studier, F.W., Rosenberg, A.H., Dunn, J.J. and Dubendorff, J.W. (1990) *Methods Enzymol.*, **185**, 60.

45. Arnold, F.H. (1991) *Bio/Technology*, **9**, 151.

46. Smith, M.C., Furman, T.C., Ingolia, T.D. and Pidgeon, C. (1988) *J. Biol. Chem.*, **263**, 7211.

47. Petty, K.J. (1994) in *Current Protocols in Molecular Biology* (F.M. Ausebel *et al.*, eds). Greene Associates/Wiley Interscience, New York, Vol. 2, p. 10.11.8.

48. Hochuli, E. 1990) in *Genetic Engineering, Principles and Practice* (J. Setlow, ed.). Plenum Press, New York, Vol. 12, p. 87.

49. Nikolov, D., Hu, S., Lin, J., Gasch, A., Hoffmann, A., Horikoshi, M., Chua, N., Roeder, R. and Burley, S. (1992) *Nature*, **360**, 40.

50. Hoffmann, A. and Roeder, R. (1991) *Nucleic Acids Res.*, **19**, 6337.

51. Schmidt, T.G.M. and Skerra, A. (1993) *Prot. Engineering*, **6**, 109.

52. Schein, C.H. (1989) *Bio/Technology*, **7**, 1141.

53. Schein, C.H. and Noteborn, M.H.M. (1988) *Bio/Technology*, **6**, 291.

54. Schein, C.H. (1990) *Bio/Technology*, **8**, 308.

CHAPTER 12
INCLUSION BODIES AND REFOLDING
D.R. Thatcher, P. Wilks and J. Chaudhuri

1. INTRODUCTION

1.1. Origin of inclusion bodies and their advantage in the purification of recombinant proteins

A major impact of molecular biology has been the development of methods to express foreign genes in potentially any organism. One common outcome of the expression of recombinant genes in foreign host cells is the accumulation of insoluble aggregated protein in what are now well recognized morphological structures called inclusion or refractile bodies. Why should this phenomenon occur? First, the recombinant protein which is synthesized, although possessing the correct amino acid sequence, may not possess native covalent structure. For full biological activity, foreign proteins may require specialized post-translational modification by enzyme systems which are lacking in the recombinant host cell (e.g. proteases, glycosyl transferases, etc.). Second, the natural conformational structure of the recombinant protein may not be adapted to the particular conditions prevailing in the host cell cytosol (e.g. extracellular secreted proteins expressed as intracellular recombinant protein) or may be overproduced at such a rate that the physiological solubility limit for that protein is exceeded. Also, *in vivo*, there are a variety of additional protein factors (1, 2) which can assist in protein folding and the level and specificity of these 'molecular chaperones' may differ in the host cell and thereby affect refolding and solubility of the recombinant protein.

Recovery of functionally active protein usually requires that these inclusion bodies are solubilized using highly denaturing solvent followed by refolding under artificial conditions. Although inclusion bodies have now been observed in all the major host vector systems (3), the main body of literature relates to *Escherichia coli* (3–6) and this chapter will be confined to discussion of methods for the isolation of inclusion body protein from this organism. The principles are generic and are applicable to inclusion bodies isolated from any species.

The production of recombinant protein as inclusion bodies has a number of advantages: high levels of protein can accumulate; simple primary recovery steps result in rapid purification at high yield and the route enables the isolation of protease-sensitive proteins. The disadvantage of this route is that refolding yields determine overall productivity and these can often be extremely poor. Operationally a good refolding yield has been defined as >20% recovery of activity (7).

1.2. Structure of inclusion bodies

The recombinant expression systems which facilitate inclusion body formation have been reviewed by Marston (8). Inclusion bodies are visible by light microscopy and are of the order of 0.2–1.5 μm in size and usually present as a single structure per cell (9). Often the inclusion bodies are so large that they distort the wall of the cell. Physicochemical analysis of isolated inclusion bodies has shown them to consist almost entirely of recombinant protein and packed at densities equivalent to those observed for protein crystals (10). Electron microscopic evidence shows that they are amorphous.

Inclusion bodies are usually assumed to consist of densely packed aggregates of misfolded protein (11). Evidence, such as the solubilization of some inclusion bodies under nondenaturing conditions (12), the detection of almost native secondary structure in interleukin-1β inclusion bodies by attenuated total reflectance Fourier transform infrared (FTIR) spectroscopy (13) and the detection of significant biological activity in intact inclusion bodies of recombinant enzymes (14, 15), suggests that this view may be an oversimplification and that significant native secondary and tertiary structure may be present.

1.3. Factors affecting inclusion body formation

There is no simple generic relationship between the formation of inclusion bodies and any one factor, genetic, physiological or chemical. In terms of the structural parameters of the recombinant protein itself, major correlations were found (16) between the probability of inclusion body formation and the average net charge of the protein and also the predicted number of reverse turns in the protein secondary structure. Solubility is an intrinsic property of a protein. It is sequence-dependent and is clearly a key factor influencing the formation of inclusion bodies, being particularly relevant to expression of recombinant proteins which are normally secreted and do not naturally accumulate to any significant level in the cytosol. The solubility and rate of folding of folding intermediates could also be intrinsic factors influencing susceptibility to inclusion body formation (17, 18). For example, although recombinant human interleukin-1β accumulates as a soluble cytosolic protein, a number of mutations have been isolated which form inclusion bodies. Some of these mutants after isolation and renaturation have similar solubilities to the wild-type protein. The conclusion from these observations is that the propensity towards inclusion body formation lies in the solubility and rate of folding of refolding intermediates.

The physiological state of the host cell can frequently outweigh any effects intrinsic to the recombinant protein structure. For instance, under varying fermentation conditions, the same gene product can be located in either the solid inclusion body phase or as soluble protein in the cytosol (19). Growth rate, expression level and in particular temperature (20, 21) appear to be the key physiological variables. In addition, pH, osmolarity and the presence of high levels of sugars in the medium can affect inclusion body formation (22).

1.4. When should the inclusion body route be chosen?

Advances in molecular biology have opened up a variety of different options for the expression of recombinant proteins: secreted versus intracellular or periplasmic accumulation; fungal versus mammalian cell versus bacterial expression. Inclusion bodies are usually associated with high level intracellular expression in bacteria.

The major advantage of the inclusion body route is an outcome of their size and density which enables rapid isolation from other cellular components. In addition, as inclusion bodies are usually composed of >70% recombinant protein, the route provides a rapid method and in many cases an extremely productive strategy for protein purification. Also, proteins bound up in inclusion bodies tend to be inactive, enabling the purification of proteins which would normally be toxic to the cell (e.g. ribonuclease). Finally, inclusion bodies are resistant to proteolytic degradation, providing a route to the isolation of proteins which would normally not accumulate in the cytoplasm because of protease sensitivity.

The major disadvantages of the inclusion body expression strategy center on the difficulties of obtaining an acceptable refolding yield. No hard and fast predictions can be made about the probability of success in refolding. Monomeric proteins with minimal post-translational modification and molecular weights of less than 30 kDa can usually be purified at extremely high yield. Many proteins are difficult to refold and this limits the general applicability of the approach. Proteins whose stability is dependent on interaction with a protein subunit(s) of different structure (hetero-oligomers such as hemoglobin) and proteins which usually have significant post-translational modification (e.g. proteins which have >3 disulfide bonds, or disulfide bonds in the presence of free thiol groups, or multiple glycosylation sites such as tissue plasminogen activator) do not refold well. Also proteins which have complex multidomain structures, integral membrane proteins and in general proteins with a molecular mass of >40 kDa are difficult to refold.

2. METHODS FOR THE ISOLATION AND SOLUBILIZATION OF INCLUSION BODIES

Once a stable and reproducible fermentation system has been established (either shake flask procedure or fermentation process), several strategies are available for the recovery of protein from the inclusion body state (*Figure 1*).

If expression and accumulation levels are very high (>50% total cell protein) and no cell breakage equipment is available, then direct extraction of the cells (*Figure 1*) with a suitable solubilization agent such as 8 M guanidine hydrochloride (GdnHCl) (23) is possible. Normally the inclusion body fraction is enriched prior to solubilization, renaturation and purification (*Figure 1*).

Figure 1. Recovery of proteins from inclusion bodies.

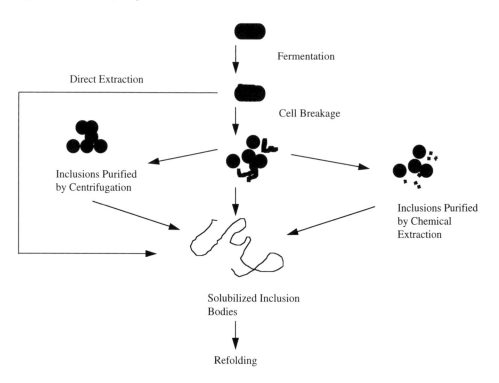

Fermentation

Direct Extraction

Cell Breakage

Inclusions Purified
by Centrifugation

Inclusions Purified
by Chemical
Extraction

Solubilized Inclusion
Bodies

Refolding

2.1. Methods for the isolation of inclusion bodies

Methods of cell breakage
Methods of inclusion body recovery rely on efficient cell breakage (see Chapter 2) to release the inclusion body from the cell debris and the soluble components of the cell. As inclusion bodies are highly resistant to shear forces compared with other cellular structures, excellent recovery without damage can be achieved using mechanical cell disruption techniques such as high pressure homogenization or ultrasonication. These are the cell breakage methods of choice. The presence of large inclusion bodies seems to improve cell rupture by these physical methods through increased cell fragility in the recombinant cells (24). Pre-incubation of the cells with lysozyme may also be used to weaken cell walls and increase the efficiency of cell breakage (6). If downstream purification options are limited, it may be important to achieve >95% cell breakage as unbroken cells will tend to copurify with the inclusion body fraction and lead to higher levels of contamination after solubilization.

Methods for the preparation of inclusion bodies
Inclusion bodies may be separated from the soluble components of the cell by centrifugation or microporous membrane filtration. Because inclusion bodies are of such a high density, careful optimization of centrifugation conditions (g force, viscosity and density of the suspension buffer) can be used to pellet inclusion bodies, whilst leaving soluble cytoplasmic protein and membrane debris in suspension. By repeated resuspension of this pellet under the same conditions, large inclusion bodies can be recovered at > 90% purity as judged by sodium dodecyl sulfate–polyacrylamide gel electrophoresis (SDS–PAGE) or electron microscopy. This strategy can easily be scaled to commercial applications using large-scale continuous centrifuges (25). At the laboratory scale, a high level of purification can be achieved using sucrose density centrifugation.

If only a conventional laboratory centrifuge is available, the cell debris can be depleted of contaminating nonrecombinant protein by resuspending the pellet in buffers which are known to solubilize *E. coli* membrane proteins (0.05–3.0 M GdnHCl or 0-4 M urea in the presence of 0.05 M Tris–HCl, pH 7.4, 0.02 M EDTA and 1% Triton X-100). These buffers usually do not solubilize inclusion body protein (26).

The quality of an inclusion body preparation can be assessed by SDS–PAGE and light or electron microscopy. The soluble fractions of all separations should also be checked by SDS–PAGE for slow solubilization of the inclusion body protein and if necessary the separation buffers modified to decrease this loss, thereby increasing overall process yield. It is also important to have determined the concentration of soluble recombinant protein at the time the cells are harvested. In many cases the protein partitions into both cytosolic and inclusion body phases and small changes in fermentation conditions can result in poor yields. Extraction for this analysis is best done by lysozyme treatment and sonication of a small sample of the harvest, followed by high speed (100 000 g) centrifugation. SDS–PAGE or high-pressure liquid chromatography (HPLC) of both solubilized pellet and supernatant should be performed to establish the ratio of cytosolic to inclusion body protein.

2.2. Solubilization

The exact conformation or subset of conformations that a particular protein occupies when in the inclusion body state is not known. There is no strong evidence for any intermolecular disulfide cross-linking of molecules within the inclusion body. Solubilization usually necessitates the complete disruption of tertiary structure. Common denaturing solvents are based on:

(i)   chaotropes;
(ii) anionic or cationic detergents; or
(iii) exposure to extremes of pH.

The major chaotropic solvents used are GdnHCl and urea, although relative solubility often follows the Hofmeister series (27) Gdn > Li > K >Na for cations and SCN > I > Br >Cl for anions. The concentration at which an inclusion body dissolves is usually related to the concentration at which the native structure unfolds. Inclusion bodies formed from recombinant proteins which normally unfold at relatively low concentrations of chaotrope (28) tend to be solubilized at lower concentrations than the inclusion bodies of more stable proteins. GdnHCl is relatively expensive when used at scale, whereas urea is much cheaper. However, care must be taken with concentrated urea solutions as cyanate can build up spontaneously and modify amino groups at basic pH. Urea solutions should, therefore, be de-ionized before use, in order to minimize this risk.

Detergents provide the most economical means of solubilizing inclusion bodies and detergent solubilization is the basis of the most commonly used analytical method, namely SDS–PAGE. A major drawback to the preparative use of detergents is the reduction in the number of downstream purification options available. Detergents usually bind with extremely high avidity to proteins and are difficult to remove; they interfere with hydrophilic and ion-exchange separations and, because of their micellar nature, are not easily removed by cross-flow filtration or dialysis. Commercial scale processes have, however, been established using solubilization in the anionic detergent SDS (29, 30) and the cationic detergent cetyltrimethylammonium bromide (CTAB) (31). The mild detergent lauryl sarcosine has also found useful application in inclusion body processing (32, 33). This detergent does not solubilize all inclusion body protein and may selectively solubilize those inclusion bodies made up of protein which retain some kind of near native conformation.

Proteins that are resistant to extremes of pH may be solubilized and renatured from either organic acids at high concentration (34) or dilute alkaline buffers (35).

3. RENATURATION OF SOLUBILIZED INCLUSION BODIES

The refolding of denatured protein is usually much less than 100% efficient. To a greater or lesser degree competing side reactions such as aggregation and proteolytic degradation reduce the overall yield.

3.1. Refolding

Effective renaturation requires the removal of the solubilizing buffer under conditions where the major fraction of the desired protein can refold into the active native conformation. Strategies for maximizing recovery are depicted in *Figure 2* and may be classified as:

(i)   single-step dilution of the solubilized inclusion body into a nondenaturing buffer;
(ii) staged dilution from solubilized fully denatured inclusion body protein to an intermediate denaturant concentration (this denaturant not necessarily being the same as the solubilization buffer) prior to buffer exchange into a nondenaturing buffer;
(iii) slow feeding of denatured protein solutions into a stream of renaturation buffer;
(iv) reversible immobilization of the denatured protein followed by buffer exchange into a renaturation buffer and then elution of the folded renatured protein from the support (refolding of spatially isolated molecules).

Single-step dilution is usually only effective if the protein in the solubilization solution is extremely dilute (10–100 μg ml^{-1}). This low concentration is required because protein folding is an intramolecular reaction which follows first order reaction kinetics (36). On the other hand, aggregation is a higher-order reaction which dominates at higher protein concentrations. As the solubilities of denatured and refolding intermediates of some proteins are extremely low in

Figure 2. Refolding strategies.

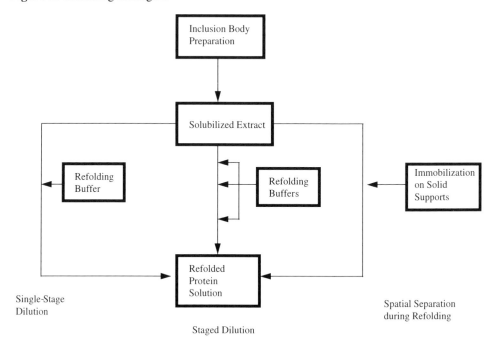

aqueous buffers, single-stage dilution may prove impracticable. One solution to the obvious practical constraint on renaturation buffer volume, is to add a small amount of the solubilized inclusion body protein to a fixed volume of buffer, allow folding to occur and then add a further pulse of denatured protein to the same renaturation solution (37). By repeating this process, a much larger concentration of fully renatured protein can be obtained. Another alternative to mix a stream of denatured protein solution continuously with a stream of renaturation buffer (38).

Staged dilution involves the dilution of a denatured protein to a denaturant concentration which just allows the formation of native tertiary structure. At this concentration the diluted solvent often has sufficient solubilizing power to prevent the aggregation of folding intermediates and incubation of the refolding mixture at this intermediate denaturation concentration allows time for the denatured states to refold (39, 40). When the tertiary structure is fully formed the denaturant may be removed. Usually these conditions have to be determined empirically by a series of small-scale optimization experiments.

Unfolded proteins, when solubilized in the nonionic chaotrope urea (41) or dilute alkali (42), can be adsorbed to an ion exchange matrix and renatured *in situ* by reducing the concentration of denaturant without eluting the protein from the column. Yields from this method vary depending on the protein in question. Thogersen (43) has shown that cycling pulses of denaturant followed by renaturing buffer, can result in increased recovery on elution. The maximum concentration of the denaturing solvent is gradually reduced in each cycle.

If a protein contains a cofactor, then it may be necessary to refold the protein in the presence of that cofactor (44). Alternatively, if the protein is known to interact strongly with another protein, for example a subunit of a hetero-oligomeric protein, then refolding may proceed more effectively in the presence of the second protein (45, 46).

Refolding yield may be increased by the addition of additives which suppress aggregation, such as arginine (47, 48), glycine (49) or polyethylene glycol (50).

3.2. Disulfide bond formation

Many recombinant proteins which form inclusion bodies are by their nature extracellular proteins and a large number of extracellular proteins contain disulfide bonds. Under the reducing conditions prevailing in the cytosol, these bonds are unable to form and intracellular inclusion bodies contain no disulfide bonds, although processing in the absence of reductant can lead to oxidation (51). It is, therefore, advisable to incorporate a low molecular weight reductant such as 2-mercaptoethanol or dithiothreitol (DTT) in the solubilization buffer.

The timing and control of disulfide bond oxidation during refolding is critical. Some proteins such as bovine pancreatic trypsin inhibitor have little or no stable conformation in the reduced state (52), whereas other proteins such as bovine growth hormone have a near-native structure under the same conditions (53). Such stability considerations are an important factor in determining refolding strategy: proteins that are poorly soluble and unstable in the reduced state have to be oxidized prior to refolding; reduced proteins that adopt a near-native structure may be oxidized after removal of denaturing buffers (e.g. α-interferon) (19).

Two types of oxidation reaction can be used to form disulfide bonds:

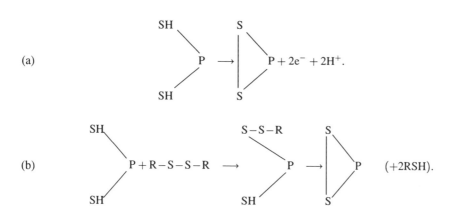

Electron acceptors for (a) are air/oxygen, with or without catalysis, iodosobenzoic acid and hydrogen peroxide. Such reactions have the disadvantage that cysteine residues in the recombinant protein sequence can be overoxidized to cysteic acid and methionine to methionine sulfone. The method has the advantage of being extremely inexpensive. Reaction (b) involves disulfide exchange and is the route through which disulfide bond formation occurs in nature (54). Generally this approach involves treatment of the reduced denatured protein with a redox buffer (e.g. glutathione and reduced glutathione). Disulfide bonds can often form under slightly reducing conditions, presumably driven by the free energy generated by the preference of the refolding protein to adopt a particular near-native conformation. In this case, the thiol component of the redox buffer should be in approximately fivefold molar excess, which corresponds to the ratio found in the endoplasmic reticulum (54). Native disulfide bond formation is thought to be catalyzed by the enzyme protein disulfide isomerase (55). Addition of this enzyme to refolding protein preparations has been shown to result in increased yield (56).

Table 1. Proteins purified from inclusion bodies

Protein	Ref.	Protein	Ref.
Albumin, bovine serum	57	Interleukin-6	69
Antibody, scFv	58	β-Lactamase	70
Antibody, F_{ab} fragment (MAK 33)	59	Lysozyme, hen egg white	71, 72
Aldolase	60	Papain	73
Citrate synthase	61	Papilloma virus HPV16 E7MS2	74
DNase	62	fusion protein	
α-Glucosidase	63	Prochymosin	75
Growth hormone, carp	64	Pre-prothrombin	76
Hemoglobin	65	Pro-urokinase	77
Insulin-like growth factor	66	Rhodanese	78
Insulin-like growth factor 1	67	tPA	62, 79
Interferon-γ	62, 68	Tumor necrosis factor-β	80

The rate of thiol oxidation by either route (a) or (b) is dependent on the refolding conditions used and the properties of the protein in question.

Table 1 is a list of the major papers and patents covering the purification of proteins from inclusion bodies. These examples illustrate the general principles outlined in this chapter for solubilization of inclusion bodies and subsequent renaturation of proteins. The references given should be consulted for further details.

4. DESIGNING A PROCEDURE *DE NOVO* FOR REFOLDING AND PURIFICATION OF AN INCLUSION BODY PROTEIN

Rational design of a purification procedure requires consideration of the following properties of the recombinant protein:

(i) expression level;
(ii) resistance to acid and alkali;
(iii) number of disulfides;
(iv) molecular mass;
(v) quaternary structure;
(vi) denaturation/renaturation conditions.

At the beginning of a project, little information may be available on the physicochemical behavior of the native fully folded protein. If only the molecular mass and behavior on SDS–PAGE is known, the following strategy should be considered (*Figure 3*).

The first experiments should be designed to establish a small-scale method for isolating inclusion bodies: either by differential centrifugation or chemical extraction with urea or GdnHCl/Triton X-100 buffers. The solubilization conditions must then be identified. If no information on the renaturation behavior of the protein is available, the solubility has to be determined empirically by suspending the purified inclusion bodies in buffers containing increasing concentrations of

Figure 3. Generic purification approach for inclusion body protein when no or limited information on the properties of the protein is available. (Suitable for molecular mass 12–35 kDa.)

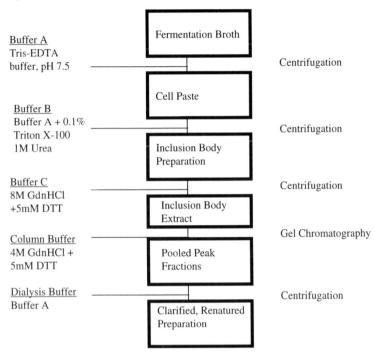

GdnHCl or urea. The extract should then be clarified by high-speed centrifugation and purified by gel filtration in the solubilization buffer. Gel filtration removes low molecular mass components (lipid, endotoxin) and simplifies spectrophotometric analysis during refolding studies.

The pooled material after gel filtration, should be adjusted to between 0.1 and 1.0 mg ml^{-1} and then diluted 10-fold with renaturation buffer. Success in refolding is dependent on such variables as ionic strength and pH of the refolding buffer, the temperature at which refolding occurs and the time for which refolding is allowed to proceed. Small-scale experiments varying these parameters may allow a favorable set of conditions to be determined empirically. After clarification the level of renaturation, as based on formation of soluble biologically active protein, should be determined. If yields are unacceptable, the solubilized inclusion bodies should be dialyzed stepwise against buffers of decreasing denaturant concentration under oxidizing conditions. The level of renaturation should then be determined.

Once the protein has been refolded, ion-exchange chromatography or hydrophobic chromatography may be necessary to remove soluble misfolded protein and other contaminants.

The properties of the purified protein fraction may then be studied with a view to obtaining information which can be used in process optimization. Most importantly the refolding behavior of the purified protein should be studied and the conditions for correct and complete disulfide bond formation defined. The purification route can then be redesigned taking these variables into account. A number of protocols of generic value have been published recently (81).

5. REFERENCES

1. Gething, M.J. and Sambrook, J. (1992) *Nature*, **355**, 33.

2. Georgopoulos, C. and Welch, W.J. (1993) *Annu. Rev. Cell. Biol.*, **9**, 601.

3. Thatcher, D.R. and Hitchcock, A.G. (1994) in *Mechanisms of Protein Folding* (R.H. Pain, ed.). IRL Press, Oxford University Press, Oxford, p. 229.

4. Hlodan, R., Craig, S. and Pain, (1991) *Biotechnol. Genet. Eng. Rev.*, **9**, 47.

5. Wetzel, R. (1992) in *Stability of Protein Pharmaceuticals* (T. Ahern and M.C. Manning, eds), Part B. Plenum Press, New York, p. 43.

6. Fischer, B., Sumner, I. and Goodenough, P. (1992) *Arzneimittelforsch./Drug Res.*, **42**, 1512.

7. Datar, R.V., Cartwright, T. and Rosen, C.G. (1993) *Bio/Technology*, **11**, 349.

8. Marston, F.A.O. (1986) *Biochem. J.*, **240**, 1.

9. Bowden, G.A., Paredes, A.M. and Georgiou, G. (1991) *Bio/Technology*, **9**, 725.

10. Taylor, G., Hoare, M., Gray, D.R. and Marston, F.A.O. (1986) *Bio/Technology*, **4**, 553.

11. Kane, J.F. and Hartley, D.L. (1988) *Trends Biotechnol.*, **6**, 95.

12. Chang, J., McFarland, N.C. and Swartz, J.R. (1993) *World Patent Application 93/11240.*

13. Ober, K., Chrunyk, B.A., Wetzel, R. and Fink, A.C. (1994) *Biochemistry*, **33**, 2628.

14. Worrall, D.M. and Goss, N.H. (1989) *Aust. J. Biotechnol.*, **1**, 28.

15. Tokatlidis, K., Dhiurjati, P., Millet, J., Begun, P., and Aubert, J.P. (1991) *FEBS Lett.*, **282**, 205.

16. Wilkinson, D.L. and Harrison, R.G. (1991) *Bio/Technology*, **9**, 443.

17. Wetzel, R. and Chrunyk, B.A. (1994) *FEBS Lett.*, **350**, 245.

18. Mitraki, A. and King, J. (1992) *FEBS Lett.*, **307**, 20.

19. Thatcher D.R. and Panayotatos, N. (1986) *Methods Enzymol.*, **119**, 16.

20. Schein, C.H. and Noteborn, M.H.M. (1988) *Bio/Technology*, **6**, 291.

21. Kopeczki, E., Schumacher, G., and Buckel, P. (1989) *Mol. Gen. Genet.*, **216**, 149.

22. Bowden, G.A. and Georgiou, G. (1988) *Biotechnol. Prog.*, **4**, 97.

23. Naglack, T.J. and Wang, H. (1992) *Biotechnol. Bioeng.*, **39**, 732.

24. Middleberg, A.P., O'Neill, B.K., Bogle, I.D.C. and Snowwell, M.A. (1991) *Biotechnol. Bioeng.*, **38**, 363.

25. Dorin, G., Lin, L.S.-L., and Hamish, W.H. (1986) *Eur. Patent 0206828.*

26. Babbitt, P.C., West, B.L., Buechter, D.D., Kuntz, I.D. and Kenyon, G.L. (1990) *Bio/Technology*, **8**, 945.

27. Lim, W.K., Smith-Sommerville, H.E. and Hardman, J.K. (1989) *Appl. Environ. Microbiol.*, **55**, 1106.

28. Hart, R. and Bailey, J.E. (1992) *Biotech. Bioeng.*, **39**, 1112.

29. Rausch, S.K. and Meng, H. (1987) *World Patent Application 87/102800/15.*

30. Koths, K., Thompson, J., Kunitani, M., Wilson, K. and Hanisch, W. (1986) *US Patent 4569790.*

31. Cadamone, M., Puri, N., Sawyer, W.H., Capon, R.J. and Brandon, H.R. (1994) *Biochim. Biophys. Acta*, **1206**, 71.

32. Fraenkel, S.F., Sohn, R. and Leinwand, K. (1991) *Proc. Natl Acad. Sci. USA*, **88**, 1192.

33. Evans, T.W. and Knuth, M.W. (1986) *Eur. Patent 0263902.*

34. Huij, O., Tomasselli, A.G., Readon, I.M., Lull, J.M., Brunner, D.P., Tomisch, C-S.C. and Heinrickson, R.L. (1993) *J. Prot. Chem.*, **12**, 323.

35. McCoy, K.M. and Frost, R.A. (1990) *Eur. Patent Application 0373325.*

36. Kiefhaber, T., Rudolf, R., Kohler, H.H. and Buchner, J. (1991) *Bio/Technology*, **9**, 825.

37. Fischer, B., Perry, B., Sumner, I. and Goodenough, P. (1992) *Protein Eng.*, **5**, 593.

38. Ambrosius, D. and Rudoph, R. (1992) *World Patent Application 92/09622.*

39. Orsini, G., Skyrzynia, C., and Goldberg, M.E. (1975) *Eur.J. Biochem.*, **5**, 433.

40. Orsini, G. and Goldberg, M.E. (1978) *J. Biol. Chem.*, **253**, 3453.

41. Creighton, T.E. (ed.) (1986) in *Protein Structure, Folding and Design.*, Alan Liss, New York, p. 249.

42. Suttnar, J., Dyr, J.E., Hamsikova, E., Novak, J. and Vonka, V. (1994) *J. Chromatog. B Biomed. Applications*, **656**, 123.

43. Etzerodt, M., Hotet, T.L. and Thogersen, H.C. (1994) *World Patent Application 94/18227.*

44. Smith, A.T., Santama, N., Daceys, S., Edwards, M., Bray, R.C., Thornley, R.N.F. and Burke, J.F. (1990) *J. Biol. Chem.*, **265**, 1335.

45. Tam, J., Shih, D.T.-B., Pagnier, J., Fermi, G. and Nagai, K. (1991) *J. Mol. Biol.*, **218**, 761.

46. Lu, W.P., Schiau, I., Cunningham, J.R. and Ragsdale, S.W. (1993) *J. Biol. Chem.*, **268**, 5605.

47. Winkler, M.E. and Blaber, M. (1986) *Biochemistry*, **25**, 4041.

48. Rudolf, R., Buchner, J., and Lenz, H. (1991) *US Patent 5077392.*

49. Burnier, J.P., and Johnston, P.D. (1988) *Eur. Patent. Application 0251615.*

50. Cleland, J.L. and Wang, D.I.C. (1990) *Bio/Technology*, **8**, 1274.

51. Schoemaker, J.M., Brasnett, A.H. and Marston, F.A.O. (1985) *EMBO J.*, **4**, 775.

52. Creighton, T.E. (1990) *Biochem. J.*, **270**, 1.

53. Holzman, T.F., Brems, D.N. and Dougherty, J.J. (1986) *Biochemistry*, **25**, 6907.

54. Gilbert, H.F. (1994) in *Mechanisms of Protein Folding* (R.H. Pain, ed.). IRL Press, Oxford University Press, Oxford, p. 104.

55. Freedman, R. (1989) *Cell*, **57**, 1069.

56. Schonbrunner, R.R. and Schmid, F.X. (1992) *Proc. Natl Acad. Sci. USA*, **59**, 4510.

57. Yoshii, H., Furuta, T., Yasunishi, A. and Kojima, T. (1994) *Biotechnol. Bioeng.*, **43**, 57.

58. Pantoliano, M.W., Bird, R.E., Johnson, S., Asel, E.D., Dodd, S.W., Wood, J.F. and Hardman, K.D. (1991) *Biochemistry*, **30**, 10117.

59. Buchner, J. and Rudolph, R. (1991) *Bio/Technology*, **9**, 157.

60. Rudolph, R., Siebendritt, R. and Kiefhaber, T. (1992) *Prot. Sci.*, **1**, 654.

61. Buchner, J., Schmidt, M., Fuchs, M., Jaenicke, R., Rudolph, R., Schmid, F.X. and Kiefhaber, T. (1991) *Biochemistry*, **30**, 1586.

62. Cleland, J.L., Builder, S.E., Swartz, J.R., Winkler, M., Chang, J.Y. and Wang, D.I.C. (1992) *Bio/Technology*, **10**, 1013.

63. Höll-Neugebauer, B., Rudolph, R., Schmid, F.X. and Buchner, J. (1991) *Biochemistry*, **30**, 11609.

64. Fine, M., Daniel, V., Levanon, A. and Gertler, A. (1993) *Biotech. Tech.*, **7**, 769.

65. Hart, R.A. and Bailey, J.E. (1992) *Biotechnol. Bioeng.*, **39**, 1112.

66. Kotlarski, N., Yeates, R.S., Milner, S.J., Francis, G.L., O'Neill, B.K. and Middelberg, A.P.J. (1995) *Trans. IChemE. (C)*, **73**, 27.

67. Chang, J.Y. and Swartz, J.R. (1993) in *Protein Folding In Vivo and In Vitro* (J.L. Cleland, ed.). ACS Symposium Series, Washington DC, Vol. 526, p. 178.

68. Vandenbroeck, K., Martens, E., D'Andrea, S. and Billiau, A. (1993) *Eur. J. Biochem.*, **273**, 481.

69. Breton, J., La Fiura, A., Bertelo, F., Orsinin, G., Valsasina, B., Ziliotto, R., De Fillipis, V., Polverino De Laureto, P. and Fonatan, A. (1995) *Eur. J. Biochem.*, **227**, 573.

70. Valax, P. and Georgiou, G. (1993) in *Biocatalyst Design for Stability and Specificity* (M.E. Himmel and G. Georgiou, eds). ACS Symposium Series, Washington DC, Vol. 516, p. 126.

71. Goldberg, M.E., Rudolph, R. and Jaenicke, R. (1991) *Biochemistry*, **30**, 2790.

72. Fischer, B., Sumner, I. and Goodenough, P. (1993) *Arch. Biochem. Biophys.*, **306**, 183.

73. Taylor, M.A.J., Pratt, K.A., Revell, D.F., Baker, K.C., Sumner, I.G. and Goodenough, P.W. (1992) *Prot. Eng.*, **5**, 455.

74. Suttnar, J., Dyr, J.E., Hamiskova, E., Novak, J. and Vonka, V. (1994) *J. Chromat. B*, **656**, 123.

75. Tang, B., Zhang, S. and Yang, K. (1994) *Biochem. J.*, **301**, 17.

76. Dibella, E.E., Maurer, M.C. and Scherage, H.A. (1995) *J. Biol. Chem.*, **270**, 163.

77. Orsini, G., Brandazza, A., Sarmientos, P., Molinari, A., Lansen, J. and Cauet, G. (1991) *Eur. J. Biochem.*, **195**, 691.

Inclusion Bodies

78. Zardeneta, G. and Horowitz, P.M. (1992) *J. Biol. Chem.*, **267**, 5811.

79. Rudolph, R., Opitz, U., Hesse, F., Riessland, R. and Fischer, S. (1992) *Biotech. Intl,* 321.

80. Jin, H., Uddin, M.S., Huang, Y.L. and Teo, W.K. (1994) J. *Chem. Tech. Biotechnol.*, **59**, 67.

81. Wingfield, P.T. (1995) in *Current Protocols in Protein Science* (J.E. Coligan, B.M. Dunn, H.L. Ploegh, D.W. Speicher and P.T. Wingfield, eds). John Wiley & Sons, Vol. 1, p. 6.1.1.

CHAPTER 13
INDUSTRIAL SCALE PURIFICATION OF PROTEINS

D.R. Thatcher

1. INTRODUCTION

1.1. Scope

Protein products now form a small but significant part of the western economy. The market for protein therapeutic products alone is now a multibillion dollar business. Industrial scale, in the context of protein production, can be any scale which generates products for commercial sale. This scale can vary from tonnes (e.g. industrial enzymes such as chymosin and glucose oxidase) to gram levels for some monoclonal antibody products used in medical diagnostic applications and the reagent business sector. The level of purification required for each individual product can also vary from almost no purification at all (1) (simple extraction of biotransformation catalysts), to processes which achieve levels of purity well beyond those required for even the most demanding research applications (e.g. *in vivo* protein therapeutics such as α-interferon and erythropoeitin). All commercially sold protein products have one common feature—a quality standard. The purification method, whatever the scale, must deliver a given level of quality during regular batch production. This method is the purification process and it differs from the research scale procedure in that it is usually performed by nonprotein chemists by a precisely documented sequence of unit operations. Unlike the research procedure, the industrial process must be sufficiently robust to absorb natural biological variability in the feed stock and still deliver product of acceptable quality every time. Whatever the application or scale the major challenges of developing industrial processes are always a combination of the technical problems of achieving process reproducibility at the desired scale and process economics. In the case of very high value added protein therapeutics quality assurance issues override production cost. In the case of low value bulk products yield and process efficiency are usually paramount. In both types of process intense commercial competition results in very little information being published on production processes. Even the patent literature is not helpful as patents are always filed well before scale up process development begins. This chapter is restricted to key features of process development and scale up. Several useful in depth reviews in this area have been published (2–15) and some of the individual unit operations are described in Chapter 5 (salting out), Chapter 6 (chromatographic procedures), Chapter 7 (covalent chromatography) and Chapter 12 (inclusion bodies).

1.2. Process scale up

A number of approaches to scale up have been defined (16). In practice, only a combination of experience guided by the general principles of biochemical engineering and a step-by-step empirical approach is likely to be successful. Our poor knowledge of the physical behavior of biological polymers, compounded by the fact that all biological materials are complex mixtures, dictates that precise mathematical descriptions of simple unit operations are not possible. Scale up using theory alone is therefore bound to fail. Nevertheless some approximations have been developed and these are the basis of a number of 'expert systems'. Improvement of these systems is an area of active research and in the future the utility of such systems will be a great boon to the rapid scale up of industrial purification processes (17–20).

Integration of the unit operations involved in a production process is important from the outset (21–24). The first step in scale up process development is therefore the design of an integrated 'scale down process' in which the purification procedure is defined entirely in terms of scaleable unit operations. The process is then scaled by a step-by-step empirical approach, collecting empirical data at each stage and employing such process correlations as are available to help predict process optimization at the next successive step of scale up.

2. DEVELOPING A SCALE DOWN PROCESS

The starting point of process development must be a well defined research scale procedure which delivers purified protein of the required quality and is preferably robust and reproducible. Careful definition of the scale down process is the key to success in scale up. All the parameters which will influence the industrial scale process need to be considered (25–26):

(i)   batch size/yield;
(ii) purity/reproducibility;
(iii) cost.

All the nonscaleable steps in the procedure need to be replaced by scaleable steps (e.g. dialysis supplanted by cross flow filtration, sonication by high pressure homogenization, etc.).

Wheelwright (27) has defined five 'rule of thumb' or 'heuristics' for the design of a good process:

(i)   choose a combination of separation processes that are based on different physical properties;
(ii) choose processes that preferentially exploit the greatest differences in physical properties between the product and the impurities;
(iii) carry out the step with the greatest separation between product and impurities first;
(iv) carry out the most expensive unit operation last;
(v)   keep the process as simple as possible.

Wheelwright (3) and Seader and Westerberg (28) point out that the development of large scale processes is an evolutionary process in itself.

The initial stage in the development of a process is the definition of a process flow sheet. This is usually based on a research purification procedure. Using experience from the operation of this research procedure, various separation principles, information from available expert systems, heuristics and chemical engineering correlations, the flow sheet is modified until an operable 'scale down' process has been developed.

3. BLOCK DIAGRAMS AND PROCESS FLOW SHEETING

The formal conventions of block diagrams and process flow sheeting are more than just process summaries, they are working documents which are an essential tool in the design of effective processes (8, 29). A flow sheet is really a detailed road map of the process which graphically identifies each unit operation in its proper sequence. These flow sheets are usually presented in the form of detailed diagrams that include information on input parameters of each step (volumes, masses, pH, buffer compositions, etc.) and characteristics of the separation step (e.g. in the case

Figure 1. Block diagrams and process flow sheeting.

(a) Block Diagram   (b) Process Flow Sheet

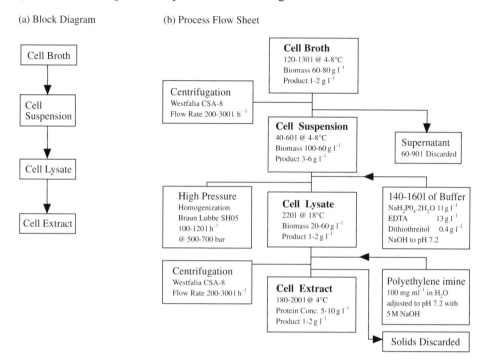

of centrifugation, centrifuge type and speed; in the case of chromatographic steps, matrices and volumes). The final process flow sheet should enable the complete mass and energy balances for the process to be estimated. The first stage in the development of a process flow sheet is the definition of a 'block diagram'. Block diagrams illustrate the basic design concepts behind the overall process strategy. A good starting point for virtually all protein purification processes is the block diagram described by Belter (30). Such a block diagram and an example of an accompanying flow sheet for a recombinant protein are shown in *Figure 1*.

4. SELECTION OF UNIT OPERATIONS

Figure 2 shows the options available for the industrial scale purification of most protein products. Protein processes, compared with most other industrial processes, are dilute aqueous processes and the initial stages are often concerned with dewatering the biomass or initial extract. The next stages are nominally concerned with removing the bulk of the nonproteinaceous biomass from the process stream (such materials such as lipid, glycolipids and nucleic acids interfere with subsequent purification steps). These steps are collectively called primary recovery. For high value added products subsequent steps are concerned with the removal of contaminating proteins (high resolution steps) and in some cases removal of modified forms of the product (polishing).

4.1. Primary recovery
Extraction
Dewatering of extracellular microbial biomass is usually achieved through solid/liquid separations. In the case of intracellular products, cell disruption is required for product extraction

Figure 2. Downstream processing technologies used in the industrial production of proteins.

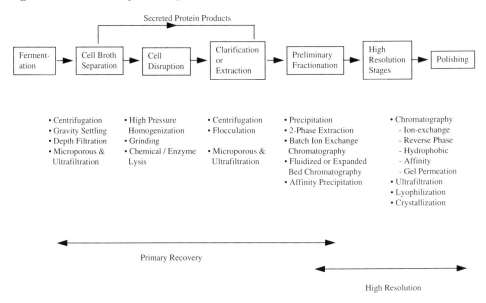

(31). Only high-pressure homogenizers and some ball mills are available in a range of sizes that allows modular scale up. Cell disruption in ball mills (32) and high-pressure homogenizers (33, 34) can be reliably approximated by pseudo first order equations, and scale up behavior can be predicted relatively accurately from measurements made on scale down equipment. A range of equipment is then available, capable of processing up to several cubic meters of biomass per hour (35).

Clarification

Centrifugation. Removal of biomass from the product stream is an essential stage of primary recovery. Usually the solids are discarded. In the case of inclusion bodies, the solids are collected (see Chapter 12). Crude biological solids are usually highly compressible gelatinous solids which are not easily filtered. Consequently centrifugation has become an almost universal primary recovery method. Scale up from conventional laboratory batch centrifuges is difficult and process development usually requires a scale down continuous centrifuge, normally of either a disk stack or tubular bowl design. These machines can handle a minimum of 25–50 l. However good scale up correlations are available as is a range of equipment that allows reliable scale up. Performance of particular continuous centrifuges of different size can be related using the Sigma Factor Σ (36, 37).

Cell debris has a very light density and is difficult to remove by centrifugation. One option is to add flocculating agents to the cell extract to increase the size of the particles to be centrifuged. Flocculating agents such as polyethylene imine (38) are useful and also have the added bonus of quantitatively precipitating nucleic acid. In the case of inclusion body processing the centrifugation step can be carefully optimized to separate cell debris from the more dense inclusion body fraction (see Chapter 12).

Filtration. Centrifugation at scale usually does not result in a completely clarified process stream. Residual solid material may be removed by filtration. Filters used at industrial scale are either of a depth or membrane type. The major issue with all filtration unit operations is nonspecific adsorption of product to the filter matrix and this must be quantified at reduced scale before filter selection is made.

Depth filters are usually thick pads of fibrous material (cellulose, nylon, diatomaceous earth, glass fiber, asbestos). When their capacity is reached, depth filters begin to shed particles back into the process stream and the process plant is usually designed so that filters are placed in parallel so that one filter may be renewed without interrupting processing.

Membrane filters are classified as microporous (which remove colloidal particles: these filters are usually defined in terms of the size of particle they are guaranteed to retain, e.g. 0.22 μm) or ultrafilters (which fractionate molecules of differing molecular weight: these filters are usually defined in terms of their nominal molecular weight cut off, NMWCO). The capacity of membrane filters to retain solids is relatively poor and all display rapidly decreasing flux rates as the membrane clogs with solids. This process of fouling can be reduced by the use of in-line depth filters or by directing a recirculating flow of process feed across the surface of the membrane (tangential or cross-flow filtration).

Microporous membrane filters may be used as dead end filters (always preceded by a depth filter) to clarify or sterilize soluble products in the process stream. Alternatively microporous membrane filters can be used to concentrate particulate products (cells, inclusion bodies or other protein precipitates) in tangential flow configuration (39, 40).

Ultrafilters are always used in tangential flow mode. Ultrafiltration is used almost exclusively to concentrate protein prior to formulation or gel filtration or to exchange buffer in diafiltration mode.

Scale up of all filtration steps is relatively straightforward, filter area being the primary variable that can be scaled arithmetically. In tangential flow configurations the recirculation rate/unit area, transmembrane pressure and flux rates per unit area must be kept constant.

Fractional precipitation. Precipitation is an easily and directly scaleable unit operation which can be used in combination with centrifugation to dewater the process stream, remove much of the nonproteinaceous contamination, soluble lipid, carbohydrate and low molecular weight metabolites. These methods are described in Chapter 5. At scale the procedure is limited by the physical properties of the floc (size density and stability). These properties determine which separation method is appropriate. Floc size becomes a key variable during harvesting by centrifugation and may be increased and stabilized by careful control of mixing conditions (41–43).

Aqueous two phase systems. Organic solvent extraction is not widely used in large-scale protein purification due to the denaturing effect of these solvents, particularly at the solvent interface. Several effective aqueous two phase systems have been developed for the primary recovery of protein products (44). Although these systems have aroused a great deal of interest because of the ease of scale up, most systems are expensive and have not been used extensively in the industrial scale separation of protein products.

4.2. High-resolution purification steps

Liquid chromatography is the only widely applicable, scaleable, high-resolution purification step (45–48). Although isoelectric focusing and electrophoretic methods give similar, if not greater, resolution than liquid chromatographic separations at the analytical scale, this resolution is diminished at preparative scale. Despite many attempts to engineer efficient preparative scale equipment, degradation of the separation zones by convective heating is a design-limiting constraint on electric field-dependent separations which has yet to be overcome.

As described in Chapter 6, chromatography systems exploit differences in overall net charge (ion-exchange chromatography) (49), relative surface hydrophobicity (hydrophobic and reverse-phase chromatography) (50, 51), molecular size (gel filtration) (52, 53) and specific molecular recognition (affinity chromatography) (14, 54–58). Chromatography may be used in a subtractive mode, where the process stream is passed through a bed of matrix without adsorption of the product but with adsorption of contaminants; in batch mode, where the matrix is suspended in a tank and the product adsorbed and eluted from the suspended and agitated matrix; and by column chromatography, using isocratic partition elution, step elution or gradient elution.

The major issues facing development of a chromatographic process to industrial scale operation are (59–63):

(i)   the order of the chromatographic steps;
(ii) yield;
(iii) the cost of the matrices.

These issues may require redesign of the research scale process; for example, affinity chromatography steps such as protein A agarose in immunoglobulin G purification have superb separation efficiencies and theoretically would be better placed as early as possible in the process sequence. However, such matrices are extremely expensive, easily fouled and difficult to clean and sanitize. Similarly high-pressure liquid chromatography (HPLC) methods have enormous separating power but are easily fouled. In these cases the process stream must be conditioned so that nonproteinaceous polymers such as nucleic acids and lipids are removed before the high resolution steps. Although this can often be accomplished by the use of anion-exchange chromatography, HPLC and affinity separation steps tend to be used later in the separation sequence. The high resolution steps employed depend on the properties of the product and those of the contaminating proteins.

The scale down chromatographic process should have defined packing materials, column sizes, flow velocities and buffer compositions. Scale up of column separations generally involves increasing column diameter whilst holding the relative volumes of feed, buffer washes and elution buffers constant. Gel filtration is a poorly scaleable step as the separation is dependent on the feed volume being <5% of the column volume. Consequently this step, if used, tends to be employed at the end of a chromatographic sequence.

Polishing step
For high value added products where high quality assurance is required, the final steps are usually called polishing steps. These serve to remove any potentially toxic residual chemicals added to the process stream during earlier steps and also to remove potentially antigenic degraded or modified forms (e.g. aggregates) of the product. These can be any of the above chromatographic steps although reverse-phase (RP-HPLC) and gel filtration have found favor.

Formulation
All industrial scale products need to be formulated in a form that preserves activity. This may vary from spray drying, for a low value added industrial enzyme, to carefully controlled lyophilization of a biotherapeutic. Normally the formulation buffer is used during the final polishing step.

5. REFERENCES

1. Roffler, S.R., Blanch, H.W. and Wilke, C.R. (1984) *Trends Biotechnol.*, **2**, 129.

2. Ho, S.V. (1990) *ACS Symp. Series*, **427**, 14.

3. Wheelwright, S. (1991) in *Protein Purification*. Hanser Publishers, Munich.

4. Naveh, D. (1988) *BioPharm.*, **1**, 34.

5. Becker, T., Ogez, J.R. and Builder, S. (1983) *Biotechnol. Advs*, **1**, 247.

6. Harrison, R.G. (1993) in *Bioprocess Technol.*, Vol. 18. Marcel Dekker, New York.

7. Willson, R.C. and Ladisch, M.R. (1990) in *ACS Symp. Series*, **427**, 1.

8. Giorgio, R.J. (1988) *BioPharma*, **1**, 38.

9. Lee, S.M. (1989) *J. Biotechnol.*, **11**, 103.

10. Asenjo, J.A. and Patrick, I. (1989) in *Protein Purification Methods: A Practical Approach* (E.L.V. Harris and S. Angal, eds). IRL Press, Oxford.

11. Glad, M. and Larsson, P.O. (1991) *Curr. Opin. Biotechnol.*, **2**, 413.

12. Gray, D. (1992) in *Polymer Applications for Biotechnology* (D.S. Soane, ed.). Prentice Hall, Englewood Cliffs, NJ.

13. Headon, D.R. and Walsh, G. (1994) *Biotechnol. Advs*, **12**, 635.

14. Labrou, N. and Clonis, Y.D. (1994) *J. Biotechnol.*, **36**, 95.

15. Flaschel, E. and Friehs, K. (1993) *Biotechnol. Advs*, **11**, 31.

16. Kline, P.E., Vogel, A.J., Young, A.E., Townsend, D.I., Moyer, M.P. and Aerstin, P.G. (1974) *Chem. Eng. Progress*, **70**, 67.

17. Bryant, C.H., Adam, A., Taylor, D.R. and Rowe, R.C. (1994) *Anal. Chem. Acta*, **297**, 317.

18. Asenjo, J.A., Herrara, L. and Byrne, B. (1989) *J. Biotechnol.*, **11**, 275.

19. Cooney, C.L., Petrides, D., Barrera, M. and Evans, L. (1988) *ACS Symp. Series*, **362**, 39.

20. Forkslund, G. (1995) *Expert Systems*, **12**, 27.

21. Dauglis, A.J. (1991) *Curr. Opin. Biotechnol.*, **2**, 408.

22. Szathmany, S. and Grandics, P. (1990) *Bio/Technol.*, **8**, 924.

23. Clarkson, A.I., Lefeivre, P. and Tichner-Hooker, N.J. (1993) *Biotechnol. Prog.*, **9**, 462.

24. Schugerl, K., Kretzmer, G., Freitag, R. and Scheper, T. (1994) *Chem. Ingen. Technik.*, **66**, 1585.

25. Remer, D.S. and Idrovo, J.H. (1990) *Biopharm.*, **3**, 37.

26. Datar, R.V., Cartwright, T. and Rosen, G.G. (1993) *Bio/Technology*, **11**, 349.

27. Wheelwright, S.M. (1989) *J. Biotechnol.*, **11**, 89.

28. Seader, J.D. and Westerberg, A.W. (1977) *Am. Inst. Chem. Eng. J.*, **23**, 951.

29. Jackson, A.T. and DeSilva, R.L. (1985) *Process Biochemistry*, December, 185.

30. Belter, P.A., Cussler, E.L. and Hu, W.S. (1988) in *Bioseparations: Downstream Processing for Biotechnology*. John Wiley & Sons, New York.

31. Kula, M.R. and Schutte, H. (1987) *Biotechnol. Progress*, **3**, 31.

32. Currie, J.A., Dunnhil, P. and Lilly, M.D. (1972) *Biotech. Bioeng.*, **14**, 725.

33. Middelberg, A.P.J., O'Neill, B.K. and Bogle, I.D.L. (1991) *Biotech. Bioeng.*, **38**, 363.

34. Sauer, T., Robinson, C.W. and Glick, B.R. (1989) *Biotech. Bioeng.*, **33**, 1330.

35. Schutte, H. and Kula, M.R. (1990) *Biotechnol. Appl. Biochem.*, **12**, 599.

36. Ambler, C.M. (1959) *J. Biochem. Microbiol. Technol. Eng.*, **1**, 185.

37. Brunner, K.H. and Hemfort, H. (1988) in *Downstream Processing; Equipment and Techniques* (A. Mizrahi, ed.). Alan Liss Inc., New York, p. 1.

38. Atkinson, A. and Jack, G.W. (1973) *Biochim. Biophys. Acta*, **308**, 41.

39. Belfort, G., Davis, R.H. and Zydney, A.L. (1994) *J. Membr. Sci.*, **96**, 1.

40. Bowen, R. (1993) *Trends Biotechnol.*, **11**, 451.

41. Bell, D.J., Hoare, M. and Dunnhil, P. (1983) *Adv. Biochem. Eng.*, **26**, 2.

42. Chan, M.Y.Y., Hoare, M. and Dunnhil, P. (1986) *Biotechnol. Bioeng.*, **28**, 387.

43. Richardson, P., Hoare, M. and Dunnhil, P. (1989) *Chem. Eng. Res. Des.*, **67**, 273.

44. Kohler, K., Ljungquist, C., Kondo, A., Veide, A. and Nilsson, B. (1991) *Bio/Technology*, **9**, 642.

45. Hammond, P.M. and Scawen, M.D. (1989) *J. Biotechnol.*, **11**, 119.

46. Prouty, W.F. (1993) *ACS Symp. Series*, **529**, 43.

47. Boschetti, E. (1994) *J. Chromatog.*, **658**, 207.

48. Frez, J. (1993) *ACS Symp. Series*, **529**, 1.

49. Chase, H.A. (1984) in *Ionexchange Technology* (D. Nordern and M. Streat, eds). Ellis Horwood, Chichester, p. 400.

50. Cretier, G. and Rocca, J.L. (1994) *J. Chromatog.*, **658**, 195.

51. Cox, G.B. (1993) *ACS Symp. Series*, **529**, 165.

52. Lesec, J. (1985) *J. Liquid Chromatog.*, **8**, 875.

53. Charm, S.E., Matteo, C.C. and Carbon, R. (1969) *Anal. Biochem.*, **30**, 1.

54. Lowe, C.R., Burton, S.J., Burton, N.P., Alderton, W.K., Pitts, J.M. and Thomas, J.A. (1992) *Trends Biotechnol.*, **10**, 442.

55. Arnold, F.H. (1991) *Bio/Technology*, **9**, 151.

56. Clonis, Y.D. (1987) *Bio/Technology*, **5**, 1290.

57. Narayanan, S.R. (1994) *J. Chromatog.*, **658**, 237.

58. Chase, H.A. and Draeger, N.M. (1992) *J. Chromatog.*, **597**, 129.

59. Janson, J.C. and Hedman, L. (1982) *Adv. Biochem. Eng.*, **25**, 44.

60. Cramer, S.M. and Jayaraman, G. (1993) *Curr. Opin. Biotechnol.*, **4**, 217.

61. Christie, Y. and Moo-Young, H. (1990) *Biotechnol. Advs*, **8**, 699.

62. Spears, R.M. (1991) *ACS Symp. Series*, **460**, 169.

63. Jungbauer, A. (1993) *J. Chromatog.*, **639**, 3.

CHAPTER 14
PEPTIDE SYNTHESIS

B.J. McGinn

1. INTRODUCTION

The development of theories on protein structure in the early 1900s was the driving force leading organic chemists to attempt the synthesis of peptides. The first insights into protein structural features began in 1902 when Hofmeister (1) remarked upon the regularly recurring amide linkage which was defined as the 'peptide' bond by Emil Fischer (2) in 1906. The chemical synthesis of peptides was a significant challenge and little progress was made in the next fifty years. A comprehensive review (3) entitled 'Naturally occurring peptides' in 1953 reported the confirmed structures of only six small peptides. However, the same year signaled the beginning of modern peptide chemistry with the synthesis of oxytocin by du Vigneaud *et al.* (4), at a time when it was recognized that many biologically active molecules had relatively short and simple amino acid sequences compared with large proteins. Since that time over five hundred analogs of oxytocin have been synthesized and several peptide pharmaceuticals have been developed and are in production. The development of peptide analogs containing amino acid residues not found in natural proteins has been a particular success of chemical methods, many analogs exhibiting modified or higher biological activity than the parent sequence.

This chapter presents the main issues in modern peptide chemistry, with particular reference to solid-phase methods which have revolutionized the field since the 1960s. The general aspects of Merrifield's methods (5) are shown to explain the basic principles. The choice of solid support is related to the apparatus and mode of synthesis to be used. A protection group strategy is essential and two major approaches are described. Merrifield's original strategy is still widely used but the modern Fmoc techniques have become popular. The formation of the peptide bond needs to be a mild yet quantitative step and modern coupling agents are discussed. Solid-phase chemistry tends to be blind, in that the growing chain remains attached until the end of synthesis. However, methods for monitoring and control of the progress of synthesis are described, followed by the techniques for final analysis, purification and handling of synthetic peptides. Finally, the design and preparation of peptides for antibody production is considered, including prediction of antigenic sites, choice of carrier and suitable conjugation methods.

Useful reviews on peptide synthesis include established reference works of Merrifield (5), Meinhofer (6), Sheppard (7), and Stewart and Young (8) in addition to more recent reviews; on the Fmoc method by Atherton and Sheppard (9); a peptide users guide by Grant (10) covering design, synthesis, evaluation and applications of synthetic peptides in addition to advice on the set-up of a core facility; on practical peptide synthesis protocols by Pennington and Dunn (11) containing information on phosphopeptides, amide bond reduced peptides, and the synthesis of unusual amino acids; their companion volume (12) details peptide analysis protocols together with information on conjugation techniques, epitope prediction, epitope mapping and the recent developments in multiple and combinatorial peptide library synthesis.

2. GENERAL ASPECTS OF SOLID-PHASE PEPTIDE SYNTHESIS

The advent of polymer-supported synthesis by Merrifield was a key development in peptide chemistry. Following unfruitful attempts with cellulose derivatives as supports, he examined cross-linked polystyrene and, in 1963, successfully demonstrated his ideas with the synthesis of the tetrapeptide Leu-Ala-Gly-Val (13, 14). This technique, which earned Merrifield the 1984 Nobel prize in chemistry, has since been widely accepted as the most promising route to synthetic peptides and has been used successfully in countless laboratories worldwide.

The key feature of the solid-phase principle is the covalent attachment of the growing peptide chain to an insoluble polymeric support. Reaction byproducts are removed readily by filtration and washing of the support matrix. This ease of separation allows the use of excess reagents, forcing reactions to completion. The polymer-supported peptide is retained in a single reaction vessel throughout the synthesis, thereby avoiding mechanical losses, simplifying manipulations and facilitating automation.

In the general procedure, a pendant functional group on the polymeric support is utilized as the anchor point for the first *N*-protected amino acid. The main requirement for the peptide–polymer bond is its stability in all subsequent reaction conditions, but it should be cleaved easily at the end of the synthetic procedures. A suitable *N*-terminal protecting group should be stable during the anchoring and coupling stages. It should be quantitatively removed under mild conditions during deprotection to expose the free amino terminus in readiness for the coupling of the next *N*-protected amino acid. The coupling step must be quantitative and the amino acid's chiral integrity must be maintained. The deprotection and coupling cycles are repeated until the required peptide sequence has been assembled. Detachment from the polymer may yield the peptide in a fully protected form, but usually any peptide side-chain functional groups which have been protected from the outset, are also deprotected at this point to give the free peptide.

3. SOLID-PHASE SUPPORTS, MODES OF SYNTHESIS AND RELATED APPARATUS

A potential support for peptide synthesis must satisfy a number of prerequisites:

(i)   it must be chemically stable under all the reaction conditions encountered;
(ii) it must contain enough reactive sites for peptide chain attachment to furnish a useful yield of product;
(iii) the microstructure of the polymer must allow rapid and unhindered contact between growing peptide chains and soluble reagents;
(iv) finally, it is essential that the support is easily separable from the liquid phase at every stage of the synthesis, while being physically stable during all manipulations.

The degree of peptide loading is typically $0.5–1.0$ mmol g^{-1} and is essentially a compromise. The general assumption is that too high a peptide loading will eventually lead to decreased reaction efficiencies. Large target peptides further exacerbate the problem and lower loaded supports are commonly used. Steric crowding, intersite side reactions, decreased diffusion of reagents and changes in the overall solvation properties of the system have all been held responsible for low yields of product.

Various materials have been evaluated and used as supports (*Table 1*). The traditional supports fall into two broad classes, microporous gel beads and macroporous rigid or supported materials. This distinction is independent of the peptide chemistry to be used in subsequent synthesis, but refers to the physical properties and therefore the required handling of the support material. All copolymer beaded supports are somewhat fragile and it is important to avoid grinding actions of magnetic stirrers, since this causes fragmentation and filter or frit blockage. It is also important to understand that microporous gel beads must not be used in continuous flow synthesizers since these supports are soft and compressible, quickly blocking the reagent flow through pumped columns. In general, microporous supports are used 'batchwise' in manual or automatic synthesizers which have shaker or bubbler reaction vessels, whereas macroporous supports are used in synthesizers with pumped columns under 'continuous flow' conditions. Almost any solid surface can be derivatized to enable peptide synthesis and many unusual support systems have been proposed. Examples are shown in *Table 1*, and some have specific advantages and applications, but few have found widespread use.

The original supports used by Merrifield were soft gel beads of 50–100 μm diameter. They were weakly cross-linked copolymers of styrene and 1% divinylbenzene functionalized by chloromethylation to 0.1 mmol g^{-1}. They typically swell to volumes of 6–8 ml g^{-1} in nonpolar solvents producing large internal volumes for reagents to permeate. These have proved extremely successful and many of the supports for peptide synthesis available today are based upon the same basic polystyrene network.

New microporous gel supports were developed following a concern about problems arising from the dissimilar solvation properties of the support matrix and the growing peptide chain (15). These 'polyamide' supports were intended to have solvation properties comparable to those of the peptide chain and were reported to yield purer products in some cases.

Macroporous supports were developed to improve the practical aspects of synthesis. It had long been accepted that batchwise washing of beads was inefficient and costly on solvents. The new macroporous resins allowed pumped columns to be used in place of shakers and bubblers. Washing efficiencies were improved, solvent consumption was reduced and reaction cycle times were shortened. A wide range of continuous flow synthesizers are now available which also offer continuous reaction monitoring via the pumped reagents.

A simple manual apparatus for batchwise synthesis has been described by Corley (24). The resin is contained within a reaction vessel fitted with a glass sinter. The four-way tap has two positions. In one position, the reaction is stirred by bubbling nitrogen from below and the vapor drawn from above by the vacuum. In the other position the solution is removed by vacuum into the waste flask while nitrogen is blanketed from above. All reagents and wash solvents are added manually into the reaction vessel.

A simple manual continuous flow apparatus has also been developed (9, p. 90). The macroporous rigid resin is contained in the column and reagents are circulated by the pump. The system has reservoirs for deprotection and wash solutions, while activated amino acid solutions are drawn in via the syringe barrel. The reagent flow may be monitored by the UV cell where UV absorbing species are used as in the 9-fluorenylmethoxycarbonyl (Fmoc) strategy.

4. PEPTIDE–RESIN LINKAGE AND PROTECTING GROUP STRATEGY

The need for protecting groups becomes immediately apparent from the simplest of cases where two free amino acids (A and B) are coupled to give the intended dipeptide (AB). Since each contains an amino and a carboxylate group, they may combine to form a variety of products. Dipeptides AA, BB, and BA will be formed in addition to AB. In addition, the reaction will progress to the tripeptide stage (eight possible combinations) and beyond. The only way to ensure

Table 1. Materials for peptide synthesis supports

Material	Characteristics
Microporous gel supports	Soft compressible gels. Must be used in 'batchwise' mode with 'shaker' or 'bubbler' type apparatus. Cannot be packed in columns. Inexpensive support material
Polystyrene 1% cross-linked (referred to as 'Merrifield' resins)	Merrifield's original support (13, 14), soft microporous gel, generally low derivatization, inexpensive, widely available as underivatized polymer or with a range of linkers and initial amino acids attached
Polydimethylacrylamide 1–2% cross-linked (referred to as 'polyamide' resins)	Improved solvation properties (15, 16), soft microporous gel, widely available as underivatized polymer or with a range of linkers and initial amino acids attached
Polyhydroxyphenylethylacrylamide 5% cross-linked	Ultra-high load support (17). Good solvation properties. Soft microporous gel. At 5 mmol g^{-1} peptide loading is near theoretical maximum. Useful for scale-up synthesis
Macroporous rigid supports	Rigid incompressible supports. Designed for 'continuous flow' mode of synthesis on apparatus using packed columns of support material. More costly than gel type, but savings made due to reduced solvent consumption. Can also be used in 'batch' mode if required
Kieselguhr-supported polyamide composite (known as 'Macrosorb') (18)	Polyamide gel supported in the pores of a rigid Kieselguhr matrix. The initial type of support for continuous flow synthesis. Commercially available as 'Macrosorb'(Sterling Organics), 'Novasyn K'(Novabiochem), and 'Ultrasyn'(Pharmacia-LKB)
Polystyrene-supported polyamide composite (known as 'Polyhype') (19)	Polyamide gel supported in the pores of a highly cross-linked rigid polystyrene matrix. Known generally as 'Polyhype' and commercially available as 'Novasyn P'(Novabiochem)
Polyoxyethylene grafted on to polystyrene (known as Rapp resin) (20)	Recent promising support with polyoxyethylene grafted on to a rigid highly cross-linked polystyrene matrix
Unusual support materials Polypropylene Pins (21)	The ends of small polypropylene pins are derivatized for peptide synthesis. Peptides generally remain attached to the pins following synthesis and are used as reusable probes for epitope mapping, etc. Moderately expensive
Cotton	Cotton has been proposed as a simple and inexpensive carrier for multiple peptide synthesis (22)
Cellulose	Cellulose paper has been used as the basis of a peptide 'spot synthesis' for use in a direct solid phase ELISA (23)

ELISA = enzyme-linked immunosorbent assay.

that the intended product is obtained is to use protected amino acids, PGn-A and B-PGc, where PGn and PGc are amino and carboxyl protecting groups respectively. The product (PGn-AB-PGc) is then deprotected to give AB.

The basic principle behind the use of protection groups is that for any given coupling reaction only one amino and one carboxyl are available for reaction, thus avoiding ambiguous products. The polymer support acts as the carboxyl protection and an *N*-protected amino acid is coupled. The *N*-terminal protecting group is removed and another *N*-protected amino acid is introduced. The cycles of *N*-deprotection and coupling are repeated to complete the sequence. Detachment from the support is in fact also the deprotection of the peptide C-terminus. Of course, the reactive side-chains of certain amino acids must also be protected until assembly is complete.

The overall protection strategy for peptide synthesis must therefore include the *N*-protection for the incoming amino acids, the side-chain protection and the link to the support. The *N*-terminal protection must be easily and quantitatively removed at each cycle, whereas the side chain protection and the resin link must be stable to this step, but removable at the end of the synthesis without damaging the peptide sequence.

Table 2 shows commonly used linkers and protecting groups for the two most popular strategies in solid-phase peptide synthesis. The standard Merrifield approach uses the tert-butyloxycarbonyl (t-Boc) amino temporary protection with benzyl-type side-chain groups and a benzyl ester link to the support. Removal of the t-Boc is achieved in each cycle with neat trifluoroacetic acid (TFA) and final detachment/deprotection with stronger acid, typically liquid HF. Both reagents are acids and the selectivity is due simply to the relative reactivities of t-butyl and benzyl groups. This may cause problems with longer sequences when side-chain protection groups must withstand many cycles of TFA treatment. A further problem in the strategy is the use of the hazardous reagent HF. These difficulties are avoided by the more modern 'Fmoc' strategy where the Fmoc group is used for *N*-terminus protection and t-butyl based groups now protect the side-chains. The Fmoc group is rapidly removed under mildly basic conditions and final detachment/deprotection may be achieved with TFA. The resin link must also be susceptible to TFA (a condition which could not be used in the Merrifield strategy).

Scavenging agents are often employed to reduce damaging side reactions during peptide deprotection and detachment. During acidolic deprotection, the chemical structures which provide ready cleavage of benzyl and t-butyl groups are also responsible for the stabilization of corresponding carbocations. These may react with electron-rich amino acid side-chains causing irreversible acylation and even reattachment to the support (from linking agent cations). Scavenging agents are themselves electron-rich carbocation acceptors and are most commonly used in mixtures of 2–5%. Deprotection mixtures are generally unpleasant, volatile cocktails of corrosive acids and pungent thiols. They must always be handled in a well-ventilated fume hood. During post-deprotection work-up, the scavengers are easily removed from the free peptide by ether precipitation or ether/water extraction. *Table 3* shows commonly used deprotection mixtures containing scavenging agents for deprotection following synthesis by the Fmoc strategy.

5. CARBOXYL ACTIVATION AND COUPLING: PEPTIDE BOND FORMATION

Formation of the peptide bond in modern peptide chemistry involves chemical activation of the carboxyl group of the incoming *N*-protected amino acid followed by coupling to the free amino group of the polymer-supported growing peptide chain. The activation needs to be good enough to produce high coupling yields, but not so active (e.g. acid chlorides) as to encourage side reactions. This is particularly important in solid-phase methods, where side reaction products and 'deletion' sequences are retained on the solid support until the detachment step, where they are released, contaminating the required product. In practice, three- to fourfold excesses of *N*-

Table 2. Protection strategies in solid-phase peptide synthesis

Strategy/functions	Linkers and protection groups	Amino acids and notes
Merrifield Strategy		
N-Terminal protection	tert-Butyloxycarbonyl (t-Boc) (25)	Removed by TFA
Resin links	Benzyl esters (14)	Detached by HF, gives C-terminal free acids
Side-chain protection	Nitro (26)	Arginine
	2,4-Dinitrophenyl (Dnp) (27)	Histidine, removed by thiophenol
	Benzyl ethers (OBz) (28)	Serine and threonine
	Benzyl ethers (OBz) (29)	Tyrosine
	Benzyl thioether (SBz) (30)	Cysteine
	Benzyl esters (OBz) (31)	Aspartic and glutamic acids
Fmoc strategy		
N-Terminal protection	9-Fluorenylmethoxycarbonyl (Fmoc) (32,33)	Removed by 20% piperidine in DMF, 7 min
Resin links	p-Benzyloxybenzyl ester (Wang linker) (34)	Detached by TFA, gives C-terminal free acids
	Hydroxymethylphenoxy ester (35)	Detached by TFA, gives C-terminal free acids
	Dimethoxyphenyl-Fmoc-amino-methyl-phenoxy (Rink linker) (36)	Detached by TFA, gives C-terminal amides
Side-chain protection	4-Methoxy-2,3,6-trimethyl-benzenesulfonyl- (Mtr) (37)	Arginine
	2,2,5,7-Pentamethylchroman-6-sulfonyl (Pmc) (38)	Arginine
	Trityl (Trt) (39)	Asparagine and glutamine
	tert-Butyl esters (tBu) (40)	Aspartic and glutamic acids
	Trityl (Trt) (41)	Cysteine
	tert-Butyloxymethyl (Bum) (42)	Histidine
	tert-Butyloxycarbonyl (t-Boc) (25)	Lysine
	tert-Butyl ethers (43)	Serine, threonine and tyrosine
	tert-Butyloxycarbonyl (t-Boc) (44)	Tryptophan

Table 3. Deprotection/scavenger mixtures suitable for the Fmoc strategy

Deprotection mixture with scavengers	Notes
TFA/water 95/5	Peptide has no Trp, Tyr, Arg, Cys or Met
TFA/phenol 95/5	Peptide contains Arg/Tyr
TFA/anisole/ethanedithiol 95/2.5/2.5	Peptide contains Met/Cys/Trp
TFA/phenol/anisole/ethanedithiol 94/2/2/2	Peptide contains Arg/Tyr and Met/Cys/Trp
Triethylsilane (45)	Peptide contains Arg/Tyr and Met/Cys/Trp. *Care*, can reduce indole ring in Trp

protected C-activated amino acids are generally used to push the coupling to completion and to help shorten coupling times to around thirty minutes. Of the many chemical procedures available in synthetic chemistry for carboxyl activation only a few have found widespread use in solid-phase peptide chemistry. *Table 4* shows commonly used 'active species' which couple with the free amino group and are formed using one of a range of 'coupling reagents'.

Symmetrical anhydrides have been used in peptide chemistry for many years. They are highly active and couple to available amino groups rapidly. They are usually formed prior to coupling by the action of a carbodiimide on the carboxylate. They suffer from a disadvantage in that it takes two amino acid molecules to form one anhydride, resulting in considerable waste of expensive reagents.

The 'active ester' is the more recent and popular form of carboxyl activation in peptide chemistry. Phenyl esters with electronegative substituents make good acylating agents and more recently esters of *N*-hydroxylamine derivatives have shown promise. Of these families, the pentafluorophenyl esters, the 1-hydroxybenzotriazole esters and the related *N*-hydroxy-oxo-

Table 4. Active species and coupling reagents

	Characteristics
Active species	
Symmetrical anhydrides (46)	Formed prior to coupling generally with diimide. Wasteful of amino acid derivative
Pentafluorophenyl (Pfp) active esters (47)	Stable crystalline solids. May be obtained pre-formed
1-Hydroxybenzotriazole (HOBt) active esters (48)	Formed *in situ* or immediately prior to coupling reaction
Coupling reagents	
Dicyclohexylcarbodiimide (DCC) (49)	Dicyclohexylurea precipitates and must be filtered off before coupling in solid-phase synthesis to prevent blockages
Diisopropylcarbodiimide (DIC) (50)	The diimide of choice for automated solid-phase work. Diisopropyl urea byproduct is soluble and need not be removed before coupling
Benzotriazolyloxy-trisdimethylaminophosphonium hexafluorophosphate (BOP) (51)	Generates *in situ* benzotriazole active esters. *Hazard*, withdrawn due to the carcinogenicity and respiratory toxicity of the byproduct hexamethylphosphotriamide (HMPA)
2-(1H-Benzotriazole-1-yl)-1,1,3,3-tetramethyluronium hexafluorophosphate (HBTU) (52)	Generates *in situ* benzotriazole active esters. Effective coupling agent and replacement for BOP
2-(1H-Benzotriazole-1-yl)-1,1,3,3-tetramethyluronium tetrafluoroborate (TBTU) (52)	As for HBTU above
Bromo-tris-pyrrolidino-phosphonium hexafluorophosphate (PyBroP) (53)	Powerful coupling agent. Will couple to hindered (e.g. *N*-methyl) amino acids when other reagents are ineffective

Peptide Synthesis

dihydrobenzotriazine esters have been widely adopted. In addition, 1-hydroxybenzotriazole is often recommended as a catalyst in coupling reactions. Its effect is probably due to the formation of *in situ* benzotriazole ester. Some active esters (e.g. pentafluorophenyl esters) are stable crystalline solids and may be purchased pre-formed ready for coupling.

Coupling reagents are responsible for the activation of amino acid carboxyls irrespective of the final activated species. Dicyclohexylcarbodiimide (DCC) has been used since the 1950s and the closely related diisopropylcarbodiimide is used commonly for solid-phase synthesis due to the higher solubility of the byproduct diisopropyl urea. It acts upon N-protected amino acids to form symmetrical anhydrides, or active esters if suitable hydroxylated compounds are also present. Benzotriazolyloxy-trisdimethylaminophosphonium hexafluorophosphate (BOP) proved to be a successful coupling reagent but was withdrawn in recent years from most peptide laboratories due to the carcinogenicity and respiratory toxicity of the byproduct hexamethylphosphotriamide (HMPA). It has been replaced by the equally effective and closely related reagents, 2-(1H-benzotriazole-1-yl)-1,1,3,3-tetramethyluronium hexafluorophosphate (HBTU) and 2-(1H-benzotriazole-1-yl)-1,1,3,3-tetramethyluronium tetrafluoroborate (TBTU). Both reagents act upon N-protected amino acids to form benzotriazole active esters. A further variation is bromo-tris-pyrrolidino-phosphonium hexafluorophosphate (PyBroP), a powerfully activating reagent which can effect couplings to sterically hindered residues such as N-methyl amino acids.

6. SYNTHESIS MONITORING AND CONTROL

A long-standing criticism of solid-phase peptide synthesis has been the inability to isolate, analyze and, if necessary, purify intermediates due to the immobilization to the solid support. There are, however, a number of techniques to monitor the progress of the synthesis and allow some control should difficulties occur. The simplest and most widely used methods are the qualitative color tests (*Table 5*) which may be used in batchwise or continuous flow approaches. These aim to imply completion of the coupling reaction by showing an absence of resin-bound free amine groups. While these tests involve removal and destructive testing of small resin samples, continuous flow methods provide an opportunity to monitor the reagent flow nondestructively for UV absorbing species or by using conductivity profiles. In addition, the observation of resin color changes during coupling with some additives has been developed successfully as a further non-destructive monitoring technique (*Table 6*). Other methods available include amino acid analysis and high-pressure liquid chromatography (HPLC; *Table 7*). These methods are not used routinely but give useful information in the event of synthesis problems.

7. SYNTHETIC PEPTIDE ANALYSIS, PURIFICATION AND HANDLING

Crude isolated synthetic peptides are likely to have a number of impurities present. Attachment to a solid support implies that products of incomplete coupling or deprotection and of side reactions are retained and accumulated during the synthesis to contaminate the final deprotected free peptide. Analysis and purification of large sequences is hampered by the close structural similarity between target peptide and its impurities. Chromatographic methods (see Chapter 6) are widely used for both analysis and purification (12). Although gel filtration and ion-exchange chromatography have been used, by far the most widely employed technique is reverse-phase HPLC which is capable of great resolution. Soft ionization mass spectrometry (MS) (see Chapter 16) has proved immensely valuable in establishing the purity and identity of the reaction product. Amino acid analysis and

Table 5. Qualitative color tests on resins

Method and notes	Reagents and procedures
Ninhydrin (54) Test for primary amines. Will not work for proline. Sensitivity to 5 μmol g^{-1} of resin, equivalent to 99% coupling on 0.5 mmol g^{-1} starting resin. Can be variable depending on resin type and amino acid sequence	*Reagent 1.* Ninhydrin (0.5 g) dissolved in ethanol (10 ml) *Reagent 2.* Phenol (80 g) in ethanol (20 ml) *Reagent 3.* Potassium cyanide TOXIC! (2 ml of 0.01 M stock solution, made from 33 mg KCN in 50 ml water) diluted to 100 ml with pyridine *Procedure.* A few beads of washed and dried resin from a coupling reaction are placed in a small test tube and 2–3 drops of each reagent are added. The tube is heated at 100°C for 5 min and viewed against a white background. A negative test (no primary amines, i.e. complete coupling) is indicated by a straw-yellow colored solution and no coloration of the beads. A positive control sample (withdrawn after deprotection and before coupling) should be strong dark blue/purple
Isatin (55) Useful test for secondary amino groups. Used when testing amino acid couplings to *N*-terminal proline. Sensitivity approaching that of ninhydrin	*Reagent.* Add isatin (2 g) to benzyl alcohol (60 ml) and stir for 2 h. Filter and add Boc-Phenylalanine (2.5 g) to 50 ml of the filtrate *Procedure.* A few beads of washed and dried resin from a coupling reaction are placed in a small test tube and 2-3 drops of isatin reagent are added with 2–3 drops of each of the three ninhydrin reagents. The tube is heated at 100°C for 5 min and on cooling the beads are washed by decantation with acetone. Blue to red beads indicate a positive test, colorless beads are negative
Trinitrobenzene sulfonic acid (TNBS) (56) Test for primary amines. Will not work for proline. Sensitivity to 3 μmol g^{-1} of resin, equivalent to 99.5% coupling on 0.5 mmol g^{-1} starting resin	*Reagent 1.* 10% diisopropylamine in DMF (made fresh daily) *Reagent 2.* TNBS (5 mg) in above solution (0.5 ml) (make immediately before test) *Procedure.* A few beads of washed and dried resin from a coupling reaction are placed in a small test tube and 2-3 drops of each reagent are added. A positive test is indicated by strong orange-colored beads, colorless beads being negative
Fluorescamine (57) Highly sensitive test for primary amines. Will not work for proline. Sensitivity to 0.6 μmol g^{-1} of resin, equivalent to 99.9% coupling on 0.5 mmol g^{-1} starting resin. Beware false positives, particularly with polyamide resins	*Reagent 1.* 10% triethylamine in chloroform *Reagent 2.* Fluorescamine (10 mg) in chloroform (1 ml) *Procedure.* A few beads of washed and dried resin from a coupling reaction are placed in a small test tube and 2–3 drops of Reagent 1 are added. The beads are then washed four times with ethanol and chloroform. A further 2–3 drops of Reagent 1 are added, followed by 2–3 drops of Reagent 2 and the beads examined under UV light (360 nm). A positive test is indicated by yellow-green fluorescence (475 nm emission), no fluorescence being negative

Table 6. Continuous flow monitoring systems

Method	Procedures and notes
Reagent flow – UV absorbance (58) Requires UV absorbing amino protecting groups. Mainly used for Fmoc strategies	Widely used in automatic and many manual continuous flow synthesizers. Fmoc groups may be monitored at 310–320 nm, using a UV spectrometer (1 mm pathlength cell) reading the absorbance of the reagent flow. Circulating coupling solutions contain 3–4 equiv. of Fmoc amino acid and produce an oscillating set of peaks which spread out to a flat baseline. The uptake of one equivalent on to the resin can be determined only semi-quantitatively. The deprotection step involves the release of one equivalent of Fmoc groups, producing a sharp peak on the spectrometer output
Reagent flow – conductivity (59)	The change in electrical conductivity of the reagent flow may be monitored using a reaction vessel fitted with conductivity electrodes. Similar information to that obtained by UV monitoring may be collected and used for feedback control of automatic synthesizers
Resin coloration – Dhbt esters (60)	When Dhbt esters are used, the resin becomes bright yellow due to ionic binding of the weakly acidic Dhbt to terminal free amino groups. When coupling nears completion the Dhbt is liberated into solution and the resin loses its bright yellow color. This has been used as a visual check of coupling and as the basis of feedback control of automatic synthesizers
Resin coloration – bromophenyl blue (61)	Bromophenyl blue has been used as an additive in coupling reactions to act as an indicator. The reagent forms an ion pair with the resin-bound amino (as in the case of Dhbt) turning the support deep blue. Complete coupling liberates the bromophenyl blue returning the resin to its original color. This is an excellent visual check of coupling, but tertiary amines must not be used in the reaction

sequencing (see Chapter 23) can also provide important information, but are of limited use in establishing peptide purity. There is no one best analytical method since each generates different information about the synthetic peptide. A useful combination is the application of HPLC to establish purity, coupled with MS data to confirm structural identity.

Problems in the handling and use of peptides are due to their special properties and may be physical, chemical or biological in nature (8, p. 49). An awareness of such problems and approaches to minimize difficulties in practice (*Table 8*) is invaluable to those preparing or handling synthetic peptides. Peptides vary considerably in their stability and shelf-life. Sensitive sequences can degrade faster than they can be purified, while others may remain stable for years. Examination of the sequence will allow the anticipation of particular problems, while the deep frozen storage of peptide lyophilizates will help to prolong the long-term shelf-life of all peptides.

Table 7. Other methods for monitoring peptide synthesis

Method	Procedures and notes
Amino acid analysis of peptide-resins (9, p. 110)	Amino acid analysis of the growing peptide at intermediate steps is generally too slow to be of use in monitoring the progress of peptide synthesis. It is useful for the determination of first amino acid attachment or when resin-bound assembly is complete, particularly where a reference amino acid has been incorporated between the polymer and the reversible linking agent
HPLC (62)	A small sample of resin may be withdrawn at intermediate stages of synthesis and the peptide detached and deprotected for HPLC analysis. In practice this is complex, with the crude material containing partially side-chain protected peptides, protecting group fragments and deprotection scavengers. Only with experience can this technique be valuable, and is generally too slow to be of routine use in monitoring the progress of synthesis

8. SYNTHETIC PEPTIDES FOR ANTIBODY PRODUCTION

The use of synthetic peptides to generate antibodies is based on two properties of the immune system. First, the size of antibody binding pockets corresponds to around six amino acids (1.6–2.0 nm in diameter). Small synthetic peptides are easily prepared and can be shown to act as antigens by binding to anti-protein antibodies. However, these antigenic peptides alone do not generate anti-peptide antibodies, that is, they are not immunogenic. This is due to the second feature of the immune system in that immunogenic molecules need to be much larger than small peptides. They are generally above 5 kDa and highly immunogenic molecules are over 100 kDa (63). The solution to this dilemma is to attach short synthetic peptides to large immunogenic 'carrier' proteins. The techniques and relevant issues have been widely reviewed (10, 64–66).

The choice of peptide sequence to act as synthetic epitope is the initial problem to be faced. A number of predictive methods have been used (*Table 9*). They all attempt to identify protein surface regions possessing a degree of mobility. In this way the synthetic construct is likely to resemble closely the conformational structure in the native protein. It has been accepted that such conformational variations are partly responsible for the often poor cross-reactivity of anti-peptide antibodies with their native protein (can be 200 times less effective than anti-protein antibodies). In addition, the 'continuous' epitope is usually part of a 'discontinuous' antigenic region. The N or C terminus of a protein is likely to be a good choice, often resulting in antibodies binding strongly with the native protein, the chain representing a flexible continuous epitope. Other exposed features such as loops or random coils are also favored. In the absence of firm evidence of protein structure from X-ray or nuclear magnetic resonance (NMR) data, the best results may be obtained by applying a range of predictive methods and using the best consensus result. To assist the adoption of a suitable conformation, a few amino acids either side of the identified antigenic site are normally included. In practice, peptides of 10–15 amino acids are used as potential synthetic antigens.

The choice of carrier protein (*Table 10*) is generally related to the animal of choice, and should be suitably 'foreign' to that species. KLH and BSA are the most widely used carriers and a review

Table 8. Problems in handling and use of peptides (8, p. 49)

Problem	Notes and actions
Hydrolysis of peptide	Usually by associated solvent molecules, e.g. acetic acid Bonds containing aspartate are acid sensitive Histidine imidazole ring catalyzes hydrolysis Keep peptide dry and deep frozen. Lyophilized material is best Allow containers to warm to room temperature before opening, thereby avoiding condensation from atmosphere Solutions lead to faster degradation–deep freeze/avoid long storage
Glass surfaces causing hydrolysis and adsorption	Glass surfaces are basic which leads to alkaline hydrolysis Old glass is especially bad Basic peptides are adsorbed on to negatively charged glass surfaces Hydrophobic peptides also adsorb Problem is serious with very dilute solutions Wash glassware and rinse in acetic acid solution before carefully drying Silanization of glass surfaces will help Use polyethylene or polypropylene containers (not polystyrene)
Insolubility of hydrophobic peptides	Peptide may dissolve better in pure water rather than a buffer solution Dissolve peptide in minimum solvent such as DMSO, acetic acid solutions or aqueous methanol, then dilute carefully with water or buffer
Oxidation of cysteine and methionine sulfur atoms	Especially bad in very dilute solutions, or when adsorbed to a surface Keep stock solutions concentrated. Divide into small aliquots/freeze Avoid repetitive freeze/thaw actions Bubble solutions with nitrogen to exclude dissolved oxygen Rapid disulfide exchange occurs in alkaline solutions, so keep solutions slightly acidic if possible
Degradation of peptides by micro-organisms on gel filtration and ion-exchange columns	Colonies of microorganisms can easily become established on dextran-based columns and peptides can be severely degraded upon chromatography Store columns in solutions containing sodium azide

(75) of anti-peptide antibodies listing around 200 examples clearly shows KLH to be the immunologist's carrier of choice. Problems may arise with weakly immunogenic peptides, when only anti-carrier antibodies are produced. In this case the inert carriers (liposomes or MAPs) provide promising alternatives. Carrier proteins are also thought to influence the conformational

Table 9. Prediction of immunogenic peptide sequences (12, Ch.11, 67)

Prediction method	Notes
Identify N- and C-terminal sequences	The ends of a protein are often exposed and flexible. Synthetic peptides have similar conformations and mobilities to these regions and produce good cross-reactive antibodies
Hydrophilicity indices of Hopp and Woods (68) or of Kyte and Doolittle (69)	Identifies surface regions from calculated hydrophilicity averages over five or seven residues. Success rate 65–75%. Does not identify amphipathic regions
Segmental mobility indices of Karplus and Schultz (70)	Identifies mobile and accessible surface regions from predictions of mobility using a factor scale calculated using data obtained from 31 protein X-ray structures
Hydrophilicity indices of De Lisi *et al.* (71) and of Sette *et al.* (72)	Incorporates prediction of amphipathic helices
Protrusion indices of Thornton *et al.* (73)	Identifies regions that protrude from the surface of the protein and are therefore accessible to act as immunogens
Turn prediction of Schulze-Gahmen *et al.* (74)	Mathematical calculation and prediction of protein turns, often immunogenic

Table 10. Carriers for synthetic peptide immunogens

Carrier	Notes
Keyhole limpet hemocyanin (KLH) (76)	Most popular carrier protein in use. Generally used in rabbits. KLH has around 60 surface lysine side-chains for coupling antigens
Bovine serum albumin (BSA) (77)	Popular and inexpensive carrier protein. 30–35 surface lysine side-chains available for coupling antigen
Ovalbumin (78)	Contains around 20 surface lysine side-chains for coupling antigens. Often used in immunoassays as nonrelevant carrier where KLH has been used as the carrier for immunization
Tetanus toxoid (79)	May be used for immunization in humans. Often used where work is directed towards development of synthetic vaccines
Liposomes (80)	Liposomes may act as both carriers of antigen and as adjuvants
Polymer-bound peptides from synthesis (81)	The assembled peptide from solid-phase synthesis remains attached to the support which is ground to a slurry and used directly for immunization
Multiple antigen peptide system (MAP) (82)	The peptide is built on a pre-formed lysine core. Structures contain four or eight peptide antigens. The core is generally non-immunogenic avoiding carrier response problems. However, the peptide conformation may not mimic that of the protein

structure of the attached synthetic peptide which may adopt a more favorable shape along the carrier surface. A change of carrier may give widely different antibody responses which may be impossible to predict.

Finally, a suitable method for attaching the synthetic peptide to its carrier is required (*Table 11*). Reagents such as glutaraldehyde and carbodiimide work by coupling together amino groups or

Table 11. Conjugation methods to link peptides to carriers

Conjugation method	Notes
Glutaraldehyde (83, 64, p. 394)	Couples amine groups to amine groups. General nonspecific cross-linker. Widely used method, but gives poorly defined products
1-Ethyl-3-(-3-dimethyl-aminopropyl)carbodiimide (EDIC) (84)	Activates carboxyl groups which subsequently couple with amino groups. General cross-linker which usually gives poorly defined products.
Maleimidobenzoic acid *N*-hydroxysuccinimide ester (MBS) (85)	Succinimide ester reacts with amines (on carrier protein) and maleimide couples with thiol groups (on cysteine of synthetic peptide). Specific reactions give well-defined products. Water-soluble (sulfo-MBS) also available
Bis-Diazobenzidine (86)	Reacts with tyrosine residues. Not widely used method

amino and carboxyl groups respectively. Peptide and carrier polymeric structures are unavoidable and many researchers are concerned with the resulting ambiguous mix of products. However, the immune system is obviously less concerned and the method has been applied successfully for many years. Heterobifunctional reagents give well-defined products since they utilize different functional groups from the peptide and protein. In a first step, MBS reacts with lysine side-chain amino groups on a carrier resulting in many pendant maleimide groups. In a second step, these capture a free thiol on the synthetic peptide from a cysteine added to the N or C terminus. In this way, peptide to peptide or protein to protein couplings are avoided.

The peptide–protein conjugates are usually made in milligram quantities and immunizations are often on the microgram scale with suitable adjuvants. The resulting antibodies are best tested for anti-peptide activity using the peptide coupled to a different carrier to eliminate anti-carrier activity in the assay. Recovered antibodies may be used for immunoassays or for affinity purification of native proteins.

9. REFERENCES

1. Hofmeister, F. (1902) *Ergeb. Physiol., Biol. Chem. Exp. Pharmacol.*, **1**, 759.

2. Fischer, E. (1906) *Ber. Dtsch. Chem. Ges.*, **39**, 530.

3. Bricas, E. and Fromegot, C. (1953) *Adv. Protein Chem.*, **8**, 1.

4. du Vigneaud, V., Ressler, C., Swan, J.M., Roberts, C.W., Katsoyannis, P.G., and Gordon, S. (1953) *J. Am. Chem. Soc.*, **75**, 4879.

5. Merrifield, R.B. (1973) in *The Chemistry of Polypeptides* (P.G. Katsoyannis, ed.). Plenum, New York, p. 335.

6. Meinhofer, J. (1973) in *Hormonal Proteins and Peptides* (C.H. Li, ed.). Academic Press, New York, p. 45.

7. Sheppard, R.C. (1977) in *Molecular Endocrinology* (I. MacIntyre and M. Szelke, eds). Elsevier/North Holland Biomedical Press, p. 43.

8. Stewart, J.M. and Young, J.D. (1984) *Solid Phase Peptide Synthesis (2nd edn)*. Pierce Chemical Company, Rockford, IL.

9. Atherton, E., and Sheppard, R.C. (1989) *Solid Phase Peptide Synthesis: A practical approach*. IRL Press, Oxford.

10. Grant G.A., ed. (1992) *Synthetic Peptides, A Users Guide.* W.H. Freeman and Company, New York.

11. Pennington, M.W. and Dunn, B.M. (1994) *Methods in Molecular Biology – 35: Peptide Synthesis Protocols.* Humana Press, Totowa, New Jersey.

12. Dunn, B.M. and Pennington, M.W. (eds) (1994) *Methods in Molecular Biology – 36: Peptide Analysis Protocols.* Humana Press, Totowa, New Jersey.

13. Merrifield, R.B. (1962) *Fed. Proc., Fed. Amer. Soc. Exp. Biol.,* **21**, 412.

14. Merrifield, R.B. (1963) *J. Am. Chem. Soc.,* **85**, 2149.

15. Atherton, E., Gait, M.J., Sheppard, R.C., and Williams, B.J. (1979) *Bioorganic Chem.,* **8**, 351.

16. Atherton, E. and Sheppard, R.C. (1975) in *Peptides 1974* (Y. Wolman, ed.). John Wiley, New York, p. 123.

17. Epton, R., Marr, G., McGinn, B.J., Small, P.W., Wellings, D.A. and Williams, A. (1985) *Int. J. Pept. Prot. Res.,* **7**, 289.

18. Miles, B.J. (1976) *British Patent 1586364.*

19. Small, P.W. and Sherrington, D.C. (1989) *J. Chem. Soc., Chem. Commun.,* 1589.

20. Bayer, E. and Rapp, W. (1986) *Chemistry of Peptides and Proteins.* Walter de Gruyter & Co., Berlin, Vol. 3, p. 3.

21. Geysen, H.M. (1987) *J. Immunol. Methods,* **102**, 259.

22. Eichler, J., Bienert, M., Stierandova, A. and Lebl, M. (1991) *Peptide Res.,* **4**, 296.

23. Frank, R. (1992) *Tetrahedron,* **48**, 9217.

24. Corley, L., Sachs, D.H. and Anfinsen, C.B. (1972) *Biochem. Biophys. Res. Commun.,* **47**, 1353.

25. Carpino, L.A. (1957) *J. Am. Chem. Soc.,* **79**, 4427.

26. Bergmann, M., Zervas, L. and Rinke, H. (1934) *Hoppe-Seyler's Z. Physiol. Chem.,* **224**, 40.

27. Shaltiel, S. (1967) *Biochem. Biophys. Res. Commun.,* **29**, 178.

28. Okawa, K. (1956) *Bull. Chem. Soc. Jpn,* **29**, 486.

29. Wunsch, E., Fries, G. and Zwick, A. (1958) *Chem. Ber.,* **91**, 542.

30. du Vigneaud, V., Audrieth, L.F. and Loring, H.S. (1930) *J. Am. Chem. Soc.,* **52**, 4500.

31. Marshall, G.R. and Merrifield, R.B. (1965) *Biochemistry,* **4**, 2394.

32. Carpino, L.A. and Han, G.Y. (1972) *J. Org. Chem.,* **37**, 3404.

33. Atherton, E., Logan, C.J. and Sheppard, R.C. (1981) *J. Chem. Soc., Perkin Trans.,* **1**, 538.

34. Wang, S.-W. (1973) *J. Am. Chem. Soc.,* **95**, 1328.

35. Sheppard, R.C. and Williams, B.J. (1982) *Int. J. Pept. Prot. Res.,* **20**, 451.

36. Rink, H. (1987) *Tetrahedron Lett.,* **28**, 3787.

37. Atherton, E., Sheppard, R.C. and Ward, P. (1985) *J. Chem. Soc. Perkin Trans.,* **1**, 2065.

38. Ramage, R., Green, J. and Blake, A.J. (1991) *Tetrahedron Lett.,* **47**, 6353.

39. Sieber, P. and Riniker, B. (1991) *Tetrahedron Lett.,* **32**, 739.

40. Visser, S., Roeloffs, J., Kerling, K.E.T. and Havinga, E. (1968) *Recl. Trav. Chim. Pays-Bas,* **87**, 559.

41. McCurdy, S.N. (1989) *Pept. Res.,* **2**, 147.

42. Colombo, R. (1984) *J. Chem. Soc., Chem. Commun.,* 292.

43. Beyerman, H.C. and Bontekoe, J.S. (1962) *Recl. Trav. Chim. Pays-Bas,* **81**, 699.

44. White, P. (1992) in *Peptides, Chemistry and Biology; Proc. 12th American Peptide Symposium* (J.A. Smith and J.A. Rivier, eds). ESCOM, Leiden, p. 537.

45. Pearson, D.A., Blanchette, M., Baker, M.L. and Guindon, C.A. (1989) *Tetrahedron Lett.,* **30**, 2739.

46. Wieland, T., Kern, W. and Sehring, R. (1950) *Justus. Liebigs. Ann. Chem.* **569**, 117.

47. Kisfaludy, L. and Schon, I. (1983) *Synthesis,* 325.

48. Konig, W. and Geiger, R. (1970) *Chem. Berichte,* **103**, 2024.

49. Sheehan, J.C. and Hess, G.P. (1955) *J. Am. Chem. Soc.,* **77**, 1067.

Peptide Synthesis

50. Sarantakis, D. (1976) *Biochem. Biophys. Res. Commun.*, **73**, 336.

51. Castro, B., Domoy, J.R., Evin, G. and Selve, C. (1975) *Tetrahedron Lett.*, **14**, 1219.

52. Knorr, R., Trzeciak, A., Bannwarth, W. and Gillessen, D. (1989) *Tetrahedron Lett.*, **30**, 1927.

53. Coste, J., Dufour, M.-N., Pantaloni, A. and Castro, B. (1990) *Tetrahedron Lett.*, **31**, 669.

54. Kaiser, E., Colescott, R.L., Bossinger, C.D. and Cook, P.I. (1970) *Anal. Biochem.*, **84**, 595.

55. Kaiser, E., Colescott, R.L., Bossinger, C.D. and Olser, D.D. (1980) *Anal. Chemicta Acta*, **118**, 149.

56. Hancock, W.S. and Battersby, J.E. (1976) *Anal. Biochem.*, **71**, 261.

57. Felix, A.M. and Jimenez, M.H. (1973) *Anal. Biochem.*, **52**, 377.

58. Sheppard, R.C. (1988) *Chem. Brit.*, 557.

59. Fox, J., Newton, R., Heegard, P. and Schafer-Nielsen, C. (1990) *Innovation and Perspectives in Solid Phase Synthesis* (R. Epton, ed.). Pub. SPCC (UK) Ltd, Birmingham, UK, p. 144.

60. Atherton, E., Holder, J., Meldal, M., Sheppard, R.C. and Valerio, R.M. (1988) *J. Chem. Soc., Perkin Trans.*, **1**, 2887.

61. Krchnak, V., Vagner, J., Safar, P. and Lebl, M. (1988) *Coll. Czech. Chem. Commun.*, **53**, 2542.

62. Atherton, E., Hubscher, W., Sheppard, R.C. and Woolley, V. (1981) *Z. Physiol. Chem.*, **362**, 833.

63. Catty, D., ed. (1988) *Antibodies Vol 1, A Practical Approach.* IRL Press, Oxford.

64. Walker, J.M., ed. (1994) *Methods in Molecular Biology – 32: Basic Protein and Peptide Protocols.* Humana Press, Totowa, New Jersey.

65. Porter, R. and Whelan, J., eds (1986) *Synthetic Peptides as Antigens, Ciba Foundation Symposium 119.* John Wiley and Sons, Chichester.

66. Friede, M., Muller, S., Briand, J.P., Schuber, F. and van Regenmortel, M.H.V. (1993) in *Immunotechnology* (J.P. Gosling and D.J. Reen, eds). Portland Press Proceedings, London, p. 1.

67. van Regenmortel, M.H.V., Briand, J.P., Muller, S. and Plaue, S. (1988) *Synthetic Polypeptides as Antigens, in Techniques in Biochemistry and Molecular Biology* (R.H. Burdon and P.H. von Knippenberg, eds). Elsevier Press, Amsterdam.

68. Hopp, T.P. and Woods, K.R. (1981) *Proc. Natl Acad. Sci. USA*, **78**, 3824.

69. Kyte, J. and Doolittle, R.F. (1982) *J. Mol. Biol.*, **157**, 105.

70. Karplus, P.A. and Schultz, G.E. (1985) *Naturwissenschaften*, **72**, 212.

71. De Lisi, C. and Berzofsky, J.A. (1986) *Proc. Natl Acad. Sci. USA*, **82**, 7048.

72. Sette, A., Doria, G. and Adauin, L. (1986) *Mol. Immunol.*, **23**, 807.

73. Thornton, J.M., Edward, M.S., Taylor, W.R. and Barlow, D.J. (1986) *EMBO J.*, **5**, 409.

74. Schulze-Gahmen, U., Prinz, H., Glatter, U. and Beyreuther, K. (1985) *EMBO J.*, **4**, 1731.

75. Palfreyman, J.W., Aitcheson, T.C. and Taylor, P. (1984) *J. Immunol. Meth.* **75**, 383.

76. Malley, A., Saka, A. and Halliday, W.J. (1965) *J. Immunol.*, **95**, 141.

77. Dayhoff, M.O. (1976) *Atlas of Protein Sequence and Structure*, Vol. 5, Suppl. 2. NRB Foundation, Washington, DC.

78. Nisbet, A.D., Saundry, R.H., Moir, A.J.G., Fothergill, L.A. and Fothergill, J.E. (1981) *Eur. J. Biochem.*, **155**, 335.

79. Bizzini, B., Blass, J., Turpin, A. and Raymond, M. (1970) *Eur J. Biochem.*, **17**, 100.

80. Alving, C. (1991) *J. Immunol. Methods*, **140**, 1.

81. Goddard, P., McMurray, J.S., Sheppard, R.C. and Emson, P. (1988) *J. Chem. Soc. Chem. Commun.*, 1025.

82. Tam, J.P. (1988) *Proc. Natl Acad. Sci. USA*, **85**, 5409.

83. Habeeb, A.F. and Hiramoto, R. (1968) *Arch. Biochem. Biophys.*, **126**, 16.

84. Yamada, H. (1981) *Biochemistry*, **20**, 4836.

85. Lerner, R.A. (1981) *Proc. Natl Acad. Sci. USA*, **78**, 3403.

86. Ratam, M. and Lindstrom, J. (1984) *Biochem. Biophys. Res. Comm.*, **122**, 1225.

CHAPTER 15
CRYSTALLIZATION PROCEDURES
L. Sawyer

1. INTRODUCTION

One rather trite reflection of the immense impact that detailed protein molecular structures are having on modern molecular biology, used here in the broadest sense, is that many journals, from food science to chemistry and beyond, now carry on their covers diagrams of some aspect of protein architecture. The determination of the crystal structure of a protein by X-ray crystallography (1,2) is one of the two methods available (the other is nuclear magnetic resonance (NMR), see Chapter 19, and ref. 3) for obtaining the accurate, three-dimensional arrangement of the atoms in the macromolecule. The prerequisite for such a structural study is a ready supply of good, well-formed crystals and the methods available for producing these are outlined in this chapter together with the underlying rationale. A list of more complete accounts of the techniques which can be used not only for proteins but also for nucleic acids can be found in Section 9. It should be said at the outset that a complete protein crystal structure determination is likely to take many months and to require milligram quantities of pure protein. Further, it is far more appropriate to have a protein supply of a few milligrams on a recurring basis than to be given 50 mg on a single occasion.

Methods have also been developed for crystallizing integral membrane proteins, and while there is an extra degree of complexity because of the two-phase nature of the membrane-solution system (4,5), the rewards to be gained from the successful crystallization make the attempts well worth pursuing (see Chapter 10).

2. OVERVIEW

For precipitation to occur, it is necessary for a protein solution to become saturated with respect to the protein component and this state can be achieved in a variety of ways (6–8). For crystals, as opposed to amorphous material, to appear there is a region close to the precipitation point where the solution is saturated with respect to the crystalline phase but unsaturated with respect to the amorphous one (9,10). This region can be quite small. A plot of protein solubility versus ionic strength generally rises from zero ionic strength, passes through a maximum and then falls exponentially with increasing ionic strength (salting out, see Chapter 5). The methods available seek to allow the cautious approach to the point of crystallization that is necessary for good crystals to grow (11–14).

The solubility curve implies that most proteins are less soluble in distilled water than in dilute buffer and this affords one method for their crystallization. More usually, however, the reduction in solubility is achieved by adding a precipitant which competes for the water of hydration of the protein and by its removal, hastens the aggregation of the molecules to produce a precipitate which one hopes will be crystalline. *Table 1* lists a large variety of factors which affect this process. The most important ones are listed in the upper part but each new protein must be treated as a new problem. Perhaps the best, although not always practicable, method for overcoming poor crystallization is to change the species from which the protein is obtained.

Table 1. Some of the factors which affect the crystallization of a protein

Protein purity	Critical; microheterogeneity can be a real concern: glycosylation, proteolysis, oxidation state, deamidation, genetic variants, denaturation, state of oligomerization. (see Chapters 8 and 9)
Precipitant	Commonly: $(NH_4)_2SO_4$, NaCl, PEG, MPD, propan-2-ol
Precipitant concentration	Protein dependent. Usually expressed (inadequately) as % saturation
pH	Protein dependent; typically between 4.0 and 8.0
Temperature	Typically 4, 16, 22°C but well controlled
Protein concentration	5–100 mg ml^{-1} (see Chapter 4). Solution should be monodisperse (light scattering)
Buffer type	Phosphate, citrate-phosphate, Hepes, cacodylate, Tris (pH = fn(T)!), Pipes, etc. (see Chapter 3)
Dielectric constant	Altered by adding ethanol, acetone, propan-2-ol; temperature control important
Ionic charge	For salting out, monovalent cations, polyvalent anions prove generally the most effective
Substrates	May stabilize flexible molecule
Cofactors	May be absent in recombinant protein
Inhibitors	As for substrates
Metal ions	May be essential, or highly undesirable!
Reducing agents	DTT, glutathione or 2-mercaptoethanol if free sulfydryl groups are present
Additives	Several small molecules in traces (1 mM) have been found to aid crystallization and prevent twinning: e.g. acetonitrile, dioxan, octyl-β-glucoside, PEG, benzamidine
Vibration	Consider incubators with distant compressors
Protein source	Lack of success can often be overcome by changing the species from which the protein is prepared

PEG, polyethylene glycol; MPD, 2-methyl-pentan-2, 4-diol; DTT, dithiothreitol.

Many proteins exhibit several solubility minima (there is always one at the isoelectric point) which may lead to a variety of crystal forms, the distinction between which can be subtle changes in cell dimensions or the more draconian change of space group. Occasionally, such polymorphism can produce different crystal forms within the same solution. Because there are likely to be several sets of conditions which will lead to crystals of any given protein, the first stage of the crystallization process is usually a survey of as wide a range of possibilities as the available protein will allow (15–19). Small-scale methods are therefore particularly suitable and *Table 2* lists the various methods that have been tried. Most laboratories use the hanging or the sitting drop methods for most if not all of their crystallizations, since these are both economical and easily assessed—a 2 μl drop of protein at 5 mg ml^{-1} allows 100 conditions to be tried for 1 mg of protein (11,13,14,16,18,25). *Table 3* lists some suppliers of the more specialist equipment used for protein crystallization.

3. GROWTH OF LARGE CRYSTALS

The crystallization process begins with the formation of nuclei upon which further crystal growth occurs (7,9). The nuclei may be imperfections on the surface of the container, dust or denatured protein rather than nascent protein crystals. If there are many nuclei, then it is likely that showers of small, poorly formed crystals will result. To limit the number of nuclei, ultrafiltration and/or centrifugation is recommended. It is also possible to supply nuclei deliberately to a supersaturated

Table 2. Methods for crystallizing proteins

Method	Advantages	Disadvantages
Batch (direct addition)	Simple, rapid	Too rapid precipitation forms showers of tiny crystals
Hanging drop (vapor diffusion)	Good for examining trials. Economical	Requires a steady hand. Needs treated cover slips
Sitting drop (vapor diffusion)	As above. Easier to set up	Not so easy to examine. Crystals can stick to the well bottom
Dialysis	Gentle. Reversible (?)	Wasteful of material
Microdialysis	Efficient. Can recycle	Tiresome to set up
Free interface diffusion	Versatile	May give immediate precipitate
Repeated extraction	Often provides first crystals	Not a micro method. Does not usually produce large crystals
Temperature change	Worth trying if other methods fail	Hard to control properly
Automated methods		
Autopipette	Very reproducible, even on a small scale	Expensive instrumentation. Wasteful of small amounts of protein (20,21)
Driven syringe	Very small scale	Possible handling problems (22,23)
Microgravity	Can produce near-perfect crystals	Needs a space shuttle! (24)

solution in order to generate crystals, a procedure called seeding (28,29). A seed can be anything from a whole, small crystal which by careful transfer from drop to drop will gradually grow larger, to the minute amount of the supernatant from a crushed and centrifuged crystal transferred by stroking first the supernatant and then the drop with a whisker. Alternatively, the whisker can be stroked across the surface of an existing crystal in order to pick up a few potential nuclei for transfer to a crystallizing drop.

For X-ray crystallography, a single crystal with a minimum dimension of 0.1 mm is required. The edges should be sharp and there should be no crystals growing into or out of the one chosen, although often some careful 'surgery' is appropriate to separate out a satisfactory specimen (30). A sharp and complete extinction when the crystal is viewed under crossed polarizers on the microscope is the sign of a good single crystal, but no extinction may reflect a high symmetry space group.

4. GENERAL SCHEME

The initial crystallization of a new protein is largely trial and error, although there is now a well-tried series of screening techniques (15,17–19). A general approach for a new protein might be the following.

(a) Purify and check the stability and homogeneity of the protein. This is critical and may well take a considerable time but will mostly derive from the initial study of the protein. It may be necessary to eliminate purification steps such as freeze-drying, acid precipitation or other extreme conditions (see Chapter 6). The protein is likely to have to remain stable and undegraded for days if not weeks.

Table 3. Some specialist supplies and suppliers

Bridges for Linbro® plates	Microbridges, 47 Purcell Road, Oxford OX3 0HB, UK (26)
Dialysis buttons	Cambridge Repetition Engineers, Green's Road, Cambridge CB4 3EQ, UK
Crystallization plates	Costar Europe Ltd, PO Box 94, 1170 AB Badhoevedorp, The Netherlands
Crystallization boxes and plates	Cryschem Inc., 5005 La Mart Drive, Riverside, CA 92507, USA
Crystal screening solutions	Hampton Research, 5225 Canyon Crest Drive, Suite 71–336, Riverside, CA 92507, USA (18)
Driven syringe robot	Douglas Instruments Ltd, 25J Thames House, 140 Battersea Park Road, London SW11 4NB, UK (23)
Linbro® and ACA CrystalPlates®	ICN Biomedicals Ltd., Thame Park Business Centre, Wenman Road, Thame, Oxfordshire OX9 3XA, UK; ICN Biomedicals, Inc., 3300 Hyland Avenue, Costa Mesa, CA 92626, USA
Light scattering equipment	e.g. Dynapro 801 - Protein Solutions Ltd, Hillside Centre, Upper Green Street, High Wycombe, Bucks HP11 2RB, UK
Lindemann glass capillaries	Wolfgang Muller, Reierallee 12, D-1000 Berlin 27, Germany
Crystallization database	National Institute of Standards and Technology, Bldg 221/A323, Gaithersburg, MD 20899, USA (27)
Crystallography books	Polycrystal Book Service, PO Box 3439, Dayton, OH 45401, USA

Micropipettes, high purity chemicals, biochemicals, greases, oils and general lab glass- and plastic-ware can be obtained from the usual laboratory suppliers

(b) Concentrate and ultrafilter the solution in a dilute buffer. Generally a concentration of about 5–15 mg ml^{-1} seems about the right concentration with which to begin. Unless protein stability dictates otherwise, use a 10 mM buffer solution of something (e.g. Hepes) which will not immediately produce salt crystals (e.g. phosphate). It is usual to initiate trials in the absence of protectants like glycerol.

(c) Set up a series of trials. Use the highest purity chemicals throughout. pH, type of precipitant, precipitant concentration and temperature are usually the first variables to try. Various additions, possibly of substrates, inhibitors, ions can be tried later. There are a variety of combinatorial methods of doing this with the 'Magic 50' (Hampton Research) technique being one that is commercially available.

(d) Expand the initial conditions. When success is achieved, make sure that the conditions will reproduce crystallization. Thus, try to produce (better) crystals again by using more and smaller steps in the variables found to work from the initial trials.

(e) Consider seeding.

5. A SIMPLE PROCEDURE FOR HANGING DROP CRYSTALLIZATION

Prepare 10 ml 3.5 M ammonium sulfate (AS) solution in 50 mM phosphate buffer at pH values 4.0, 5.5, 6.5 and 7.5. Spin or ultrafilter to remove any debris. Into a 24-well Linbro® tissue culture plate, pipette sufficient AS to give four rows, one at each pH, of 0.5, 1.0, 1.5, 2.0, 2.5, 3.0 M AS when diluted to 1.0 ml with 50 mM buffer. Round the top of each well, put a thin ring of petroleum jelly (Vaseline™ in a syringe with a plastic pipette tip works well). Pipette 1–5 μl (Gilson or similar autopipette). I use 5 μl drops, the less myopic generally use 1 or 2 μl) of the centrifuged (or ultrafiltered) protein solution (5 mg ml^{-1}, 10 mM Hepes buffer, pH 7.0) onto a clean, siliconized 22 mm coverslip and then carefully add an equal volume of the well solution, allowing the two drops to mix. Invert the coverslip and seat securely on the ring of grease. Repeat for all wells. Put a small piece of plasticine at each corner of the plate to keep the lid from resting on the coverslips. Store at the required temperature for vapor diffusion to establish an equilibrium—this involves water leaving the drop thereby increasing the salt and protein concentrations. It is sensible to leave the plate undisturbed for a week, although the equilibrium is established more rapidly. A dissecting microscope (magnification in the range ×20 to ×80) is ideal for examining the drops without disturbing them.

This basic recipe can be varied according to the quantity, known stability and other properties of the protein, the type of precipitant (for PEG use 10–30% saturated initially), the most suitable buffer, and so on.

6. A SIMPLE PROCEDURE FOR CRYSTALLIZATION BY SMALL-SCALE DIALYSIS

Perhaps the most convenient cells to use are 'dialysis buttons', small perspex cells with a groove to locate an O-ring for holding the membrane in place (25). Cambridge Repetition Engineers (*Table 3*) provide them in a variety of sizes. Alternatively, a short length (2.5 cm) of glass capillary (i.d. 1–2 mm) with an O-ring made of plastic tubing can be used. It is worth making the O-ring with two 'legs' to keep the membrane surface off the bottom of the tube (31). Dialysis allows recycling of protein, provided any precipitation is reversible.

6.1. Method for 'buttons'
Pipette the appropriate volume of protein solution into the central well (sizes range from 10–100 μl) and place a 20×20 mm square of dialysis membrane on top, ensuring that no air bubbles are trapped underneath. The O-ring is then pushed on, using a homemade device (25), which consists of a rod the same diameter as the button with a concave end which fits over the membrane, holding it securely. A cylinder which can be slid down the first rod then pushes the O-ring from the rod onto the button, locating it in the groove. The buttons are then placed in small bottles, such as scintillation vials, with some 5–10 ml of buffered precipitant solution. It is better to start with a dilute solution and increase the concentration, although the reduction of the concentration is possible too.

7. CONCLUSION

Despite the frustrations that are attendant on many crystallization experiments, the biological insight afforded by a high resolution crystal structure more than makes up for them. The information provided in this chapter should allow initial trials to be begun and, with patience and some luck, the crystallographic process, described in detail elsewhere (1,2), can be got underway.

8. REFERENCES

1. McRee, D.E. (1993) *Practical Protein Crystallography*. Academic Press, London.

2. Drenth, J. (1994) *Principles of Protein Crystallography*. Springer-Verlag, Berlin.

3. Groenenborn, A.M. (1993) *NMR of Proteins*. CRC Press, Boca Raton, FL.

4. Michel, H. (1991) *Crystallisation of Membrane Proteins*. CRC Press, Boca Raton, FL.

5. Garavito, M. and Picot, D. (1990) *Methods: a Companion to Methods in Enzymology*, **1**, 57.

6. Arakawa, T. and Timasheff, S.N. (1985) *Methods Enzymol.*, **114**, 49.

7. Mikol, V. and Giege, R. (1992) in *Crystallisation of Proteins and Nucleic Acids* (A. Ducruix and R. Giege, eds). IRL Press, Oxford, p. 219.

8. McPherson, A. (1991) in *Crystallisation of Membrane Proteins* (H. Michel, ed.). CRC Press, Boca Raton, FL, p. 1.

9. Feher, G. and Kam, Z. (1985) *Methods Enzymol.*, **114**, 77

10. Ries-Kautt, M. and Ducruix, A. (1992) in *Crystallisation of Proteins and Nucleic Acids* (A. Ducruix and R. Giege, eds). IRL Press, Oxford, p. 195.

11. Ducruix, A and Giege, R. (1992) *Crystallisation of Proteins and Nucleic Acids*. IRL Press, Oxford.

12. McPherson, A. (1989) *Preparation and Analysis of Protein Crystals (2nd edn)*. Krieger, Malabar.

13. Weigand, G. (1990) in *Modern Methods in Protein and Nucleic Acid Research: Review Articles* (H. Tschesche, ed.). De Gruyter, Berlin, p. 343.

14. Ollis, D. and White, S. (1990) *Methods Enzymol.*, **182**, 646.

15. Carter, C.W., Jr (1992) in *Crystallisation of Proteins and Nucleic Acids* (A. Ducruix and R. Giege, eds). IRL Press, Oxford, p. 47.

16. McPherson, A. (1990) *Eur. J. Biochem.*, **189**, 1.

17. Carter, C.W. Jr and Carter, C.W. (1979) *J. Biol. Chem.*, **254**, 12219.

18. Jancarik, J. and Kim, S.H. (1991) *J. Appl. Cryst.*, **24**, 409.

19. Kingston, R.L., Baker, H.M. and Baker, E.N. (1994) *Acta Cryst.*, **D50**, 429.

20. Ward, K.B., Perozzo, M.A. and Zuk, W.M. (1992) in *Crystallisation of Proteins and Nucleic Acids* (A. Ducruix and R. Giege, eds). IRL Press, Oxford, p. 291.

21. Oldfield, T.J., Ceska, T.A. and Brady, R.L. (1991) *J. Appl. Cryst.*, **24**, 255.

22. Chayen, N.E., Stewart, P.D., Maeder, D.L. and Blow, D.M. (1990) *J. Appl. Cryst.*, **23**, 297.

23. Chayen, N.E., Shaw Stewart, P.D. and Baldock, P. (1994) *Acta Cryst.*, **D50**, 456.

24. DeLucas, L.J. and Bugg, C.E. (1990) *Methods: a Companion to Methods in Enzymology*, **1**, 109.

25. Ducruix, A. and Giege, R. (1992) in *Crystallisation of Proteins and Nucleic Acids* (A. Ducruix and R. Giege, eds). IRL Press, Oxford, p. 73.

26. Harlos, K. (1992) *J. Appl. Cryst.*, **25**, 536.

27. Gilliland, G.L., Tung, M., Blakeslee, D.M. and Ladner, J.E. (1994) *Acta Cryst.*, **D50**, 408.

28. Stura, E.A. and Wilson, I.A. (1992) in *Crystallisation of Proteins and Nucleic Acids* (A. Ducruix and R. Giege, eds). IRL Press, Oxford, p. 99.

29. Thaller, C., Eichelle, G., Weaver, L.H., Wilson, E., Karlsson, R. and Jansonius, J.N. (1985) *Methods Enzymol.*, **114**, 132.

30. Rayment, I. (1985) *Methods Enzymol.*, **114**, 136.

31. Zeppezauer, M. (1971) *Methods Enzymol.*, **22**, 253.

32. Weber, P.C. (1991) *Adv. Prot. Chem.*, **41**, 1.

9. FURTHER READING

9.1. Sources of information on protein crystallization

Books
Most books on X-ray crystallography have sections on crystallization and crystal growth. The more recent ones also deal specifically with proteins. See refs 4, 11 and 12.

Reviews and articles
See refs 13, 16 and 32.

Methods Enzymol. (1985) **114**, 49.
(1990) **182**, 646.
Methods: a Companion to Methods in Enzymology (1990) **1**, 1.

International Conference on Crystallization of Biological Macromolecules Reports
First   – Stanford (1985) *J. Crystal Growth* (1986), **76**, 529.
Second – Bischenberg (1987) *J. Crystal Growth* (1988), **90**, 1.
Third   – Washington (1989) *J. Crystal Growth* (1991), **110**, 1.
Fourth – Freiburg (1991) *J. Crystal Growth* (1992), **122**, 1.
Fifth   – San Diego (1993) *Acta Crystallographica* (1994), **D50**, 337.

CHAPTER 16
SIZE DETERMINATION OF PROTEINS
A. HYDRODYNAMIC METHODS
O. Byron

1. INTRODUCTION

This section will concentrate on some aspects of modern analytical ultracentrifugation of relevance to size determination. The reason for undertaking analytical ultracentrifuge experiments is clear: if a protein system is to be truly understood it must at some point be characterized *in solution*. The information obtained from analytical ultracentrifugation is *operational*. It is also *absolute* and needs no comparison with molecular standards. The analytical ultracentrifuge can be operated in two modes to give data on solute molecular mass (from equilibrium studies) and shape (from velocity studies). This section should assist the reader in the effective design of an analytical ultracentrifuge experiment. Most of the text that follows will refer to absorption optics.

1.1. Data from the analytical ultracentrifuge
The data available from the analytical ultracentrifuge are summarized in *Table 1*.

Table 1. Parameters measurable in the analytical ultracentrifuge

Parameter	Mode	Use of parameter
$M_{w, app}^c$	SE	M of monodisperse system at loading concentration, c
		Point on graph (see *Figure 3*) for ultimate extrapolation to M_w^0
B	SE	Information on molecular conformation and Donnan effects
K_a	SE/SV	Strength of interaction between solutes
$s_{T, b}^c$	SV	Plot of $s_{T, b}^c$ versus concentration yields $s_{20, w}^0$
		$s_{20, w}^0$ is related, through M and $\bar{v}$, to f which gives information about shape and δ
k_s	SV	Information about molecular elongation and shape
$D_{T, b}^c$	SV	Plot of $D_{T, b}^c$ versus concentration yields $D_{20, w}^0$
		Dependent on M and f

Abbreviations: $M_{w, app}^c$, whole-cell apparent weight-average molecular mass (Da); SE, sedimentation equilibrium; M, molecular mass (Da); M_w^0, weight-average molecular mass at infinite dilution (Da); B, second virial coefficient (1 mol g^{-2}); K_a, association constant (M^{-1}); SV, sedimentation velocity; $s_{T, b}^c$, sedimentation coefficient s at solute concentration c, temperature T (K) and in buffer b; $s_{20, w}^0$, s at standard conditions (S; 1S=10^{-13}s); $\bar{v}$, partial specific volume (ml g^{-1}); f, frictional coefficient (g s^{-1}); δ, hydrodynamic hydration (g water g^{-1} protein); k_s, concentration-dependence coefficient for sedimentation velocity (ml g^{-1}); D, diffusion coefficient (cm^2 s^{-1}); $D_{T, b}^c$, D at solute concentration c, temperature T and in buffer b (cm^2 s^{-1}); $D_{20, w}^0$, D at standard conditions (cm^2 s^{-1}).

Table 2. Complementary techniques

Method	Sample required	Data generated
Classical light scattering[a] (1)	100 μl–3 ml, $M > 40$ kDa	M, mass distributions, R_g, B
Dynamic light scattering[b] (1)	200 μl, M > 15 kDa	D
Precision densimetry (2, 3)	1 ml at 1 mg ml^{-1}	$\bar{v}$, ρ
Viscosimetric analysis (2, 4)	2 ml, concentration strongly dependent on sample	$\eta_{T, b}$, $[\eta]$, k_η
X-ray scattering (SAXS, WAXS) (5, 6)	700 μl, typically 10 mg ml^{-1}	R_g, M, macromolecular conformation
Neutron scattering (SANS, WANS) (5, 6)	700 μl, typically 10 mg ml^{-1} often in D$_2$O buffer	R_g, M, macromolecular conformation, composition information for glycoproteins and lipoproteins

[a] Also known as multi-angle laser light scattering.
[b] Also known as quasi-elastic light scattering and photon correlation spectroscopy.
Abbreviations: M, molecular mass (g mol^{-1}); R_g, radius of gyration (nm); B, second virial coefficient (1 mol g^{-2}); D, diffusion coefficient (cm^2 s^{-1}); $\bar{v}$, partial specific volume (ml g^{-1}); ρ, density (mg ml^{-1}); $\eta_{T, b}$, viscosity of buffer, b, at temperature, T (Poise); $[\eta]$, intrinsic viscosity (ml g^{-1}); k_η, concentration-dependence coefficient for intrinsic viscosity (ml g^{-1}); SAXS, small-angle X-ray scattering; WAXS, wide-angle X-ray scattering; SANS, small-angle neutron scattering; WANS, wide-angle neutron scattering.

1.2. Complementary techniques

The interpretation of data from the analytical ultracentrifuge is greatly enhanced by the acquisition of complementary data (see *Table 2*). In some cases additional experiments are essential (see Section 4).

2. THEORY

The following introduction to analytical ultracentrifuge theory should be used in parallel with a basic primer on the subject (7, 8).

2.1. Sedimentation equilibrium

In sedimentation equilibrium experiments the rotor is spun at a speed high enough to generate a measurable distribution of the solute but low enough to allow the forces of sedimentation and diffusion to balance with each other. At equilibrium the distribution of the solute in the cell undergoes no further change and is (in the ideal case) characteristic only of the molecular mass of the solute.

Equation 1 describes the solute distribution at equilibrium (9, 10):

$$M^c_{w, app} = \frac{2RT}{(1 - \bar{v}\rho)\omega^2} \times \frac{d(\ln c)}{dr^2}, \qquad \text{Equation 1}$$

(see *Table 1* abbreviations; R, universal gas constant (8.314×10^7 erg mol^{-1} K^{-1}); ω, angular velocity ($= \pi$ rpm/30) (radian s^{-1}); r, radial distance from the rotor center (cm)). For an ideal

Figure 1. The diagnostic plot of ln (c) versus r^2 which, for an ideal monodisperse solute should display the straight line behavior represented by trace (a). Nonideality is indicated by the downwards curvature of trace (b), whilst trace (c) is typical of a polydisperse system. It is possible for a polydisperse, nonideal system to give data which reduce to a misleading trace (a).

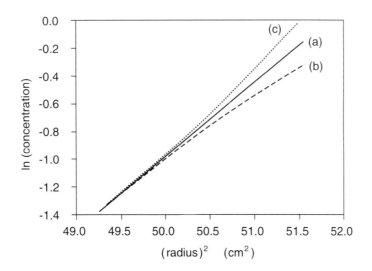

monodisperse (single component) system a plot of ln(c) as a function of r^2 should yield a straight line, the slope of which is proportional to the molecular mass. There are two diagnostic deviations from straight line behavior. If the data curve upwards away from the line (*Figure 1*) the system is displaying heterogeneity, whereas downwards curvature is symptomatic of nonideality. These phenomena do occur simultaneously for some systems (11) so a straight line plot is not conclusive evidence of a well-behaved sample.

Nonideality

If a system is monodisperse but thermodynamically nonideal, M observed at finite concentration might differ from that given by extrapolation to infinite dilution (M_w^0). This deviation can be a symptom either of molecular asymmetry or charge effects or a combination of both and is quantified by the second virial coefficient, B (see Equation 2).

$$M_{w,\,app}^c = \frac{M}{(1 + 2BMc)}.$$

Equation 2

The effect of B (normally positive for proteins) is to reduce $M_{w,\,app}^c$ (see *Figure 2*). Thus, for a nonideal system, M and B can be obtained by fitting Equation 2 to a plot of $M_{w,\,app}^c$ versus concentration.

Heterogeneity

The molecular mass obtained with absorption optics is the weight-average molecular mass, M_w, whilst the z-average parameter (M_z) can be obtained from data recorded with schlieren optics.

$$M_w = \frac{\Sigma M_i c_i}{\Sigma c_i}$$
Equation 3

and
$$M_z = \frac{\Sigma M_i^2 c_i}{\Sigma M_i c_i}$$
Equation 4

and
$$M_n = \frac{\Sigma M_i n_i}{\Sigma n_i}.$$
Equation 5

The number-average molecular mass (M_n) can be obtained from high-speed equilibrium experiments (Section 6.2). If a system is homogeneous $M_w/M_n = 1.0$, whilst this ratio will exceed unity as the degree of heterogeneity increases. Also, if $M_n = M_w = M_z$ then the sample is homogeneous, whilst polydispersity is indicated by $M_n < M_w < M_z$.

Self-association
There are several levels of complexity in the description of a self-associating system. This is well explained at a basic level in ref. 7. This equilibrium distribution can be analyzed with a number of software packages (e.g. NONLIN (33)) as described in Section 7.1.

Figure 2. A plot of $M_{w, app}^c$ versus total solute loading concentration for two hypothetical systems, both with a monomer molecular mass of 100 kDa. For the self-associating system $M_{w, app}^c$ increases, tending towards the dimer molecular mass, but being depressed with increasing concentration by the effects of nonideality. $M_{w, app}^c$ steadily decreases with increasing concentration for the monodisperse solute according to Equation 2.

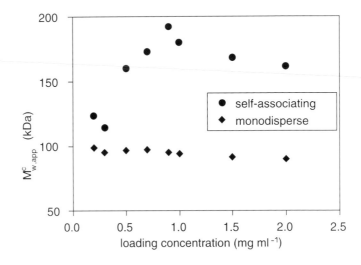

2.2. Sedimentation velocity

In a sedimentation velocity experiment the frictional behavior of solute is being explored. The rotor is run at a higher speed than in equilibrium analysis so that the process of sedimentation

Figure 3. Schematic diagram of the traces recorded with absorption optics in a typical sedimentation velocity experiment. The rate of movement of the sigmoidal boundary midpoint yields the sedimentation coefficient. The reduction in concentration apparent in the plateau region near the cell base is caused by radial dilution. The inset shows the corresponding plot of ln (r_b) versus $\omega^2 t$ from which $s^c_{T,b}$ is obtained.

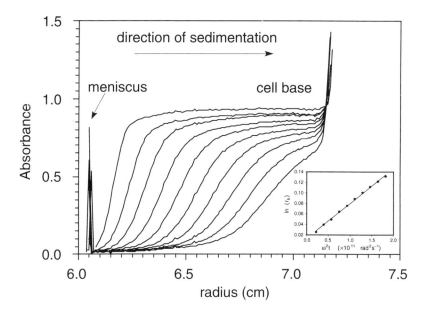

outweighs that of diffusion and the solute forms a moving boundary in the cell. Provided the solute is more dense than the solvent it will sediment. The rate of movement of the moving boundary (see *Figure 3*) gives the sedimentation coefficient $s^c_{T,b}$. This is related to other experimental parameters as follows:

$$s^c_{T,b} = \frac{1}{\omega^2} \cdot \frac{d \ln(r_b)}{dt},$$

Equation 6

where r_b is the radial position of the sedimenting boundary and t is time.

Simplistically it is the weight-average sedimentation coefficient that is measured in a sedimentation velocity experiment. This can be obtained from a plot of ln (r_b) versus $\omega^2 t$ (inset, *Figure 3*) which will have a slope of s (Section 7.2).

The sedimentation coefficient (s) is directly related to the mass (M) and frictional coefficient (f) of a particle:

$$s = \frac{M(1 - \bar{v}\rho)}{N_A f},$$

Equation 7

where N_A is Avogadro's number (mol^{-1}). Thus, if the mass of a monodisperse system is known, the frictional coefficient can be determined from a sedimentation velocity run. f is composed of two contributions: one from the deviation from sphericity of the molecular shape (f_e/f_0); the other the hydration (δ) of the particle (f/f_e):

$$\frac{f}{f_0} = \frac{f}{f_e} \times \frac{f_e}{f_0}.$$

Equation 8

Hydration

One central goal of sedimentation velocity is to gain a greater understanding of the conformation of a particle in a particular solvent. Through the frictional coefficient (f) it is possible to model this molecular shape, but the precision to which this can be done is limited by a lack of knowledge about δ (the degree of hydration of the protein). Hydrodynamic hydration (12) is difficult to quantify, although knowledge of the amino acid composition can lead to one estimate for δ. Failing this an average δ can be assigned to a protein but this average will vary according to the text consulted (9, 13). For many proteins $\delta = 0.3$ g water g^{-1} protein is a reasonable approximation. An experimental method for the determination of δ is described by (14), in which additional equilibrium and density measurements are required for the sample over a range of buffer densities.

Macromolecular shape

The frictional ratio (or Perrin ratio) can be interpreted at a number of levels of sophistication:

(i) δ can be ignored and all deviation from anhydrous sphericity be assigned to asymmetry. Reference to graphs of the Perrin ratio (9, 10) yields an axial ratio for prolate and oblate ellipsoids;
(ii) an estimated δ can be assigned and the reduced effective Perrin ratio be interpreted as above;
(iii) the molecule can be represented by an assembly of equally sized spheres for which the frictional coefficient can be quite accurately calculated using the Kirkwood equation (15);
(iv) the molecule can be represented by a more complex assembly of spheres of variable size. f and other parameters can then be very accurately calculated using the programs TRV (16) or HYDRO (17).

Concentration-dependence coefficient

The sedimentation coefficient measured at finite concentration is related to that at infinite dilution through the concentration dependence coefficient, k_s (Equation 14). A macromolecule that has a large virial coefficient (B) will usually have a large k_s (18). If the ionic strength of the solvent is moderate, the charge contribution to k_s can be neglected. So k_s can be used as a measure of asymmetry (18). If the intrinsic viscosity of the protein is known, the ratio of $k_s/[\eta]$ can be used as a probe of elongation. For spherical particles $k_s/[\eta] = 1.6$ (19). As particles become more rod-like $k_s/[\eta]$ tends towards 0.2. Such an understanding of a sedimenting species is vital in the effective interpretation of sedimentation velocity data for a self-associating system.

Self-association

Clearly because the experimental sedimentation coefficient is determined at a finite concentration it will be depressed by k_s (Equation 14). If $s^c_{T,b}$ is greater than or equal to $s^0_{T,b}$, the system is self-associating. Self-associating systems can be fully characterized by sedimentation velocity experiments, as can heterologous interactions and polydisperse noninteracting systems. These are discussed in Section 7.2. Aside from the analysis of the weight-average sedimentation coefficient, the sedimenting boundary traces themselves can also be analyzed. In self-interacting systems the normal sigmoidal traces (in absorption optics, *Figure 3*) will be replaced by those shown in *Figure 4*. The important exception to this rule is the monomer–dimer system, for which there is little deviation from the sigmoidal trace. Heterologous interactions will produce the same stepped

Figure 4. Absorption optics traces for a sedimentation velocity experiment with a heterogeneous sample in which the sedimentation coefficients of the two components differ sufficiently to give a characteristic stepped boundary. The approximate sedimentation coefficient of the slower moving component can be generated from analysis of the more compacted boundaries, whilst the more widely separated traces will yield an approximate $s_{T,b}^{c}$ for the more rapidly sedimenting solute.

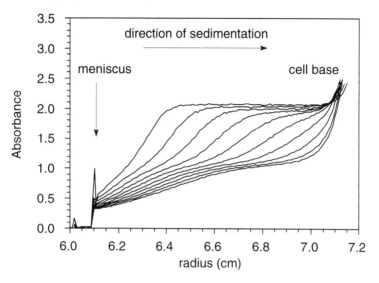

traces seen in *Figure 4* provided the difference in sedimentation coefficient between the constituent species is great enough. The rate of movement of the separately migrating regions of the boundary can be analyzed to yield approximate sedimentation coefficients for the species but a more satisfactory alternative is to use the $g(s^*)$ (sedimentation coefficient distribution function) approach (Section 7.2).

3. ADVANTAGES AND DISADVANTAGES

Table 3 lists the advantages and disadvantages of analytical ultracentrifugation.

4. ESSENTIAL ADDITIONAL INFORMATION

For the standardization of experimental parameters it is necessary to acquire additional information. Many of the calculations detailed below have been implemented in software called BIOMOLS which runs on a Macintosh and is available upon request from the Leicester laboratory of the NCMH, UK (see Section 8).

4.1. Partial specific volume

The partial specific volume ($\bar{v}$) of the protein under scrutiny is required for the conversion of buoyant properties (those actually measured in the analytical ultracentrifuge) to absolute parameters. $\bar{v}$ can be calculated with a good degree of accuracy from the amino acid composition.

Table 3. The advantages and disadvantages of analytical centrifuge experiments

Advantages	Disadvantages
Data are absolute and operational	Most instruments lack containment
Almost any solvent environment can be used	systems necessary for infectious
Proteins can be studied ±salts, ±cofactors, ±ligands,	samples
±detergents, ±nucleic acids	Proteolytic cleavage possible, can be
Any temperature between 4 and 40°C	minimized with protease inhibitors and
Experiments require a small amount of sample	low temperature
Essentially nondestructive	Protein adhesion to cell windows can
Range of optical systems	dilute sample and give high
Technique has large dynamic range	background
Well-suited to study of weak interactions	Interactions may be modified by pressure
Sedimentation experiments are self-cleaning	effects in velocity runs (20)
Experiments are rapid	Advanced data interpretation is more
Modern data acquisition systems are automatic	difficult
Data analysis is easier	Initial equipment financial outlay quite
Modern instruments are relatively easy to maintain	large
Beckman XL-A has good low-speed stability	
Wealth of literature spanning six decades	
Supported by a growing worldwide community of	
analytical ultracentrifuge specialists	
Electronic bulletin board (RASMB, see Section 8)	

The $\bar{v}$s of the constituent residues are known (21, 22), thus allowing a calculation of $\bar{v}$ for the protein by using Traube's rule:

$$\bar{v} = \frac{\Sigma n_i M_i \bar{v}_i}{\Sigma n_i M_i}$$

Equation 9

where n_i is the number of molecules, M_i is the molecular mass and $\bar{v}_i$ is the partial specific volume of component i in a multi-solute system. If the amino acid composition is unknown, $\bar{v}$ can be measured via precision densimetry (2, 3). A high level of accuracy is vital because an error of 1% in $\bar{v}$ will lead to an error in the region of 3% in $(1 - \bar{v}\rho)$ and hence in M. If the milligram quantities of sample required for densimetry are unavailable, it is not unreasonable too>approximate $\bar{v}$ as listed in *Table 4*. $\bar{v}$ is a temperature-dependent parameter. Any calculated or estimated values will require correction to the temperature of the sedimentation experiment (21).

Table 4. Approximate partial specific volumes for common biological macromolecular constituents

Constituent	$\bar{v}$ (ml g^{-1})
Protein	0.73
Carbohydrate	0.61
RNA	0.53
DNA	0.58

4.2. Buffer density

The density of the sample buffer (ideally the dialysate (see Section 6.1)) needs to be known for the interpretation of sedimentation data. This can either be measured directly via precision densimetry (2, 3) at the appropriate temperature or it can be estimated by calculation from standard data tables. Laue *et al.* (21) include a user-friendly account of density determination and how densities can be corrected to a given temperature using the Kell equation. A good source of actual density values for standard buffers is Weast (23, or later editions).

4.3. Buffer viscosity

For the conversion of sedimentation coefficients to standard conditions it is vital to know the viscosity of the buffer at the temperature of the analytical ultracentrifugation run. Viscosity can either be measured in a viscometry system (2, 4, 24) or estimated from data tables (23).

5. INSTRUMENTATION

There are few modern analytical ultracentrifuges: the Beckman Optima XL-A (scanning absorption optics (25, 26)) and the MOM 3180 (Rayleigh interference optics plus a system for the simultaneous acquisition of schlieren and interference data (27)). Some laboratories are equipped with older instruments.

5.1. Optics

The Beckman Optima XL-A is currently equipped with scanning absorption optics covering wavelengths between either 190–400 nm or 400–800 nm in a given run. Thus intrinsic chromophores can be followed. If the system lacks an absorbing group, Rayleigh interference optics can be used (if available). Here the difference in refractive index between the solution and sample is exploited to generate an interference pattern between light passing through the two sectors of a cell. This very sensitive method is suited to moderate-to-low concentration samples. The schlieren optical system (again dependent on refractive index) is suited to higher concentration systems lacking an absorbing chromophore. It should also be used for high concentration absorbing systems when the use of 'off-peak' wavelengths with absorption optics can yield nonlinear and unreliable data.

The availability of far-ultraviolet (UV) wavelength optics facilitates the characterization of poorly expressed or rare proteins and also enables the probing of strong interactions (with K_a in the region of 10^7 M^{-1}). Data obtained in the far-UV can be prone to errors induced by elevated background absorption of organic buffer components. This can be accounted for by overspeeding after an equilibrium run and is not such a cause for concern with sedimentation velocity data.

Proteins in *ultra*-low concentration regimes can be analyzed using fluorescence detection. There are, as yet, no commercially available systems but Schmidt and Riesner (28) have replaced the schlieren optics of a Beckman Model E with a fluorescence detection system which is used in tandem with UV absorption optics. The authors were able to determine the sedimentation coefficient of bovine serum albumin (BSA) at concentrations down to 0.1 nM (6.5 ng ml^{-1}).

6. MATERIALS AND METHODS

6.1. Sample preparation

After exhaustive dialysis against the required buffer the concentration of the sample should be re-

SIZE DETERMINATION OF PROTEINS

169

evaluated and the sample concentrated if necessary. About 400 μl of solution are needed for a sedimentation velocity run, whilst an equilibrium run requires only 80 μl (14 μl if ultra-short columns are being used (29)). The concentration used very much depends on the data being sought. It is useful to run a concentration series in a sedimentation velocity experiment to check for association or asymmetry (Section 7.2), but if material is scarce then it is reasonable to run samples at very low concentration and assume that $s^c_{20,\,w}$ approximates to $s^0_{20,\,w}$. The same logic applies to samples intended for study via sedimentation equilibrium. If self-interaction is being monitored, some idea of the association constant is required so that the concentrations can be appropriate to the observation of both monomer and oligomer.

6.2. Sedimentation equilibrium

The design of a sedimentation equilibrium experiment depends critically on the data being sought. Is the system under scrutiny likely to consist of a monodisperse, ideal solute? If so $M^c_{w,\,app}$ will be an invariant function of solute loading concentration and rotor velocity. If $M^c_{w,\,app}$ increases as a function of increasing loading concentration, self-association is being observed (*Figure 3*). $M^c_{w,\,app}$ will not increase as a function of loading concentration for a polydisperse, noninteracting system as the molecular mass distribution remains constant, regardless of finite concentration. If $M^c_{w,\,app}$ decreases as a function of increasing rotor velocity, the system is likely to be polydisperse, provided the rotor velocity is not high enough to sediment higher order self-associated oligomers to the cell base. Ref. 7 should be referred to for an expanded account of these diagnoses.

Rotor speed
There are two speed regimes for equilibrium experiments: low speed sedimentation equilibrium (LSSE) and high speed sedimentation equilibrium (HSSE) (see ref. 30 for an accessible account of their significances). HSSE (or meniscus depletion) is useful when an accurate value for the number-average molecular mass is being sought. But the high speed employed can lead the novice user to believe a system to be polydisperse when in fact it consists of self-interacting species with a high stoichiometry. There are also limitations to the precision achieved with HSSE (30).

Lower rotor speeds (those that ensure that at equilibrium the ratio of the cell-base solute concentration to that at the meniscus is less than 3–4) overcome the problems identified with HSSE but the equilibrium takes longer to attain. It is also true that HSSE provides an automatic baseline whilst, if there is any doubt about the integrity of the net background absorbance in a LSSE run, it is necessary to overspeed the rotor after the equilibrium distribution has been recorded so that a true baseline can be observed (see Section 6.1).

Time to equilibrium
For LSSE the approximate time for a system to reach equilibrium to within 0.1% (30) may be estimated from:

$$t_{0.1} = \frac{0.7(r_b - r_m)^2}{D} \qquad\qquad \text{Equation 10}$$

and

$$D \cong \frac{2.7 \times 10^{-5}}{(M\bar{v})^{1/3}} \qquad\qquad \text{Equation 11}$$

(where r_m is the radial position of the meniscus), but this is only approximate and Equation 11 overestimates D for asymmetric solutes (30). In LSSE experiments the time to reach equilibrium is appreciably longer and is harder to estimate but the equations of van Holde and Baldwin (31) can be used. Clearly the column length is the universal variable in tailoring the experimental time for either HSSE or LSSE. If a sample is unstable, it is appropriate to use very short columns (29).

6.3. Sedimentation velocity

Rotor speed
If the molecular mass is known with reasonable certainty, a rotor speed can be selected so that the sedimenting boundaries move a measurable distance during the scanning interval. An increased rotor speed is necessary in denaturing conditions where the frictional coefficient is much higher than for compact monomer, and in low-temperature runs where the increased buffer viscosity reduces the effective sedimentation rate. Similarly, rotor speed estimations can be inaccurate when the molecular species is highly asymmetrical. Use of too low a speed will give broad boundaries which are difficult to analyze.

Scan intervals
The scan interval will often be limited by the instrument. In the case of the Beckman Optima XL-A it can take 5 or 6 min to acquire one full-cell velocity trace. If three cells are being scanned, the minimum scan interval may be in the order of 20–25 min. The scan interval is a much lesser consideration in nonscanning optical systems, that is in photographic or direct digital camera systems which have the potential to generate far more data for a given run. Scans that are not fully resolved from the meniscus (i.e. early in the velocity experiment) should not be included in the determination of sedimentation coefficients. Scans should, however, be acquired from the start of the run as they contain valuable information about the system in the first stages of sedimentation and can reveal the presence of heavier species or aggregates.

7. DATA ANALYSIS

7.1. Sedimentation equilibrium
Use of raw data: molecular mass, distributions, interaction constants and virial coefficients
M can be obtained from the raw data via a plot of ln (c) versus r^2 (*Figure 2*) which should, for a monodisperse, ideal system, yield a straight line whose slope is proportional to M as defined in Equation 1. Alternatively, the raw data can be fitted via nonlinear least squares fitting algorithms, for example those implemented in software supplied with the Beckman Optima XL-A (XLA-SINGLE). These algorithms enable the user to select one of four models with which to analyze the data (*Table 5*).

Table 5. Models embodied in Beckman Optima XL-A analysis software for equilibrium data

Model	System	Useful variables
ideal 1	Single ideal species	M
ideal 2	Two noninteracting ideal species	M_1, M_2
nonideal	Nonideal single solute	M, B
assoc 4	Up to four associating species	M, $N_{2,3,4}$, $K_{a\ 2,\ 3,\ 4}$

Caution should be exercised with nonlinear least-squares fitting: there may be several models that will provide an acceptable fit to experimental data. This is illustrated by McRorie and Voelker (7) and is the subject of more critical analysis by Johnson and Straume (32). Self-interacting systems can also be modelled with the program NONLIN (33) and with the OMEGA function (34) which can be used to model ideal self- and hetero-association and has the facility to model nonideality.

$M^0_{w, app}$ (and M_z together with other molecular mass distributions) can also be determined using the program MSTAR (35). MSTAR differs from other programs in one important respect – it does not assume any model in its analysis. It has been written for both absorption and interferometric data and is used at the NCMH (Universities of Nottingham and Leicester, UK (Section 8)).

Use of reduced data: molecular mass, interaction constants and virial coefficients
By plotting $M^c_{w, app}$ as a function of concentration (*Figure 3*) it is possible to obtain association (or dissociation) constants for a self-interacting system. For a monomer–dimer system the monomer molecular mass (M) and $M^c_{w, app}$ are related through the equilibrium constant (K) as a function of total solute concentration (c):

$$M^c_{w, app} = M \frac{2(1 + 8Kc)^{1/2}}{1 + (1 + 8Kc)^{1/2}}.$$ Equation 12

This can be modeled in a standard curve-fitting package. Modeling becomes more complex when the solute exhibits nonideality, especially if the second virial coefficient undergoes a marked alteration upon dimerization. This can be accounted for by combining Equation 12 with Equation 2 and making B a variable.

7.2. Sedimentation velocity

Sedimentation coefficients
How the apparent sedimentation coefficient $s^c_{T, b}$ is generated from the raw data depends very much on how they were acquired in the first instance, but essentially a plot of $\ln(r_b)$ versus time will have a slope of $\omega^2 s^c_{T, b}$. If the data have been recorded digitally, $s^c_{T, b}$ can be extracted using purpose-written software. The Beckman Optima XL-A analysis package (XLA-VELOC) enables the user to analyze data in three distinct ways (*Table 6*).

Any value of $s^c_{T, b}$ requires correction for buffer density, viscosity and temperature as in Equation 13:

$$s^c_{20, w} = \frac{(1 - \bar{v}\rho)_{20, w}}{(1 - \bar{v}\rho)_{T, b}} \times \frac{\eta_{T, w}}{\eta_{20, w}} \times \left(\frac{\eta_b}{\eta_w}\right)_T \times s^c_{T, b}.$$ Equation 13

Table 6. Models embodied in Beckman Optima XL-A analysis software for velocity data

Method	Output
Transport	Weight-average $s^c_{T, b}$
Second moment/ boundary spreading	Weight-average $s^c_{T, b}$, D
g(s*) (36)	$s^c_{T, b}$ distribution

So it is necessary to measure (or estimate) the density of solvent and its viscosity relative to water at 20°C. The remaining factors can be found in data tables (23).

Important information can be gleaned from the concentration dependence of $s^c_{20, w}$. Spherical, noninteracting particles will exhibit a slight decrease in $s^c_{20, w}$ (approximately a 1% change in $s^c_{20, w}$ per mg ml^{-1}) with increasing concentration, according to Equation 14. If there is much less dependence, or if $s^c_{20, w}$ actually increases with concentration, self- or hetero-interaction is indicated. If the decrease in $s^c_{20, w}$ with concentration is much greater, the molecule is likely to be asymmetric:

$$s^c_{20, w} = s^0_{20, w}(1 - k_s c).$$

Equation 14

Some software routines automatically correct for radial dilution (e.g. the transport method routine of XLA-VELOC) but otherwise the multiplicative correction factor can found from the following equation:

$$\frac{c_p}{c_0} = \left(\frac{r_m}{r_b}\right)^2$$

Equation 15

where c_p is solute concentration at plateau of sedimentation velocity system and c_0 is solute concentration at meniscus of sedimentation velocity system.

Sedimentation coefficient distributions
Sedimentation boundary analysis (36, 37) can be used to obtain apparent (uncorrected for diffusion) sedimentation coefficient distributions ($g(s^*)$) and equilibrium constants for polydisperse systems (38) and for associating systems (39). Because the *time* derivative of concentration is used in this method the original data are corrected for time-independent noise.

Interaction constants
Interaction constants for velocity systems can also be estimated by nonlinear least-squares analysis of the concentration dependence of $s^c_{20, w}$ (18, 40). More complex analyses including the modeling of monomer-n-mer, isodesmic and ligand-induced self-association are described in ref. 41.

8. NATIONAL ANALYTICAL ULTRACENTRIFUGE FACILITIES

At the time of writing there are two national facilities for analytical ultracentrifugation, one in the UK (the National Centre for Macromolecular Hydrodynamics (NCMH), which has two laboratories) and one in the USA (the National Analytical Ultracentrifugation Facility (NAUF)). Their addresses are as follows.

NCMH, Department of Biochemistry, University of Leicester, University Road, Leicester LE1 7RH, UK. Tel: +116 252 3448/3452; fax: +116 252 3369 email: *ajr@le.ac.uk or ob1@le.ac.uk.*

NCMH, Department of Applied Biochemistry & Food Science, University of Nottingham, Sutton Bonington Campus, Sutton Bonington LE12 5RD, UK. Tel: +115 951 6148; fax: +115 951 6141 email: *sczsteve@szn1.nott.ac.uk.*

NAUF, Biotechnology Center, University of Connecticut, Storrs, Connecticut 06269-3149, USA. Tel: +203 486 4462/5011; fax: +203 486 5005; email: *brass@uconnvm.uconn.edu.*

8.1. RASMB

RASMB (Reversible Associations in Structural and Molecular Biology) was devised at a meeting of the same name held in February 1994 at the University of Melbourne, Australia. It is an electronic bulletin board that acts as a forum for open discussion on analytical ultracentrifugation and related issues and has amongst its regular contributors most of the world experts in the field. Usefully it is also a database of 'free' software and recently the XL-A bibliography (compiled by Beckman Instruments) has been made available through RASMB. RASMB is administered by Dr Walter Stafford of the Boston Biomedical Research Institute (BBRI) and is actually housed at the BBRI. Prospective users can subscribe by e-mailing *rasmb-manager@bbri.harvard.edu*. The XL-A bibliography, which is a survey of literature including work performed on the XL-A on biological macromolecules, assemblies and synthetic polymers, can be accessed through Netscape by typing *http://bbri-www.harvard.edu* or via a file transfer protocol command: *ftp://bbri.harvard.edu/rasmb/xla_refs.txt*. This downloads the database which can then be examined in a word processing program.

B. MASS SPECTROMETRIC METHODS

A.R. Pitt

9. INTRODUCTION

Mass spectrometry (MS) has come of age for the protein scientist. Since electrospray ionization (ESI) and matrix-assisted laser desorption ionization (MALDI) first became available in the late 1980s (42, 43) they have found uses in a wide variety of applications. They are both 'soft' ionization techniques, capable of producing ions of low energy, and therefore have the power to generate ions from biological macromolecules of molecular mass in excess of 100 000 Da (100 kDa), and in many cases measure their mass with accuracies better than 0.01%.

Mass spectrometry relies on two key processes, the generation of gas phase ions of the molecule of interest, a process that occurs within the 'source' area of the mass spectrometer, and the analysis of the mass to charge ratio of these ions using a 'mass analyzer'. Many combinations of sources and analyzers have been used (see *Table 7*), but this section will concentrate on the two most widely available and extensively used, MALDI time of flight (MALDITOF) and ESI using a quadrupole mass analyzer (referred to as ESIMS in this chapter). A large number of reviews have been published comparing the two techniques (44–49), so only the most important points will be dealt with here. As with any rapidly growing area of science, new techniques are currently being developed with even higher performance. Fourier transform ion cyclotron resonance (FTICR) mass spectrometers (50–52) with mass accuracies of <1 p.p.m. and exceptional resolution and sensitivity are now becoming commercially available, as are quadrupole ion traps (53, 54). These exciting new techniques are, without doubt, set to become the standards of the future.

Table 7. Mass spectrometric techniques for the determination of protein molecular mass

Technique	Mass range (Da)	Advantages	Disadvantages
ESI, quadrupole detection	>100 000	Robust and easy to use. Femtomole to picomole sensitivity. Resolution up to 2500[a]. Good accuracy 0.01–0.001%. Observation of noncovalent complexes from water. On line HPLC, CE capability. MS–MS[b] on top range machines	Limited efficacy for mixtures. Confusion sometimes arises due to multiple charging. Intolerant of presence of salts, especially alkali metals. Not necessarily quantitative. Often significant suppression of some components of a mixture
MALDITOF	>950 000	Robust and easy to use. Attomole to femtomole sensitivity. Tolerant of mM concentration of salts. Useful for mixtures. Little sample processing needed. Can analyze membranes from blotted gels directly	Generally lower resolution[a] than quadrupole (500–1000). Accuracy 0.1–0.01%. Choice of matrix important. Not easy to interface on line. Some suppression of components in complex mixtures. Not good for noncovalent complexes
Sector detection (mainly ESI but now being interfaced to MALDI)	>100 000	Resolution up to 50 000[a] (10 000–15 000 for routine analysis). Excellent accuracy (±5 p.p.m.)	Very poor sensitivity. Expert operation advisable. Time-consuming optimization
Quadrupole ion trap	>100 000	Excellent sensitivity (attomoles of compound with ion accumulation). Can act as analyzer (resonance ion ejection) or as an intermediate ion accumulator. Good accuracy and resolution[a]	Still in development. Only just becoming commercially available

[a] There are some discrepancies in the use of the term 'resolution' in the literature. It is usually one of the following:

(i) the '10% valley' definition: the mass at which there is a valley between the two ions differing in mass by 1 Da that is 10% of the height of the largest peak;

(ii) the 'full width at half maximum' (FWHM) definition: this uses the formula, Resolution $= M/\Delta M$, where M is the mass in Da and ΔM is the width of the peak in Da at 50% of its height.

The second definition is perhaps a more apposite test for the applications discussed and is used throughout this section.

[b] MS–MS = tandem mass spectrometry.

SIZE DETERMINATION OF PROTEINS

10. MALDITOF (55)

MALDI is an 'energy-sudden' technique, where ions are formed by an intense pulsed UV laser beam. The energy of the laser pulse is absorbed by an excess of an appropriate organic molecule, referred to as the matrix, cocrystallized with the analyte onto the target area. The matrix is chosen to absorb the energy of the laser light efficiently and vaporize rapidly to generate a plume of matrix and analyte ions which can then be sampled into the mass analyzer. Matrices must be volatile and are chosen to absorb strongly at the wavelength of the laser. The two most commonly utilized lasers are nitrogen (337 nm) or frequency tripled Nd:YAG (355 nm). The choice of matrix is relatively important, and is dependent on the type of molecule to be analyzed. The most commonly used matrices are dihydroxybenzoic acid (56) and a variety of cinnamic acid derivatives (57), although a wide range have been examined (58). The sample of interest is adjusted to 1–10 pmol μl^{-1} (i.e. 1–10 μM) and a small volume (a few microliters) is mixed with a similar volume of saturated matrix solution. The resultant mixture is placed on a stainless steel target area and allowed to dry, before being mounted in the source.

The pulsed energy-sudden nature of MALDI makes it ideal for analysis using a time of flight mass analyzer. In its simplest form, the ions produced by the laser pulse are accelerated to a constant kinetic energy, whereby high masses will have a low velocity and vice versa. The ions then enter a high vacuum field free drift region (i.e. a hollow tube) of 0.5–1.5 m in length and are detected as they reach the end of this region. The length of time between the laser pulse and the arrival of the ion at the detector is related to its mass to charge ratio. Hence, after calibration of the analyzer in the required mass range, the mass of the individual species can be calculated. The use of a reflectron (59), which doubles the length of the flight tube and reduces the spread of kinetic energies of ions of the same mass, has resulted in significant improvements in the resolution (at the expense of sensitivity and accuracy). Time of flight analyzers can cope with very large masses, with the largest so far reported being close to 1 000 000 Da! The lower mass range, below 500 Da, is usually obscured by ions from the matrix.

MALDITOF has a wide range of applications (see *Table 8*). Most importantly, it can cope with relatively complex mixtures of proteins (60), quite high concentrations of salts and glycosylated proteins (61). Most of the ions produced at the target are analyzed, making it significantly more sensitive than ESIMS. In many respects it is complementary to ESIMS.

11. ESIMS (62–64)

A wide range of mass spectrometers are available with electrospray sources (65). ESI is an atmospheric pressure ionization technique, where ions are formed continuously in the source at, or near, atmospheric pressure. The pressure is then reduced in stages through the source by high-capacity vacuum pumps, to reach the 10^{-5} torr or less needed in the analyzer. A continuous flow of solvent (at flow rates ranging from 1 μl min^{-1} up to 1 ml min^{-1}) is pumped into the source through a fine capillary needle which is held at a high potential (2–4 kV) relative to an adjacent sampling plate. This forms a fine spray of droplets. The formation of gas-phase ions takes place in four steps (63, 64):

(i) the formation of a fine spray of droplets with relatively high surface charge densities due to the high potential on the needle;
(ii) evaporation of the carrier solvent molecules from the droplets;
(iii) explosive fragmentation of the droplets;
(iv) eventually the formation of gas-phase ions of individual molecules.

Table 8. Applications of MALDITOF and ESIMS

Application[a]	MALDITOF	ESIMS
C-terminal sequencing (protein ladders)	OK for small peptides (83, 84)	Often spectra too complex
Characterization of proteins	Good (85, 86)	Good (86)
Characterization of covalently modified proteins	Good (87)	Good (57, 88, 89)
Comparison to expected mass from gene sequence	Internal mass standard needed	Good, masses usually sufficiently accurate
Complex mixtures	Good (60)	Limited
Determination of glycosylation patterns	Good (matrix choice important)	OK (90). Not for highly glycosylated forms
Identification of noncovalent complexes	No	In many cases good (73–76)
Identification of post-translational modifications	Good (91)	Good (92–94)
High molecular masses (Da)	Good >250 000 (47, 95–97) (poor mass accuracy)	0.02% accuracy to 150 000 (97)
Monitoring of purifications and digestions	Salt removal unnecessary	Poor (and slow)
MS–MS sequencing	Not yet available (being developed)	Good on more advanced machines (98–100)
On line CE-MS	Not easy	Good (69)
On line HPLC-MS	No	Good (70)
Proteolytic mapping	Good	Needs HPLC or CE
Quantification	Good (with care)	OK
Screening for heterogeneity	Resolution often limiting	Good if not too complex (101)
Screening for proteins of interest	Best sensitivity	Best mass accuracy
Suitable for direct use of blots from gels	Yes (102, 103)	No

[a] Assuming protein of interest is 6000 Da or above. MALDITOF has wider applicability if molecular masses are below 6000.
CE = capillary electrophoresis.

The formation of the stable spray of fine droplets can be aided at higher flow rates by the addition of a sheath of N_2 gas around the electrospray needle to promote nebulization (pneumatically assisted electrospray) or by replacing the capillary needle with an ultrasonic nebulizer. The evaporation of solvent from the droplets is usually promoted by heating the source (to 40–150°C), and in a number of designs by the use of a flow of nitrogen gas through the source. The formation of charged ions can be assisted by the addition of volatile acids or bases to the solvent. In many cases only a proportion of the analyte produces ions and the sampling of the ions is relatively inefficient. Hence, electrospray sources have limited sensitivity, especially at higher solvent flow rates.

Electrospray can generate either positive or negative ions allowing analysis of proteins with a wide range of pK_as. The sample of interest is usually dissolved at 5–100 pmol μl^{-1} (i.e. 5–100 μM) in a mixture of water, a small amount (0.2–0.5% v/v) of organic acid (usually methanoic or ethanoic acid) for positive ion or a base (either a volatile organic base, such as triethylamine or a volatile ammonium base, such as ammonium acetate) for negative ion electrospray and a volatile organic solvent (usually methanol or acetonitrile). This is injected directly into the solvent stream entering the source using a high-pressure liquid chromatography (HPLC) injection valve. It is usual to use 5–100 μl of the solution for each analysis.

The nature of the electrospray source means that it can be coupled to a number of analyzers, but most commonly a quadrupole mass analyzer is used. Quadrupole analyzers have a reasonable mass to charge range (up to 4000) and are robust, easy to tune and do not require a very high vacuum (10^{-5} torr is sufficient). The electrospray source produces ions carrying a number of charges such that the mass to charge ratio (m/z) of the ions, which is the parameter that is measured, usually lies in the region 500–2000 (although under some conditions this may increase to >4000). These are ideal for detection by a quadrupole analyzer. The quadrupole analyzer is a scanning analyzer, meaning that it scans across a mass range only allowing those ions of the correct mass to reach the detector. This means that only a fraction of the ions produced reach the detector, hence the lower sensitivity compared with MALDITOF.

Figure 5 shows a typical electrospray mass spectrum for a protein. It is usual to see a range of charged states adopting a Gaussian distribution, each of the peaks in the spectrum differing by the addition of one charge caused by the addition (positive ion electrospray) or removal (negative ion electrospray) of a proton. Under ideal conditions it should be possible to resolve species differing by 1 Da in 10 000 with a quadrupole analyzer, and 1 Da in 50 000 with a double focusing sector analyzer. Complex mixtures can be difficult to analyze due to the confusion of peaks that all appear at around the same mass. All ESIMS come with software that can deconvolute these multipeak spectra to give the molecular mass of the parent molecule, but many of these also have difficulty with multicomponent mixtures. One of the most effective is the MaxEnt (66–68) software supplied by VG Analytical, Manchester, UK. This uses a maximum entropy algorithm to identify the components from the raw data without any pre-processing, and can give exceptionally good resolution. However, in many cases, the best results have been obtained using manual picking of peaks in a series.

One of the major advantages of electrospray, apart from its accuracy and resolution, is that it can be directly interfaced to HPLC (69) and capillary electrophoresis (CE) (see Chapter 8 and ref. 70) for on-line monitoring of separated materials directly. This has proved to be a powerful technique for the identification of components of complex mixtures, especially in the use of proteolytic digests for the location of modifications (71).

ESI is a soft ionization technique that occurs from biologically compatible solvents. This allows for noncovalent and labile covalent protein–ligand, protein–protein and probably protein–DNA complexes to remain intact during the ionization process and be observed. This is useful for the identification of bound cofactors, including metal ions (72), and for the identification of enzyme–substrate complexes (73–76). Quaternary structures of proteins have also been observed (77–79).

Figure 5. ESIMS of a mixture of horse heart myoglobin (A) (expected molecular mass 16 753), and human hemoglobin α (B) and β (C) subunits (expected subunit molecular masses $\alpha = 15\ 126$, $\beta = 15\ 867$ Da) (Hum Hb HH Myo Mix). (a) A typical smoothed raw data set of multiply charged ions (the numbers above the peaks refer to the number of charges on each ion and the letters to the parent protein). (b) A MaxEnt deconvolution of spectrum (a), showing the derived molecular masses (the numbers above the peaks refer to the derived molecular mass in Da and the letters to the parent protein). The major peaks to the higher mass side of the main peaks are due to the formation of adducts with sodium.

Hum Hb HH Myo Mix

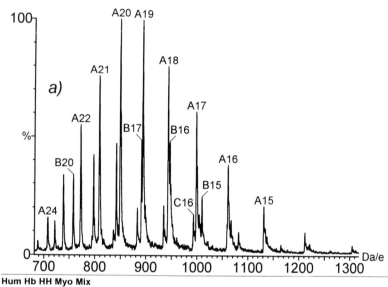

Hum Hb HH Myo Mix

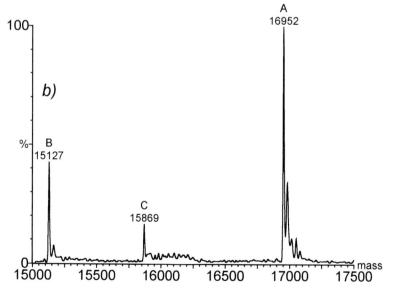

12. COMMON PROBLEMS AND LIMITATIONS

12.1. MALDITOF

MALDITOF is the more robust of the two techniques, and has fewer problems with sample quality and operation. Nonuniformity in the matrix-analyte mixture is the most common cause of both lack of signal and problems with quantitative analysis (80), so care must be taken to mix the sample well. Many systems allow the laser spot to be scanned over the surface of the target area, allowing sampling at a number of sites giving improved reproducibility. Adducts with matrix molecules are often formed, caused by reaction of the photo-excited state of the matrix molecule with the protein. These are observable as a shoulder on the higher mass side of a peak separated by approximately the mass of the matrix molecule. The main limitation of MALDITOF has been poor mass accuracies and resolution (81). Careful calibration in the mass range of interest, and preferably the addition of an internal standard to the analyte mixture, have been used to improve accuracy (82), while the use of a reflectron can improve resolution (59).

Table 9. Some do's and don'ts for ESIMS

DO (whenever possible)	DO NOT
Spend some time preparing the sample: this is *always* time well spent	Believe everything that you see
Exchange the solvent for high grade water	Use high ion concentrations (which includes the protein)
Freeze dry the sample and redissolve in the carrier solvent	Expect the technique to be quantitative, especially for components that are very different
Remove all nonvolatile buffer salts	Use phosphate buffers
Increase the needle potential, source heating and drying gas flow rate for less volatile solvents	Assume that a more concentrated sample will always give better results
Try to work in 50:50 water:acetonitrile or water:methanol (the machine often needs to be retuned for each different solvent mixture)	Assume that quoted errors are for the derived mass – they are more likely to refer to the quality of the data for each individual run
Check that your protein is soluble in the carrier solvent, especially if the solvent for your sample is different (precipitated protein is very bad news!)	Be too worried if the mass differs from that predicted from the DNA sequence. Sequencing errors and post-translational modification are common
Repeat analysis for error calculation	Use high salt steps (e.g. ion exchange) as the last step of the purification
Be on the lookout for peaks at +22.0 and +38.1 in positive ion electrospray, which correspond to sodium or potassium adducts	Use sodium or potassium salts of your buffers for positive ion electrospray (NH_4^+ salts are acceptable counterions, although H^+ is best)
Filter or centrifuge sample before use	
Use average isotopic masses in calculations	

12.2. ESIMS

Electrospray is sensitive to a number of factors, mainly the quality of the spray, the desolvation/declustering of the ions and the quality of the sample (see *Table 9* for further details).

The spray quality depends on many factors, especially the surface tension of the solvents and constant solvent delivery. The dissociation of clusters of solvent molecules and non-volatile material present in the mixture (buffer ions, etc.), known as declustering/desolvation, is markedly affected by the volatility of the solvent, the presence of other small molecules in the analyte and the thermal energy transmitted to the molecules in the source area. Generally the carrier stream

Table 10. Average isotopic masses of amino acids and protein end groups

Name	Single letter code	Molecular formula	Average mass[a]
Amino acids			
Alanine	A	C_3H_5NO	71.0788
Cysteine	C	C_3H_5NOS	103.1448
Aspartate	D	$C_4H_5NO_3$	115.0866
Glutamate	E	$C_5H_7NO_3$	129.1155
Phenylalanine	F	C_9H_9NO	147.1766
Glycine	G	C_2H_3NO	57.0520
Histidine	H	$C_6H_7N_3O$	137.1412
Isoleucine	I	$C_6H_{11}NO$	113.1595
Lysine	K	$C_6H_{12}N_2O$	128.1742
Leucine	L	$C_6H_{11}NO$	113.1595
Methionine	M	C_5H_9NOS	131.1986
Asparagine	N	$C_4H_6N_2O_2$	114.1039
Proline	P	C_5H_7NO	97.1167
Glutamine	Q	$C_5H_8N_2O_2$	128.1308
Arginine	R	$C_6H_{12}N_4O$	156.1876
Serine	S	$C_3H_5NO_2$	87.0782
Threonine	T	$C_4H_7NO_2$	101.1051
Valine	V	C_5H_9NO	99.1326
Tryptophan	W	$C_{11}H_{10}N_2O$	186.2133
Tyrosine	Y	$C_9H_9NO_2$	163.1760
N-terminal groups			
Hydrogen		H	1.0079
N-formyl		HCO	29.0183
N-acetyl		CH_3CO	43.0452
C-terminal groups			
Free acid		OH	17.0073
Amide		NH_2	16.0226

[a] To calculate the average molecular mass of the protein, sum the masses of the constituent amino acids and add the relevant N- and C-terminal groups from the table (for other post-translational modifications see *Tables 11* and *12*). Average isotopic masses are based upon the following average atomic masses for the elements: H=1.00794, C=12.011, N=14.00674, O=15.9994, P=30.97376, S=32.066.

Size Determination

Table 11. Common post-translational modifications (excluding glycosylations)

Modification	Mass change[a]
Pyroglutamic acid from glutamine	−17.0306
Disulfide bond formation[b]	−2.0159
C-terminal amide from glycine[c]	−0.9847
Deamidation of glutamine or asparagine	0.9847
Methylation	14.0269
Hydroxylation	15.9994
Oxidation	15.9994
Oxidation of methionine:	
sulfoxide	15.9994
sulfone	31.9988
Formylation	28.0104
Acetylation[b]	42.0373
Carboxylation of asparagine or glutamine	44.0098
Phosphorylation	79.9799
Sulfonation	80.0642
Cysteinylation	119.1442
Incomplete N-terminal methionine removal [b]	131.1986
Glycosylation (see *Table 12*)	
Farnesylation	204.3556
Myristoylation	210.3598
Biotinylation	226.2994
Pyridoxal phosphate (Schiff base to lysine)	231.1449
Glutathionylation	305.3117
5′-adenosylation	329.2091
4′-phosphopantotheine	339.3294

[a] Add the quoted mass to the calculated molecular mass of the unmodified sequence.
[b] Incomplete post-translational processing of overexpressed proteins is quite a common occurrence and will result in mixtures of proteins being observed, separated by the indicated masses.
The amide (which is essential for biological activity of many hormones, growth factors and neurotransmitters) is formed by hydroxyglycine of an additional glycine residue in the precursor. The hydroxyglycine dissociates to form the peptide amide and glyoxylic acid. Therefore, the overall mass change is −58. 0367.

and sample solvent are mixtures of equal volumes of water and a low-viscosity volatile organic solvent, although 100% water and 100% organic phase can be used. While it is possible to use significant concentrations of volatile buffers (5–10 mM), resolution and mass accuracy can be impaired. Nonvolatile solutes should be avoided as these associate with the surface of the protein resulting in the formation of peaks with long tails to the higher mass side leading to poor resolution, inaccuracies in mass determination, reduced sensitivity and, in serious cases, complete loss of information. Low concentrations of nonvolatile buffer salts, such as Tris–HCl, have been used (up to 5 mM can usually be tolerated) but the quality of the spectra is always compromised. The technique is *very intolerant* of alkali metal ions (Na^+ and K^+ especially). These replace hydrogens as the counterions on the surface of the protein with high affinities, resulting in a series of peaks separated by 21.982 or 38.090 Da, respectively. Careful preparation of samples is therefore essential.

Table 12. Common sugars occurring in glycosylated proteins

Monosaccharide residue	Formula	Mass[a]
Deoxypentoses (dRib)	$C_5H_{10}O_4$	116.117
Pentoses (Ara, Rib, Xyl)	$C_5H_{10}O_5$	132.116
Deoxyhexoses (Fuc, Rha)	$C_6H_{12}O_5$	146.143
Hexosamines (GalN, GlcN)	$C_6H_{13}NO_5$	161.158
Hexoses (Gal, Glc, Man)	$C_6H_{12}O_6$	162.142
Hexuronic acid (GlcA)	$C_6H_{10}O_7$	176.126
Heptose	$C_7H_{14}O_7$	192.169
N-acetylhexosamines		
(GalNAc, GlcNAc)	$C_8H_{15}NO_6$	203.195
2-keto-3-deoxyoctanoate	$C_8H_{14}O_8$	220.179
Muramic acid	$C_{11}H_{19}NO_8$	275.259
Sialic acids:		
N-acetylneuraminic acid (NeuAc)	$C_{11}H_{19}NO_9$	291.258
N-glycolylneuraminic acid (NeuGc)	$C_{11}H_{19}NO_{10}$	307.257

[a] Add the masses of the relevant sugars to the calculated molecular mass of the unmodified sequence.

13. CALCULATION OF MOLECULAR MASSES

Neither MALDITOF nor ESIMS are particularly high-resolution techniques, with isotope peaks only being visible for low molecular mass compounds. For compounds of molecular mass of greater than the resolution limit of the machine it is usual to see a broad peak that is an average of the isotope peaks. The standard practice with spectra of these higher molecular mass compounds, and especially with the multiply charged peaks in ESIMS spectra, is to apply smoothing to reduce noise levels. It is usual, therefore, to use the top of the peak in deriving the mass of the compound, which corresponds to its *average* mass. Hence, for calculating the molecular masses of the compounds *average* isotopic masses are used. Those of the most commonly occurring elements in biomolecules are given in *Table 10*, and these masses are used to derive all the masses quoted in this chapter.

It is often easier to calculate the mass based on the amino acid sequence, and the molecular masses of the amino acids are shown in *Table 10*, along with the standard end groups. The mass changes for post-translational modifications and glycosylation are shown in *Tables 11* and *12*, respectively.

14. REFERENCES

1. Harding, S.E., Satelle, D.B. and Bloomfield, V.A. (eds) (1992) *Laser Light Scattering in Biochemistry*. Royal Society of Chemistry, Cambridge, UK.

2. Freifelder, D. (1982) *Physical Biochemistry*, *2nd edn*. W. H. Freeman & Co., New York.

3. Kratky, O., Leopold, H. and Stabinger, H. (1973) *Methods Enzymol.*, **27**, 98.

4. Bradbury, J.H. (1970) in *The Principles and Techniques of Protein Chemistry (Part B)*. Academic Press Inc., New York, p. 99.

2. INSTRUMENTATION

2.1. IR spectroscopy

There are two types of IR spectrometer, the dispersive instrument and the Fourier transform (FT) instrument. The former employs a grating to scan the frequency range, much as in an ultraviolet (UV) instrument. In recent years the FT instrument has displaced the dispersive instrument since it has a very much better performance. The FT instrument employs a Michelson interferometer to modulate the whole frequency range simultaneously and so the spectrum is acquired across the whole frequency range simultaneously.

The FT instrument is capable of far better noise performance than is the dispersive instrument, for two main reasons. First all the light passes through the sample all the time; there are no slits. The spectrum is acquired 'in parallel' and so the scan rate is enormously faster. A modern research-grade FT spectrometer can acquire spectra at 2 cm^{-1} resolution at up to 30 scans s^{-1}. This allows time-resolved phenomena to be studied and very small absorbances or changes in absorbance to be measured of the order of 0.0001. The resolution is dependent upon the distance the mirror travels in each stroke, so lower resolution spectra can be scanned faster since the distance the mirror has to travel is less. FT instruments cost between £10 000 and £150 000 ($15 000–225 000), depending on the performance and flexibility required. They are extremely reliable and easy to use.

2.2. Raman spectroscopy

Raman spectrometers are somewhat similar to fluorescence instruments in that they collect and measure scattered light. A scanning spectrometer uses monochromators and a photomultiplier but the more modern multichannel instrument accumulates all frequencies simultaneously and is thus comparable with an FTIR spectrometer. For collection of the dispersed, scattered radiation an intensified diode array or cooled charge coupled device (CCD) array detector is used. Lasers that emit at wavelengths tunable throughout the UV and visible spectrum are now available and work reliably. The Raman spectrometer has not been developed as a commercial 'package' to the same extent as FTIR spectrometers, and is often built up using parts from a variety of suppliers. It is undoubtedly more difficult to use compared with FTIR spectrometers and requires experience to avoid pitfalls such as photolysis of the sample by the intense laser beam. The cost is towards the upper range of that of FTIR spectrometers.

The FT-Raman spectrometer, a recent innovation, uses an IR laser beam at 1085 nm to promote Raman scattering, which is then detected and analyzed using an FTIR spectrometer. This configuration has the advantage that fluorescence, which can plague the application of Raman spectroscopy, is eliminated. Resonance enhancement cannot however be used with this equipment which is commercially available in packaged form (1).

3. SAMPLING

3.1. IR spectroscopy

Transmission methods have traditionally been used to sample biological materials. For the study of biological materials sampling is a crucially important aspect where it is worthwhile to pay adequate attention to detail in order to get good results. In FTIR spectroscopy biological samples can be studied in a dry state, in the form of dried films or KBr discs. Lipid films are often studied in this way, coated on a sodium chloride plate. While this is easy to do, it is often unsuitable, since reactions require solution and proteins cannot fold and unfold in the dry state. Although far from ideal in view of their strong IR absorbance, aqueous solvents can be used with FTIR (1).

Attenuated total reflectance is rapidly becoming more widely used since the sample is simply placed upon the plate and the instrument purge does not have to broken. The method works by the total internal reflection of the IR beam along a crystal (see *Table 1*). Each time the light is reflected an evanescent wave penetrates a distance approximately equal to the wavelength of the IR light into the sample. It is this part of the beam which acquires information about the sample.

The most widely used sampling methods are described in *Table 1* (1).

Table 1. The most widely used methods for sampling biological material

Sample method	Sample preparation	Applications	Spectral range (cm^{-1})
Transmission: KBr disc	Grind sample with KBr and press or deposit thin layer on flat surface of crystal	Dry material only. Not usually used with proteins	40 000–400
Transmission: NaCl plate or liquid cell	Nonaqueous materials only	Lipid layers	40 000–650
Transmission: CaF$_2$ liquid cell	Liquid cell with 50 μm Teflon spacer for deuterium oxide and 5–10 μm spacer for water. 10–200 μl sample volume required	Protein secondary structure and protein folding (sample $\approx$10 μM (minimum)). Enzyme reactions and intermediates. Sample $\approx$1 mM	50 000–1100, but restricted by water and deuterium oxide absorbance to 'windows'. 5–10 μm cells for use with water have to be dismantled between each experiment, longer pathlength cells can be filled without dismantling
Attenuated total reflection: ZnSe	Sample in aqueous solution placed directly on plate	Any biological sample where toxicity of ZnSe is not a problem; avoid extremes of pH	20 000–500
Attenuated total reflection: Ge	As for ZnSe	Biocompatible, particularly when coated with collagen or gelatin; can be used with mammalian cells	5500–600

3.2. Raman spectroscopy

The wide range of sampling procedures used in Raman spectroscopy makes it difficult to summarize the information in the form of a Table. Water is an excellent solvent for Raman spectroscopy. The lack of sensitivity of Raman spectroscopy requires that the sample be as concentrated as possible. It can, for example, be presented to the spectrometer in a quartz capillary, often using a flow system to refresh the sample, which may suffer some laser damage. Resonance Raman spectroscopy requires much less sample, owing to the greatly enhanced sensitivity, and solutions in the micromolar range can often be used. Fluorescence can be a major problem in Raman spectroscopy and there are several ways that this may be alleviated. A time-honored procedure known as 'burn off' sometimes works. The sample is simply irradiated in the laser beam for an extended period before the spectrum is taken. The fluorescence decays but so may the integrity of the sample! Another method employs gating of the light scattered by the sample. The Raman signal appears instantaneously, while the fluorescence emission is usually delayed by a few nanoseconds. The scattered light is thus collected only for the first few nanoseconds after the laser is pulsed. This method cannot be used with continuous wave (cw) lasers such as Ar^+ which often give the best Raman signals. A recent innovation, said to remove the problem of fluorescence, is the FT-Raman method. Here an IR laser is used and the scattered radiation is analyzed using an FT spectrometer (1).

3.3. Microsampling

Microscope systems are available for both IR and Raman spectrometers. These enable objects as small as 1 μm to be imaged and have spectra recorded. Mammalian cells are sufficiently large that spectra can be taken of organelles within the cell. The sampling materials and optics are as described above.

4. VIBRATIONAL FREQUENCIES

4.1. Strongly absorbing bands in biological materials

Table 2 gives the IR bands which are of primary importance in studies of biological materials, together with the type of molecules which show these bands.

CaF_2 is the most popular window material for work with aqueous solutions but cuts off at around 1100 cm^{-1}; the asymmetric P=O stretch is usually observable. There is no such restriction in Raman spectra and bands can be measured down to 50 cm^{-1}. Some 'marker' bands for DNA bases occur below 1000 cm^{-1}. Raman frequencies are nominally the same as in the IR but symmetric or polarizable groups give strong bands, while asymmetric polarized species show weak scattering. The SH and –S–S– groups, for example, absorb very weakly in the IR but scatter strongly with frequencies of around 2550 and 500 cm^{-1} respectively. The C=C group also scatters strongly at 1680–1620 cm^{-1} but absorbs weakly in the IR. In dry samples there is the advantage that bands at frequencies <1000 cm^{-1} can be observed in IR spectra.

5. ASSIGNMENT OF BANDS IN COMPLEX BIOLOGICAL MOLECULES

While bands such as the amide 1 (C=O) vibration in proteins are easily observed since there are many such groups in a protein molecule, it is often desired to observe bands which arise from single groups in macromolecules. This is generally much more difficult to achieve, since there will often be a large background absorbance which arises from other groups in the molecule, although it is sometimes possible to work in a clear region of the background spectrum (1–3).

Table 2. IR bands of primary importance in the study of biological material and the molecules that show these bands

Frequency (cm^{-1})	Group vibration	Biological molecule
3700–3100	O–H stretch	Very strong in IR; only usable with nonaqueous samples
3700–3100	N–H stretch	Not generally visible in aqueous solution
3200–2800	C–H stretch	Lipids, protein, DNA
2250	C≡N stretch	Ligand binding model, e.g. for O_2
1750–1700	C=O stretch	COOH, esters, lipids
1680–1600	C=O stretch	Amide 1, proteins
1640	O–H bend	Can obscure C=O in H_2O; D_2O often used to avoid this
1610–1540	C–O asym. stretch	COO$^-$, proteins, fatty acids
1580–1500	C–N stretch, N–H bend	Amide 2, proteins
1460–1400	C–O sym. stretch	COO$^-$, proteins, fatty acids
1400–1350	CH$_2$ wagging	Lipids
1250–1200	P=O asym. stretch	DNA, phospholipids
1150–1000	P=O sym. stretch	DNA, phospholipids

5.1. Resonance Raman spectroscopy

Resonance Raman spectroscopy is a particularly effective way of enhancing the intensity of the band of interest, since the laser frequency is tuned to excite just this chromophore so that it stands out from the background (1–3).

5.2. Difference spectroscopy

The method of difference spectroscopy uses subtraction to highlight the feature of interest that differs between two related states of the system (20–23). Thus for a lipid the difference might be taken between spectra collected at two different temperatures and so reflect a phase difference between the two temperatures. The spectrum of a protein can be subtracted from that of a protein–ligand complex to give the spectrum of the bound ligand; this can then be compared with the free ligand to assess whether there is any change in the spectrum when the ligand binds to the protein. A serious problem that can arise in the use of this method is obscuration of the ligand spectral band by features which arise from perturbation of the protein (or macromolecule) structure which occurs consequent upon binding of the ligand. This can usually be resolved as below.

5.3. Heavy atom isotope editing

If a predominant natural atom in a vibrating system such as ^{1}H is changed to ^{2}H the frequency is decreased (by approximately 800 cm^{-1} for C—H $\Rightarrow$ C—^{2}H) since the reduced mass is increased. This reduction in frequency can be used to assign specific groups to specific spectral features if the isotope substitution is itself specific to the group or groups of interest. This may also serve the purpose of shifting the frequency away from the background absorbance so that the labeled group(s) can be observed in isolation. Some groups such as amide (N—H) exchange readily with deuterium in the solvent so these cannot be used for specific assignment; this process can, however, be used to assess the solvent availability of amide groups in proteins. Substitution of ^{13}C in place of ^{12}C in ^{12}C=^{16}O to give ^{13}C=^{16}O is particularly useful for the assignment of carbonyl groups; this gives a down frequency shift of about 40 cm^{-1} (21, 23). Coincidental substitution of the ^{16}O with ^{18}O to give $^{12}C={^{16}O} \rightarrow {^{13}C}={^{18}O}$ yields a shift of about 80 cm^{-1} and can be useful in the resolution of overlapping carbonyl bands (23). ^{15}N substitution has not been used much but offers an opportunity for substitution in amides and nucleic acids.

6. SOME SELECTED APPLICATIONS OF IR AND RAMAN SPECTROSCOPIES

6.1. Lipids/phospholipids
The various C—H and CH$_2$ modes can be used to study order in lipid membranes. Temperature variation is commonly employed to define phase transition temperatures. The spectroscopic parameters allow the nature of the structural transition to be defined. The P=O stretch vibration can be used to gain information concerning the environment of the head groups in phospholipids (4,5).

6.2. Bacteriorhodopsin
The bacterial membrane light-activated proton pump bacteriorhodopsin has been the subject of many excellent studies. The structural changes which occur in the rhodopsin chromophore during the photocycle have been characterized in detail, as have the proton transfer events (i.e. Schiff base protonation) that occur during pumping. Site-specific mutagenesis of Asp and Glu residues, which are buried in the transmembrane portion of the molecule, has allowed the definition of some aspects of the proton transfer pathway through the molecule. Fast time-resolved studies using both resonance Raman and step-scan IR have played an important role (6–8).

6.3. Heme proteins
Hemoglobin and myoglobin bind oxygen and act as carriers. Resonance Raman spectroscopy, and to a lesser extent IR, has been used to characterize the interaction of the heme group with ligands. Carbon monoxide has proved to be a useful model for oxygen since it binds tightly and has a convenient absorption frequency. The geometry and energetics of the ligand interaction can be elucidated. The change that the iron atom experiences on ligand binding has been described by measurement of change in the His–Fe stretch vibration at low frequency. The His represents the axial ligand to the iron that opposes the oxygen ligand on the other face of the heme group and so is very sensitive to the state of ligation of the heme group. Similar studies have involved cytochromes and oxidase enzymes (9–11).

6.4. DNA and nucleoprotein complexes
Raman spectroscopy has been used to very good effect in the determination of the structure of oligonucleotides. By using base and sugar 'marker' bands that are conformation sensitive it is often possible to assign the conformation of a given basepair in an oligonucleotide sequence and in this way to build up a knowledge of the structure. This ability has arisen from much painstaking work on the spectra of the nucleotide components of the DNA structure, which are simplified models of the whole structure. This approach has also reaped rewards in the study of the structure of viral nucleoprotein complexes (12–14).

7. DETERMINATION OF THE SECONDARY STRUCTURE OF PROTEINS

7.1. Band fitting methods
One of the most important applications of vibrational spectroscopy to the study of biological molecules has concerned the development of methods which allow the determination of the secondary structure of proteins. IR spectroscopy has been to the forefront in this work and two fundamentally different methods have been devised. Both methods rely upon the fact that the

different types of secondary structure in a protein give rise to characteristic amide C=O absorption frequencies. The mixture of overlapping amide absorption bands can be used to define the mix of secondary structure if the various components can be separated. Resolution enhancement and second derivative methods rely upon a 'sharpening' of the multiple features which lie in the broad amide band envelope, so that they may be individually distinguished and quantified. Fourier self-deconvolution is a popular resolution enhancement method which permits direct quantitative band fitting of the structural components. This technique relies absolutely upon an accurate knowledge of the frequencies of the various structure types. This information has been acquired from measurements on simple model peptides and by a correlation analysis with known X-ray structures (15–17).

7.2. Frequencies of secondary structural elements in proteins

Typical values of the frequencies (18) that relate to specific secondary structures are shown in *Table 3*.

There is discussion concerning the validity of the use of these frequencies, since each type of structure will have amides that deviate from these exact frequency values and hence there is a problem concerning the width that should be assigned to heavily overlapping bands, which arise from microscopically dispersed species (15).

Table 3. Typical values of frequencies relating to specific secondary structures of proteins

Secondary structure	Frequency (cm^{-1})
β-Sheet	1626, 1635
Irregular	1644
α-Helix	1657
Turns	1666
β-Sheet/turns	1679

7.3. Factor analysis methods

The second main method for secondary structure determination makes use of factor analysis. The complete but unprocessed amide band envelope for a series of proteins of known structure is correlated using factor analysis, with the known structural components assessed from the X-ray structures. The structure of a new protein is determined by cross-correlation with the calibration set. Both methods have disadvantages but have recently been shown to agree, provided care is taken with the frequency assignments and the way in which the secondary structures are defined in the X-ray structures (16). Both vibrational circular dichroism and rotational Raman spectroscopies are new methods that make use of polarized light to obtain additional information concerning protein (and other) structures (19). Recently the temperature-induced unfolding/folding of ribonuclease T1 has been studied using IR spectroscopy. The different structural components are seen to 'melt' at different temperatures. Good use has been made of site-specific mutants of the enzyme and this combined approach is rapidly becoming more popular, also for the study of enzyme mechanisms (18).

8. ENZYMES

8.1. Enzyme–substrate interaction

Both IR and resonance Raman spectroscopies have proved useful in the study of enzyme–

substrate interaction. Clear evidence for hydrogen bonding of the carbonyl of dihydroxyacetone phosphate to triosephosphate isomerase in the enzyme–substrate (ES) complex has been found on the basis that the frequency is lowered on binding to the enzyme (20). This is as would be expected since the C=O bond will be weakened. An extensive series of studies of the acylenzyme intermediates of the serine proteinase chymotrypsin has been carried out using both IR and resonance Raman spectroscopies (21,22). The role of the oxyanion, hole catalytic device, which polarizes the susceptible acylcarbonyl group has been verified by the finding that the specificity is correlated with the strength of the hydrogen bonding; this has shown that ground state interaction is important in enzyme catalysis. Extensive use of isotope editing has been made in these studies (23). Highly specific substrates have not yet been studied owing to their kinetic lability. There is a need for fast time-resolved methods in vibrational spectroscopy, where the reaction cannot be recycled by light excitation as can be done with, for example, bacteriorhodopsin. Methods which use light-activated caged substrates or a recently constructed stopped-flow FTIR spectrometer for use with aqueous solutions may help to address this difficulty (24).

9. REFERENCES

1. Parker, F.S. (1983) *Applications of Infrared, Raman and Resonance Raman Spectroscopy in Biochemistry.* Plenum Press, New York.

2. Clark, R.J.H. and Hester, R.E. (1986) *Spectroscopy of Biological Systems*, Vol. 13. John Wiley & Sons, Chichester.

3. Campbell, I.D. and Dwek, R.A. (1984) *Biological Spectroscopy.* Benjamin, London.

4. Senak, L., Davies, M.A. and Mendelson, R. (1991) *J. Chem. Phys.*, **95**, 2566.

5. Mantsch, H.H., Cameron, D.G., Trembley, P.A. and Kates, M. (1982) *Biochim. Biophys. Acta.*, **689**, 63.

6. Diller, R. and Stockburger, M. (1988) *Biochemistry*, **27**, 7641.

7. Nolker, K., Weidlich, O. and Siebert, F. (1992) in *Time-resolved Vibrational Spectroscopy* (H. Takahashi, ed.). Springer-Verlag, Berlin, p. 57.

8. Diller, R., Iannone, B.R., Cowen, S., Maitli, R.A. and Hochstrasser, R.M. (1992) *Biochemistry*, **31**, 5567.

9. Jung, C., Hui Bon Hoa, G., Schroder, K.L., Simon, M. and Doucet, J.-P. (1992) *Biochemistry*, **31**, 12855.

10. Bosenbeck, M., Schweitzer-Stenner, R. and Dreybrodt, W. (1992) *Biophys. J.*, **61**, 31.

11. Dyer, R.B., Einarsdottir, O., Killough, P.M., Lopez-Garrida, J.J. and Woodruff, W.H. (1989) *J. Am. Chem. Soc.*, **111**, 7657.

12. Thomas, G.J. and Wang, A.H.-J. (1988) *Nucl. Acids Mol. Biol.*, **2**, 1.

13. Aubrey, K.L., Casjens, S.R. and Thomas, G.J. (1992) *Biochemistry*, **31**, 11835.

14. Liquiers, J., Taillandier, E., Peticolas, W.L. and Thomas, G.A. (1990) *J. Biomol. Struct. Dynam.*, **8**, 295.

15. Surewicz, W.K., Mantsch, H.H. and Chapman, D. (1993) *Biochemistry*, **32**, 389.

16. Byler, D.M. and Susi, H. (1986) *Biopolymers*, **25**, 469.

17. Lee, D.C., Haris, P.I., Chapman, D. and Mitchell, R.C. (1990) *Biochemistry*, **29**, 9185.

18. Fabian, H., Schultz, C., Naumann, D., Landt, O., Hahn, U. and Saenger, W. (1993) *J. Mol. Biol.*, **232**, 967.

19. Pancoska, P., Yasui, S.C. and Keiderling, T.A. (1991) *Biochemistry*, **30**, 5089.

20. Belasco, G.J. and Knowles, J.R. (1980) *Biochemistry*, **19**, 472.

21. White, A.J. and Wharton, C.W. (1990) *Biochem. J.*, **270**, 627.

22. Tonge, P.J. and Carey, P.R. (1992) *Biochemistry*, **31**, 9122.

23. Johal, S.S., White, A.J. and Wharton, C.W. (1994) *Biochem. J.*, **297**, 281.

24. White, A.J., Drabble, K. and Wharton, C.W. (1995) *Biochem. J.*, **306**, 843.

CHAPTER 18
CIRCULAR DICHROISM
S.R. Martin

1. INTRODUCTION

Circular dichroism (CD) is a spectroscopic property which is uniquely sensitive to molecular conformation, and is widely used in the study of proteins. The near-ultraviolet (UV) CD bands of proteins (340–250 nm) arise principally from tryptophan, tyrosine, phenylalanine and cystinyl groups, and they reflect the tertiary and quaternary structure of the protein. The far-UV CD bands of proteins (260–178 nm) derive primarily from the amide chromophore and they reflect the secondary structure of the protein (α-helix, β-sheet, β-turn).

Detailed structural information is not generally available from the CD spectra of proteins. However, the secondary structure *content* of a protein may be determined with considerable accuracy from the analysis of far-UV CD spectra. Furthermore, CD bands are extremely sensitive to the environment of the chromophore and may be used in the detection of conformational changes deriving from ionization, ligand binding, temperature changes and so on. There are several recent reviews which describe the theory of CD and its application to biochemical systems (1–5).

2. APPLICATIONS

The principal applications of CD in the study of proteins and peptides are summarized in *Table 1*.

2.1. Examination of secondary and tertiary structure of proteins
Secondary structure
Proteins show characteristic far-UV CD spectra (260–178 nm) that are related to the presence of regular secondary structure. It is generally assumed that the far-UV CD spectrum of a protein is the weighted sum of the spectra of the individual secondary-structural elements: α-helix, β-sheet, β-turn, and random or unstructured. Several approaches have been employed in attempts to

Table 1. Applications of circular dichroism in the study of proteins and peptides

Application	Section
The determination of the secondary structure *content* of proteins and peptides	2.1
The study of protein denaturation induced by heating or by the addition of chemical denaturants	2.2
The study of the effects of pH, salts and organic solvents on conformation	2.3
The study of protein–protein, protein–peptide and protein–nucleic acid interactions	2.4
The study of protein–ligand interactions and the determination of association constants	2.4
The study of the kinetics of ligand binding and protein folding and unfolding	2.5

Table 2. CD characteristics of different secondary structures found in proteins in solution (7)

Structure	Wavelength range (nm)	Expected $\Delta\epsilon_{mrw}$ range
α-Helix	193	+22
	208	−10.5
	222	−11.5
β-Sheet	191–197	+2.5 to +5.0
	207–213	−1.5 to −4.0
Random coil	192–201	−11.0 to −14.5
	208–212	−1.5 to +3.3
β-Turn	198–202	+3.5 to +11.5
	222–226	+3.5 to +6.5

determine the contribution of these individual structural components to the measured spectrum and thereby to determine the secondary structure content of the protein. Early methods used reference spectra which were derived from the spectra of either model polypeptides (6) or proteins (7). The principal features of these reference spectra are given in *Table 2*.

More recently, Provencher and Glöckner (8) and Hennessey and Johnson (9) analyzed experimental CD curves as linear combinations of the spectra of proteins, the structure of which had been determined by X-ray diffraction. The original approach of Hennessey and Johnson was extended and improved with the variable selection method (10, 11). The current version of the variable selection method gives correlation coefficients of 0.97 for α-helix, 0.76 for β-sheet, 0.49 for β-turn and 0.86 for random structure, when the predicted structure from CD is compared to the X-ray structure for the 16 proteins in the database (10). These methods are now the methods of choice since they avoid the serious difficulties inherent in the selection of appropriate reference spectra. The different methods available have been discussed by Yang *et al.* (5), van Stokkum *et al.* (12), and by Sreerama and Woody (13). The validity of the various underlying assumptions in the calculation of secondary structure content from CD have been discussed by Manning (14).

Tertiary structure
Several factors influence the intensities of near-UV CD bands (340–250 nm) deriving from the aromatic amino acids. Amongst these are the extent to which the residue is mobile (rigidity of the protein) and the extent to which the aromatic ring interacts with its surroundings (15). Intense CD bands tend to be observed when the residue is immobilized and when it interacts with neighboring aromatic residues. Because CD bands from individual residues may be either positive or negative and may vary widely in intensity it is often difficult to separate out the contributions of individual aromatic residues in a protein. Serial replacement of residues by site-directed mutagenesis offers the possibility of investigating these individual contributions (16). Although it is not generally possible to use the near-UV CD spectrum of a protein to say anything in detail about tertiary structure, a knowledge of the position and intensity of CD bands expected for a particular residue may help in understanding the near-UV CD spectrum (see *Table 3*). Near-UV CD, which provides a sensitive 'fingerprint' for the native state of a protein, is discussed in excellent reviews by Strickland (15) and Woody (4).

Mutant proteins
When working with mutant proteins it is important to examine the effect of the mutation on the overall conformation of the protein. CD provides a convenient means of doing this with limited amounts of material. Differences observed in the far-UV spectra are generally an indication of significant differences in secondary structure (subject to accurate concentration determination—

Table 3. Identifying features for residues contributing to the near-UV CD spectra of proteins[a]

Residue	Identifying features	Expected $\Delta\epsilon_M$ range
Phenylalanine	Sharp fine structure in range 255–270 nm. Peaks frequently observed at 262 and 268 nm. May be obscured if other residues present	±0.3
Tyrosine	Maximum in range 275–282 nm, possibly with a shoulder some 6 nm to the red. May be obscured if other residues present	±2.0
Tryptophan	Fine structure at $\lambda > 280$ nm. Two ^1L_b bands, one at 288 to 293 nm and one some 7 nm to the blue, with same sign	±5.0
	^1L_a band (265 nm) with little fine structure	±2.5
Cystine	Begins at long wavelength (>320 nm). One or two broad peaks above 240 nm. Long wavelength peak frequently negative	±1.0

[a] Data from ref. 15.

see Section 3.5). However, differences observed in the near-UV region may derive from subtle changes in the environment of particular aromatic residues which are not necessarily associated with any major structural change.

2.2. Studying protein denaturation

Near- and far-UV CD spectroscopy offer a convenient way to monitor the effect of temperature and chemical denaturants on protein structure, and may be used to assess the effect of specific mutations on the stability of proteins (17). The effects of ligands on the stability of the protein to denaturation are also readily studied (see, for example, ref. 18).

Effect of temperature
Heating proteins to sufficiently high temperatures often dramatically reduces CD intensities in both the far-UV (loss of secondary structure) and near-UV regions (increased mobility of aromatic side-chains). Cooperative conformational changes occur over a narrow temperature range and produce a characteristic 'melting' curve. Noncooperative conformational changes tend to occur gradually over much wider temperature ranges; such transitions are frequently observed with small peptides. The advantages of following these changes by CD measurements are that the technique requires small amounts of material and that the temperature-dependence of CD outside the transition region is generally small.

Effect of denaturing agents
For most proteins the native state has much more intense CD over most of the far- and near-UV wavelength range than the unfolded state, and a large loss of intensity therefore occurs upon treatment of the protein with denaturants such as urea or guanidine hydrochloride. Studies in the two wavelength regions, combined with other spectroscopic measurements such as fluorescence, allow one to address the question of the relative stability of the tertiary and secondary structure. Loss of tertiary structure is sometimes observed to precede loss of secondary structure.

2.3. Detecting altered protein conformation

The stability of a particular secondary structure element is known to depend upon several factors; for example, the stability of an individual α-helix is determined by the number and strength of intrahelical hydrogen bonds; by the degree of repulsion between ionized side-chains with the same charge; by stabilizing interactions between side-chains; and by interactions with other secondary structural elements. These factors, and thus the structure of the protein, are frequently influenced by changes in solution conditions.

Effect of organic solvents
Alcohols such as trifluoroethanol (TFE) and hexafluoroisopropanol are widely used as structure-inducing cosolvents and can induce the formation of stable conformations in peptides which are unstructured in aqueous solution. They are known to be helix-inducing solvents; effects on other structural elements are less well characterized. The low dielectric constant of alcohols resembles that of the interior of proteins and this presumably strengthens electrostatic interactions. Such effects may, however, be offset by increased counterion binding in presence of the alcohol. The major effect is probably attributable to weaker hydrogen bonding of the amide protons to the solvent, with concomitant strengthening of intramolecular hydrogen bonds and stabilization of regular secondary structure (19). It should be noted that alcohols tend to induce α-helix only in regions which have some helical propensity (20). The results of such studies should always be interpreted with caution since there may be little correlation between the alcohol-induced and the native protein conformation.

Effect of pH, salt
Since regular secondary structures may be stabilized by favorable charge interactions or destabilized by unfavorable ones, some effect of added salts and changes in pH on the backbone conformation of polypeptides may be anticipated. Such processes, either formation or loss of structure, are readily studied by CD (21–23). In the case of added salts it may be noted that ClO_4^- sometimes has a much greater effect in promoting the formation of α-helix than smaller anions such as Cl^- (23). This has been attributed to the binding of the bulky ClO_4^- ion to guanidinium groups and ϵ-amino groups, with neutralization of destabilizing positive charge repulsions. Ionic strength and pH changes may also have a direct effect on the tertiary structure by altering the environment of aromatic side-chains and disulfide bridges.

2.4. Studying interactions between molecules

Many interactions of proteins with ligands produce easily measurable changes in the CD spectra of the protein and/or the ligand. CD bands of the ligand are generally monitored in the visible and near-UV regions. Ligand-induced changes in secondary structure are sometimes detected in the far-UV, whilst changes in the tertiary and quaternary structure may be observed in the near-UV. It should be noted that substantial changes in the near-UV region may arise from only small local changes in the environment of aromatic residues (15). There may also be interactions between the chromophores of the ligand and protein, which may alter the CD intensities of either.

Binding of nonchromophoric ligands
Near- and far-UV CD spectra of proteins frequently undergo large changes on binding small molecules; for example, extensive changes in both far- and near-UV CD occur on the binding of metal ions to calmodulin (18) and actin (24). Because the metal ion is nonchromophoric, the interpretation of such effects in terms of effects on the structure of the protein is generally rather straightforward.

Binding of chromophoric ligands
In the case of binding of chromophoric ligands one may see changes in the CD of both the protein and the ligand and they may be difficult to distinguish if their absorption bands overlap. Major changes in ligand CD are quite often observed; for example, the strong near-UV absorption of nucleotides permits electronic coupling between the ligand and the chromophores on the protein,

and these interactions frequently generate major changes in the CD spectrum of the nucleotide. Thus, binding of ligands to glutamate dehydrogenase generates a CD spectrum for the bound ligand which is much more intense than that of the free ligand (25, 26). In such cases the changes in CD signals are large enough to enable one to measure association constants for ligand binding. The use of CD in the determination of association constants has been discussed by Greenfield (27). However, CD is not generally the method of choice for such studies, owing to the relatively poor signal-to-noise (S/N) ratio compared with other optical techniques.

Interactions between macromolecules
Mixing two protein components sometimes gives a near-UV CD spectrum which is not equal to the sum of the spectra of the two components. Such changes may derive from conformational changes and/or coupling between chromophores on the two proteins and may be used to monitor the interaction. Intuitively, one expects the near-UV CD to be nonadditive when the region of contact contains aromatic side-chains. The reduced mobility of aromatic residues in the contact region frequently leads to increased CD intensity; for example, the aggregation of insulin monomers to dimers and hexamers leads to substantial increases in the intensity of the tyrosyl CD bands (28). Changes in the far-UV are less likely, although small peptides can self-associate to form discrete aggregates with increased helical content (29). CD has been used extensively to study the interactions of proteins with nucleic acids, DNA and RNA. Conformational changes in the protein and/or the nucleic acid are easily detected using CD (30, 31).

Structure comparison studies
Finally, in studies involving limited proteolytic cleavage one may use CD to see if the fragments retain the structure they had in the intact protein. This is done by comparing the spectra of the intact protein with the appropriately weighted sum of the spectra of the fragments (18).

2.5. Kinetics
In favorable cases it may be possible to use CD to study the kinetics of ligand binding (particularly slow dissociation processes), protein aggregation, protein denaturation, or protein refolding. Such studies may provide information not readily accessible with other optical techniques. Kinetic studies of protein folding reactions are particularly important in that they provide information about transient intermediates which may play a role in the folding process. Data on the formation of tertiary structure in proteins have been available for many years because the fluorescence properties of aromatic chromophores are sensitive to their environment and stopped-flow devices with fluorescence detection are widely available. However, information on the formation of secondary structure is also needed for a complete understanding of the refolding process (32). The development of stopped-flow CD (see ref. 33 and references therein) has facilitated such studies and stopped-flow attachments for CD spectrometers are now available. Stand-alone stopped-flow CD equipment is available from Aviv and Associates (Lakewood, NJ, USA) and from Applied Photophysics (Leatherhead, Surrey, UK).

3. PRACTICAL CONSIDERATIONS

3.1. Instrumentation
CD instruments are manufactured by Jasco (Jasco Incorporated, Easton, MD, USA; Jasco (UK) Limited, Great Dunmow, Essex, UK); Jobin-Yvon (Jobin-Yvon, Longjumeau, France; Instruments S.A. UK Ltd., Stanmore, Middlesex, UK); Aviv and Associates (Lakewood, NJ, USA), and On-Line Instrument Systems Inc. (Bogart, GA, USA).

The key to retaining good instrument performance is routine maintenance. Since the high intensity light source converts oxygen into ozone, which damages the optics, the instruments should be purged with pure nitrogen for at least 20 minutes before starting the light source and whilst making the measurements. Although this simple precaution will do much to maintain the instrument in good working order, the optics will eventually degrade and this combined with an aging light source will lead to poor far-UV performance. The optics should be cleaned every 2 years and the light source and mirrors replaced when necessary.

3.2. Presentation of data

CD spectrometers measure the difference in absorbance for left and right circularly polarized light, ΔA $(= A_L - A_R)$. The molar CD extinction coefficient,

$$\Delta \epsilon_M (= \epsilon_L - \epsilon_R \text{ (units : } cm^{-1}M^{-1}))$$

is calculated from the CD version of Beer's law:

$$\Delta A = \Delta \epsilon_M cl,$$

where c is the molar concentration and l the pathlength in centimeters. In order to facilitate comparison between proteins it is common practice to average far-UV intensities over the total number of amino acid residues (N):

$$\Delta \epsilon_{mrw} = \Delta \epsilon_M / N.$$

Averaging near-UV intensities in this way is not justified because only four amino acid side-chains contribute to the CD in this region and these intensities are better reported as $\Delta \epsilon_M$. CD intensities are sometimes reported as molar ellipticity, $[\theta]_M$ (units: degrees cm^2 dmol^{-1}). Values presented in this way may be converted to molar extinction coefficients using the relationship:

$$\Delta \epsilon_M = [\theta]_M / 3300.$$

3.3. Amounts of material

The far-UV CD signals from proteins (260–178 nm) are usually intense and spectra may be recorded with small quantities of material. The concentration required depends upon the cuvette used: 200 μl of a 0.05–0.1 mg ml^{-1} solution for a 1 mm pathlength cuvette or 30 μl of a 0.5–1.0 mg ml^{-1} solution for a 0.1 mm cuvette. For best penetration into the far-UV the shorter pathlength is preferable. It is sensible practice to record a normal absorption spectrum of the sample and its buffer over the same wavelength range prior to the CD measurement. An absorbance less than 1.0 is generally acceptable. If the absorbance is greater than 1.0, the CD curve is likely to be unreliable (see Section 3.4).

Near-UV CD signals from proteins (340–250 nm) are much less intense and correspondingly more material is required. The spectra are usually recorded under similar conditions to those used for measuring a conventional absorption spectrum. For a 'typical' protein one records the near-UV CD spectrum of a 1 mg ml^{-1} solution in a 10 mm pathlength cuvette. Volumes as low as 350 μl may be sufficient if 10 mm pathlength microcuvettes are used.

The spectra of turbid solutions may be distorted by optical artifacts, so that samples should be clarified by filtration or low-speed centrifugation when necessary prior to concentration determination. The samples should be of the highest possible purity.

3.4. Choice of solvent

CD machines are prone to serious artifacts when too little light reaches the photomultiplier. In

Table 4. Low wavelength cut-off ($A=1.0$) in a 1 mm pathlength cuvette for different solvents and for common salts and buffers dissolved in water

Solvent system	Cut-off (nm)	Solvent system	Cut-off (nm)
Water	<185	0.15 M $(NH_4)_2SO_4$	191
Trifluoroethanol	<185	0.15 M NaCl	196
Hexafluoroisopropanol	<185	0.15 M $NaClO_4$	<185
Acetonitrile	185	0.15 M $NaNO_3$	245
Methanol	195		
Ethanol	196	100 mM Phosphate	<185
2-propanol	196	100 mM Tris	195
Cyclohexane	<185	100 mM Pipes	215
Dimethylsulfoxide	251	100 mM Mes	205
Dioxane	232	4 M GdnHCl	210
		4 M Urea	210

Other reagents such as EGTA, dithiothreitol and some detergents (for example, Lubrol and sodium dodecyl sulfate (SDS)) may be used at reasonable concentrations. The absorption spectrum should be checked in each case.

practical terms this means that one cannot make reliable measurements on samples with an absorbance (sample plus solvent) greater than about 1. For the far-UV region the absorbance of the sample itself is generally rather small (< 0.4). The major problem derives from absorption by solvents in this region. *Table 4* lists the effective low-wavelength cut-off for a variety of common solvents in a 1 mm cuvette. These cut-offs can, of course, be lowered by using shorter pathlength cuvettes. Buffers, salts and other added reagents may also absorb in this region. Cut-offs for various common buffers and salts (in water) are also given in *Table 4*. It should be noted that distilled water kept in a polyethylene bottle for a long time can have poor transparency in the far-UV owing to the presence of eluted polymer additives.

Although the prediction of secondary structure from far-UV CD spectra is thought to require data from 250 to at least 178 nm (11), it is still possible to make useful qualitative comparisons with spectra recorded over a much more limited wavelength range (8).

3.5. Concentration

Accurate protein concentrations are essential for the analysis of far-UV CD for secondary structure content and when one wants to make meaningful comparisons between different samples. Protein concentrations are readily determined by absorption spectroscopy using known extinction coefficients for the aromatic region or for the amide absorption at around 190 nm (9). Quantitative amino acid analysis is an alternative method. Lowry or Bradford analyses may be in error by a factor of two and are not sufficiently accurate for use with CD measurements unless they have been calibrated against a careful amino acid analysis of the protein of interest. See Chapter 4 for a detailed discussion of the determination of protein concentration.

3.6. Sources of error in CD measurements

There are several potential sources of error in CD measurements. These have been carefully reviewed by Hennessey and Johnson (34) and are summarized below.

Circular Dichroism

Intensity errors

The CD instrument must be regularly calibrated by using, for example, purified *d*-10-camphorsulfonic acid and the method of calibration should be reported. The optical properties of *d*-10-camphorsulfonic acid are (11):

$$\Delta\epsilon_{290.5} = 2.36 \, \text{cm}^{-1} \, \text{M}^{-1},$$

and

$$\Delta\epsilon_{192.5} = -4.72 \, \text{cm}^{-1} \, \text{M}^{-1}.$$

The concentration should be determined by absorption spectroscopy using

$$\epsilon_{285} = 34.5 \, \text{cm}^{-1} \, \text{M}^{-1},$$

and not by weight since the solid is hygroscopic.

Cuvettes with a pathlength of 1 mm or less should be calibrated. This is easily done using a solution with known absorbance. The major source of intensity errors is probably in the estimation of protein concentration (see above). The method used to determine concentrations should always be reported in order to facilitate comparisons with data from other laboratories.

Wavelength errors

An error of 1 or 2 nm in the wavelength scale affects CD analysis for secondary structure. This can easily be checked using, for example, *d*-10-camphorsulfonic acid (see above) with extrema at 290.5 and 192.5 nm.

Scan rate errors

Long time constants are frequently used to improve the S/N ratio and spectra are sometimes scanned excessively quickly in order to reduce the data collection time. Too high scanning rates tend to lead to errors in CD band position and intensity. The product of the time constant and the scan rate should be less than 0.5 nm. A typical procedure is to use time constants in the range 0.25–2 seconds and to collect multiple scans to increase the S/N ratio to acceptable levels. S/N is proportional to the square root of the number of scans and to the square root of the time constant.

Spectral bandwidth errors

Increasing the spectral bandwidth reduces the noise by increasing the amount of light reaching the photomultiplier. Errors due to the use of an inappropriate spectral bandwidth depend on the spectrum of the protein involved (34). The bandwidth should generally be 2 nm or less. Low values of spectral bandwidth should be used to resolve fine structure in the near-UV region.

Baseline alignment errors

The baseline should always be recorded with the cuvette and solvent used for the measurement, preferably both before and after the measurement to check for baseline drift. If the baseline is incorrectly aligned, there will be a vertical displacement in the entire CD spectrum, resulting in a change in peak height ratios but not band shape. The spectra must be started at least 20 nm before the sample signal is observed and the spectra of the sample and baseline can then be aligned to give a zero signal in this region. When recording far-UV CD spectra, generally from 260 nm, it is important to check for the absence of any small contribution from aromatic residues in the region used for baseline alignment.

Wavelength range errors

Far-UV CD spectra are frequently recorded beyond the point where the instrument has ceased to function correctly. Spectra will be excessively noisy and distorted if the photomultiplier voltage rises above a certain limit, which is most likely to be a problem when making measurements in the far-UV region. The far-UV performance may be checked by recording the CD spectrum of solutions of *d*-10-camphorsulfonic acid at different concentrations and calculating the ratio $\Delta\epsilon_{192.5}/\Delta\epsilon_{290.5}$. If this ratio is less than 1.95, then the machine is not performing correctly.

Other errors

CD spectra may be distorted if the sample scatters light to a significant extent. Therefore, turbid solutions should not generally be used for CD measurements (see above). Some distortion might also occur when measuring the CD of strongly fluorescent samples, depending upon the intensity of the fluorescence. Finally, with dilute samples, especially of small, highly charged molecules, one can get loss of sample through absorption on the cuvette wall. Such effects are generally minimized by using buffers containing high concentrations of salt, though this may pose problems for measurements into the far-UV (see Section 3.4).

3.7. Low-temperature measurements

CD spectra at low temperatures may be made on glasses (frozen solutions) or by using a suitable solvent, such as water–glycerol (1:1, v:v). An essential prerequisite in such studies is to check for any effect of the solvent itself on the conformation of the protein. This may be done by measuring the CD spectrum in the solvent at room temperature and comparing it with the spectrum measured in an aqueous buffer. Where possible, low-temperature measurements should always be made over a range of concentrations, in order to demonstrate that any observed changes do not derive from aggregation effects.

3.8. Membrane proteins

Recording the CD spectra of detergent-solubilized membrane proteins presents no particular experimental difficulties, but it must be remembered that the conformation may differ from that in the native environment. The CD spectra in the native state, in membrane fragment suspensions, are generally distorted by two artifacts (35). First, the fragments may be similar in size to the wavelength of the incident light and differential scattering of left-and right-circularly polarized light will appear as CD. Distortions due to differential light scattering are generally small and can be compensated for by using a large detection angle (i.e. by moving the detector close to the sample). The second artifact is caused by the fact that the protein is concentrated in the fragments and is not present in the surrounding solvent. This gives rise to an apparent reduction in the CD in regions of high absorption. This artifact, known as absorption flattening, introduces large distortions and several methods for correcting for this have been described (36). These distortions are generally eliminated when the proteins are incorporated in small unilamellar vesicles.

4. ACKNOWLEDGMENTS

I would like to thank Drs Peter Bayley, Maria Schilstra, and Wendy Findlay for useful comments on the manuscript.

CIRCULAR DICHROISM

Circular Dichroism

5. REFERENCES

1. Bayley, P.M. (1980) in *Introduction to Spectroscopy for Biochemists* (S.B. Brown ed.). Academic Press, London, p. 148.

2. Johnson, W.C. (1985) *Methods Biochem. Analysis*, **31**, 61.

3. Johnson, W.C. (1988) *Ann. Rev. Biophys. Chem.*, **17**, 145.

4. Woody, R.W. (1985) *The Peptides*, **7**, 15.

5. Yang, J.T., Wu, C.-S.C. and Martinez, H.M. (1986) *Methods Enzymol.*, **130**, 208.

6. Brahms, S. and Brahms, J. (1980) *J. Mol. Biol.*, **138**, 149.

7. Chang, C.T., Wu, C.-S.C. and Yang, J.T. (1978) *Anal. Biochem.*, **91**, 13.

8. Provencher, S.W. and Glöckner, J. (1981) *Biochemistry*, **20**, 33.

9. Hennessey, J.P. and Johnson, W.C. (1981) *Biochemistry*, **20**, 1085.

10. Manavalan, P. and Johnson, W.C. (1987) *Anal. Biochem.*, **167**, 76.

11. Johnson, W.C. (1990) *Proteins: Struct. Funct. Genet.*, **7**, 205.

12. van Stokkum, I.H.M., Spoelder, H.J.W., Bloemendal, M., van Grondelle, R. and Groen, F.C.A. (1990) *Anal. Biochem.*, **191**, 110.

13. Sreerama, N. and Woody, R.W. (1994) *J. Mol. Biol.*, **242**, 497.

14. Manning, M.C. (1989) *J. Pharmaceut. Biomed. Anal.*, **7**, 1103.

15. Strickland, E.H. (1974) *CRC Crit. Rev. Biochem.*, **2**, 113.

16. Craig, S., Pain, R.H., Schmeissner, U., Virden, R. and Wingfield, P.T. (1989) *Int. J. Pep. Prot. Res.*, **33**, 256.

17. Wendt, B., Hofmann, T., Martin, S.R., Bayley, P.M., Brodin, P., Grunström, T., Thulin, E., Linse, S., and Forsén, S. (1988) *Eur. J. Biochem.*, **175**, 439.

18. Martin, S.R. and Bayley, P.M. (1986) *Biochem J.*, **238**, 485.

19. Sönnichsen, F.D., Van Eyk, J.E., Hodges, R.S., and Sykes, B.D. (1994) *Biochemistry*, **31**, 8790.

20. Waterhous, D.V. and Johnson, W.C. (1994) *Biochemistry*, **33**, 2121.

21. Skehel, J.J., Bayley, P.M., Brown, E.B., Martin, S.R., Waterfield, M.D., White, J.M., Wilson, I.A. and Wiley, D.C. (1982) *Proc. Natl Acad. Sci. USA*, **79**, 968.

22. Syed, S.E.-H., Engel, P.C. and Martin, S.R. (1990) *FEBS Lett.*, **262**, 176.

23. Clark, D.J., Hill, C.S., Martin, S.R. and Thomas, J.O. (1988) *EMBO J.*, **7**, 69.

24. McCubbin, W.D, Oikawa, K. and Kay, C.M. (1981) *FEBS Lett.*, **127**, 245.

25. Bayley, P.M. and O'Neill, K.T.J. (1980) *Eur. J. Biochem.*, **112**, 521.

26. Delabar, J.M., Martin, S.R., and Bayley, P.M. (1982) *Eur. J. Biochem.*, **127**, 367.

27. Greenfield, N.J. (1975) *CRC Crit. Rev. Biochem.*, **3**, 71.

28. Strickland, E.H. and Mercola, D. (1976) *Biochemistry*, **15**, 3875.

29. Hagihara, Y., Kataoka, M., Aimoto, S. and Goto, Y. (1992) *Biochemistry*, **31**, 11908.

30. Russo, E., Giancotti, V. and Crane-Robinson, C. (1983) *Int. J. Biol. Macromol.*, **5**, 366.

31. Chandler, L.R. and Lane, A.N. (1988) *Biochem. J.*, **250**, 925.

32. Kuwajima, K., Semisotnov, G.V., Finkelstein, A.V., Sugai, S. and Ptitsyn, O.G. (1993) *FEBS Lett.*, **334**, 265.

33. Bayley, P.M. (1981) *Prog. Biophys. Mol. Biol.*, **37**, 149.

34. Hennessey, J.P. and Johnson, W.C. (1982) *Anal. Biochem.*, **125**, 177.

35. Wallace, B.A. and Mao, D. (1984) *Anal. Biochem.*, **142**, 317.

36. Bustamante, C. and Maestre, M.F. (1988) *Proc. Natl Acad. Sci. USA*, **85**, 8482.

CHAPTER 19
NMR STRUCTURE DETERMINATION OF PROTEINS

B. Whitehead and J.P. Waltho

1. WHAT IS NMR?

Nuclear magnetic resonance (NMR) is one of a very limited number of techniques capable of providing structural information about proteins at the atomic level. As with other forms of spectroscopy, NMR is concerned with the interactions between radiation and the energy levels in matter—in particular the interaction between known-phase radiofrequency (rf) radiation and the nuclear energy levels of certain nuclei. Knowledge of the phase of the radiation allows the resonance of the absorption by the nuclei to be detected. In the presence of a powerful magnetic field, the nuclear energy levels of nuclei possessing spin become nondegenerate—the magnitude of the splitting E being proportional to the strength of the magnetic field and to the gyromagnetic ratio γ of the nucleus in question. The proportionality constant is Planck's constant h divided by 2π, usually denoted $\hbar$.

$$E = \hbar\omega = 2\pi\hbar f = -\gamma\hbar B_0.$$

The frequency f (angular frequency ω) at which the nuclei interact with radiation is proportional to the energy difference between the levels, and is thus also proportional to B_0 and γ. The strength of the field is commonly not expressed in SI units (Tesla), but referred to as the resonance frequency of ^1H nuclei in the field, thus for example an 11.74 T magnet (a common magnet strength for protein NMR) is referred to as a 500 MHz magnet.

The exact frequency of resonance of specific nuclei is found to vary with the electronic environment of the nuclei—for ^1H nuclei with a frequency of approximately 500 MHz, the spread of frequencies is typically of the order of 5000 Hz for proteins. This variation, termed the chemical shift, is usually expressed in parts per million (ppm), thus ^1H nuclei in proteins have a chemical shift range of around 10 ppm. Although small, it is these chemical shift differences that make NMR useful for studying molecular structures. One of the major properties that singles out NMR from most other spectroscopies is the ability to transfer the NMR signal between nuclei. This may be achieved in one of two ways: via the chemical bonding network through an interaction known as J-coupling, or through space via an interaction known as the nuclear Overhauser effect (NOE). More information about NMR theory can be found in refs 1 and 2.

In biological systems, ^1H, ^{15}N and ^{13}C (and to a lesser extent ^{19}F and ^{31}P) nuclei give a useful NMR response. The relevant properties of these nuclei are summarized in *Table 1*. The relative sensitivity of the NMR signals is proportional to γ^3, thus ^1H nuclei give the highest sensitivity. For other nuclei, the natural abundance of the NMR active isotope must also be taken into account, and the absolute sensitivity is proportional to the product of the relative sensitivity and the natural abundance.

At present, the application of NMR to the structure determination of proteins is limited to

Protein NMR

Table 1. Properties of NMR sensitive nuclei that are commonly used in the study of proteins

Nucleus	$\gamma \times 10^{-7}$ (rad s^{-1} T^{-1})	Natural abundance (%)	Relative sensitivity	Absolute sensitivity	NMR frequency (MHz) in an 11.7 T field
^{1}H	26.7522	99.98	1.00	1.00	500.0
^{13}C	6.7283	1.108	1.59×10^{-2}	1.76×10^{-4}	126.0
^{15}N	−2.7126	0.37	1.04×10^{-3}	3.85×10^{-6}	50.7
^{19}F	25.1815	100.0	0.83	0.83	470.0
^{31}P	10.8394	100.0	6.62×10^{-2}	6.62×10^{-2}	202.0

molecules of less than 30 kDa molecular mass and therefore the technique is somewhat less widely applicable than X-ray crystallography. Despite this, NMR is in many ways complementary to crystallography. Since NMR experiments are performed on the protein in solution, proteins that are difficult to crystallize, such as those that are small and/or flexible, can be studied. In cases where proteins have been studied both by crystallography and NMR, the structures have been generally compatible, indicating that structures in the crystalline state are retained in solution (3, 4). On occasions where the techniques give conflicting information, a greater understanding of the system is obtained than by use of one technique alone (5).

NMR can also be used to probe the dynamics of protein systems. The relaxation properties of individual ^{13}C and ^{15}N nuclei in proteins can be used to monitor the degree of conformational flexibility in different regions of the molecule. NMR studies of protein folding and ligand binding, often in conjunction with techniques such as CD and fluorescence, have made unique contributions to the understanding of these events. Techniques for studying dynamics, folding and binding are reviewed in ref. 6.

The relatively low sensitivity of NMR compared with other spectroscopies that utilize higher frequency electromagnetic radiation requires the minimum sample concentration for structure determination to be of the order of 1 mM (e.g. 15 mg ml^{-1} for a 15 kDa protein). The typical sample size of 0.5 ml dictates the use of relatively large protein quantities, although the technique itself is nondestructive and the sample recoverable. The solubility of the protein at such high concentrations may be a limiting factor. The widths of NMR resonances are sensitive to restrictions in molecular rotations and are thus sensitive to oligomerization and nonspecific aggregation phenomena. The purity of the sample should be greater than 95%, although high molecular mass impurities are rarely observable owing to their relative immobility. More of a problem are small molecule impurities, for example residual buffer components, such as Tris, glycine and EDTA, that have very sharp lines that often give rise to spectral artifacts.

Experiments can be performed over a wide range of temperatures. Normally, as high a temperature as possible is used to provide the sharpest resonances. Variation in temperature is a very useful source of resolving overlapping signals since there is some differential behavior of the chemical shift of individual resonances with temperature. In terms of pH, experiments are often performed in the range 4.0–5.0. This reflects the exchange rate of amide N–H protons with the solvent being minimized around pH 4.0. At higher pH values more technically demanding experiments are often required to observe solvent-exposed amide N–H proton resonances, although it is feasible to study most proteins at neutral pH. Again, variation of pH over a small range is an important means of resolving spectral overlap.

2. PROTEIN NMR ASSIGNMENT TECHNIQUES

The ^{1}H NMR spectrum of even a small protein will contain many hundreds of resonances, and before any useful structural information can be extracted it is necessary to assign the resonances to particular nuclei in particular residues. The strategies used for assignments are described briefly below, along with a discussion of their limitations.

2.1. Homonuclear assignment techniques

The majority of NMR studies are carried out using only resonances of the hydrogen nuclei. This has the advantage that unlabeled protein can be used, obtained either from natural or recombinant sources. These studies primarily utilize two-dimensional NMR experiments in order to improve the ability to resolve resonances. In these experiments, the resonance frequencies of the nuclei are measured twice: either side of an event that transfers information between nuclei via J-coupling or NOEs. The primary experiments used are shown in *Table 2*. There are numerous variations of the basic experiments designed to suit specific circumstances and hence the spectroscopist is armed with a battery of techniques. The general approach for homonuclear assignment is outlined below, and refs 7–10 should be consulted for more information.

The first step in sequential resonance assignment is the identification of individual amino acid residues according to the topology of their hydrogen network (termed *spin-systems*) as defined by

Table 2. A list of two-dimensional NMR experiments commonly used in the study of proteins without isotopic labels[a]

Experiment	Full name	Nuclei correlated
COSY	Correlated spectroscopy	Via J-coupling in a single step between nuclei two or three bonds apart
2QFCOSY	2 Quantum filtered COSY	COSY with uncoupled resonances removed
3QFCOSY	3 Quantum filtered COSY	COSY involving three or more nuclei J-coupled together
2QCOSY	2 Quantum COSY	COSY measuring the sum of frequencies of two J-coupled nuclei
3QCOSY	3 Quantum COSY	COSY measuring the sum of frequencies of three J-coupled nuclei
RELAY	Relayed COSY	Nuclei connected by two successive COSY transfers
TOCSY (HOHAHA)	Total correlation spectroscopy	Via J-coupling between any nuclei connected within two or three bonds in a contiguous network
NOESY	NOE spectroscopy	Via NOEs between nuclei less than 0.5 nm apart in space
ROESY	Rotating frame NOE spectroscopy	Via NOEs between nuclei less than 0.5 nm apart in space using an applied rf field

[a] For experimental details see refs 1, 2, 6 and 7.

Table 3. A list of H–H NOEs and their relative intensities normally observed in regular secondary structure elements[a]

Secondary structure	NOE	NOE intensity
α helix	d_{NN} (i,i+1)	Strong
	$d_{\alpha N}$ (i,i+1)	Weak
	$d_{\alpha N}$ (i,i+3)	Medium
	$d_{\beta N}$ (i,i+1)	Medium
β sheet	d_{NN} (i,i+1)	Weak
	$d_{\alpha N}$ (i,i+1)	Very strong
	d_{NN} (cross strand)	Medium
	$d_{\alpha N}$ (cross strand)	Medium
	$d_{\alpha\alpha}$ (cross strand)	Very strong in antiparallel β-sheets

[a] The NOE nomenclature is: d=distance between two nuclei in subscript; N = amide proton; α = $C^\alpha H$ proton; $\beta = C^\beta H$ proton; the relative sequence numbers of the amino acid residues to which the hydrogens belong (e.g. i or i+1) are given in parentheses.

the J-coupling; for example, in aspartic acid the amide proton is J-coupled to the α-proton which in turn is J-coupled to both β-protons. In glutamic acid residues (but not of course aspartic acid residues) this J-coupling network is extended between the β-protons and the γ-protons. These network topologies are mapped out for each residue using COSY- and TOCSY-type experiments. However, not all residues have unique topologies; for example, it is impossible to distinguish aspartic acid from phenylalanine, or glutamic acid from methionine, if, as is commonly the case, only two or three bond J-couplings may be observed. Hence residues are categorized into topology groups. Since four or more bonds separate the hydrogens of adjacent residues, no sequential information (i.e. between one residue and its neighbor) can be obtained from COSY/TOCSY experiments. Instead, the spin-systems are linked in a sequence specific manner using NOEs from NOESY spectra. The particular NOEs involved depend upon the peptide backbone dihedral angles and hence the local secondary structure. Indeed, the sequential NOEs, supported by NOEs across the turn of α-helix or across β-sheets provide a description of the secondary structure of the protein almost immediately on completion of the sequential assignment. However, the connection of the spin systems of sequential residues using NOEs is complicated by the necessity to distinguish sequential NOEs from all other NOEs. The NOEs expected from the two major secondary structure types are summarized in *Table 3*.

Homonuclear assignment strategies are not generally useful for proteins with molecular masses above 10 kDa. As the size of protein increases, the number of resonances and hence the overlap also increases, leading to ambiguity in the resonance assignments. This problem is compounded by the increased linewidths in large proteins which leads to a decrease in the amount of NMR signal transferred via three bond H-H J-couplings, and hence to low sensitivity in COSY/TOCSY experiments.

2.2. Heteronuclear assignment techniques

Due to the low natural abundance of ^{13}C and ^{15}N nuclei (1.1% and 0.37% respectively), it is almost always necessary to use isotopically enriched protein for NMR experiments involving heteronuclei. Thus a recombinant source of protein, fed with isotopically labeled food sources, must be utilized. Heteronuclear experiments for protein resonance assignments fall into two categories.

First, with fully ^{15}N labeled (U-^{15}N) protein, a strategy similar to that outlined above is used, but with the advantage of the extra resolution offered by the ^{15}N chemical shifts. The nitrogen and hydrogen resonances may be connected by either heteronuclear multiple-quantum correlation (HMQC) or heteronuclear single-quantum correlation (HSQC). Three experiments are in common use, termed TOCSY-HMQC, NOESY-HMQC and NOESY-HMQC-NOESY. The HMQC (or the related HSQC) part of these experiments transfers the NMR signal between the hydrogens and the heteronuclei. The first two of these are simply three-dimensional extended versions of the homonuclear two-dimensional TOCSY and NOESY experiments. The third dimension corresponds to the amide ^{15}N chemical shifts, and α and side-chain protons correlated to the amide proton either by J-coupling or through space (NOE) coupling appear in the spectra. The HMQC-NOESY-HMQC experiment allows some of the ambiguities from the NOESY-HMQC experiment to be resolved. The extra dimension means that the overlap present in two-dimensional spectra is dramatically reduced. However, the experiments still rely on three bond H–H J-couplings, limiting the size of protein for which they are applicable. See ref. 11 for an example of use of ^{15}N edited experiments.

Second, the use of fully ^{13}C,^{15}N labeled (U-^{13}C/^{15}N) protein allows a different assignment strategy in which the relatively large one bond C–N and C–C J-couplings are used. This strategy avoids the potential ambiguities that arise in NOE-based assignments as discussed above. Both the resolution and the sensitivity can usually be improved relative to studies of unlabeled and U-^{15}N proteins, allowing proteins as large as 30 kDa to be assigned. Essentially, most assignment strategies involve the correlation of amide ^{1}H/^{15}N chemical shift pairs with the chemical shifts of other nuclei, such as C^{α}, H^{α} and C'. By correlating both the intraresidue and the preceding residue chemical shifts with each amide chemical shift pair, a full backbone assignment is possible. A variety of experiments are then available to assign the side-chains, of which the most widely used are the HCCH-COSY and the HCCH-TOCSY. The heteronuclear experiments most commonly used in protein assignments are listed in *Table 4*. Examples of this approach are given in refs 12–14, and refs 15–17 give more detail of the use of labeled proteins in structural work.

3. PROTEIN STRUCTURE DETERMINATION

Two spectral parameters, the NOE and J-coupling constants, yield information on internuclear distance and torsional angles respectively, that may be used in structure calculations. Although neither of these parameters produce a precise measurement of distance or angle, their interpretation leads to limits for constraints that may be applied during the generation of the molecular structure. Correspondingly, the number of constraints with which the position of one atom may be defined relative to its neighbors dictates the resolution that may be determined for the structure in that region; for example, regardless of mobility (see below) surface residues are normally less well defined than interior residues in proteins since their number of neighboring residues is lower.

By far the largest number of constraints used in structure calculations result from the observation of NOEs, thus the NOESY spectra must be as well resolved as possible in order to produce sufficient constraints. Multidimensional heteronuclear edited NOESY spectra have proven very useful in this respect (17). By acquiring a series of NOESY spectra with different transfer (mixing) times, the NOE intensities are converted to distance ranges via the relationship of the

Table 4. A list of multi-dimensional NMR experiments commonly used in the study of proteins with nitrogen, carbon or both nuclei isotopically labeled; also included are the nuclei that each experiment correlates

Experiment	Nuclei correlated[a]	Reference
2D ^{15}N-^{1}H HSQC	$H^N_i\ N^H_i$	18
3D TOCSY-HMQC	$H^N_i\ N^H_i\ H^{x,y,z,\cdots}_i$	19
3D NOESY-HMQC	$H^N_i\ N^H_i\ H^{y,z,\cdots}_{i,j,\cdots}$	19
3D HMQC-NOESY-HMQC	$N^H_i\ N^H_{j,k,\cdots}\ H^N_{j,k,\cdots}$	20,21
3D HNCA	$H^N_i\ N^H_i\ C^\alpha_i\ \{C^\alpha_{i-1}\}$	22,23
3D HN(CO)CA	$H^N_i\ N^H_i\ C^\alpha_{i-1}$	24
3D H(CA)NNH, HN(CA)H, HN(CA)HA	$H^N_i\ N^H_i\ H^\alpha_i\ \{H^\alpha_{i-1}\}$	25,26,27
3D HN(COCA)HA	$H^N_i\ N^H_i\ H^\alpha_{i-1}$	28
3D HNCO	$H^N_i\ N^H_i\ C'_{i-1}$	22,23
3D HN(CA)CO	$H^N_i\ N^H_i\ C'_i$	29
3D CBCANH, HNCACB	$H^N_i\ N^H_i\ C^{\alpha,\beta}_i\ \{C^{\alpha,\beta}_{i-1}\}$	30,31
3D CBCA(CO)NH	$H^N_i\ N^H_i\ C^{\alpha,\beta}_{i-1}$	32
3D HBHA(CBCA)NH	$H^N_i\ N^H_i\ H^{\alpha,\beta}_i\ \{H^{\alpha,\beta}_{i-1}\}$	33
3D HBHA(CBCACO)NH	$H^N_i\ N^H_i\ H^{\alpha,\beta}_{i-1}$	34
3D HCACO	$H^\alpha_i\ C^\alpha_i\ C'_i$	22,23
3D HCA(CO)N	$H^\alpha_{i-1}\ C^\alpha_{i-1}\ N^H_i$	22,23
3D HN(CA)NNH	$H^N_i\ N^H_i\ N^H_{i+1}$	35
3D H(NCA)NNH	$H^N_i\ N^H_i\ H^N_{i+1}$	35
3D H(N)CACO	$H^N_i\ C^\alpha_i\ C'_i$	36
3D HA[CAN]HN	$H^\alpha_i\ [C^\alpha_i + N^H_i\ ]H^N_i$	37
4D HCANNH, HNCAHA	$H^N_i\ N^H_i\ C^\alpha_i\ H^\alpha_i\ \{C^\alpha_{i-1}\ H^\alpha_{i-1}\}$	38,39
4D HCA(CO)NNH, HN(CO)CAHA	$H^N_i\ N^H_i\ C^\alpha_{i-1}\ H^\alpha_{i-1}$	38,40
3D HCCH-COSY	$H^x_i\ C^x_i\ H^y_i$	41
3D HCCH-TOCSY	$H^x_i\ C^x_i\ H^{y,z,\cdots}_i$	42
3D HNCCH-TOCSY	$N^H_i\ C^\alpha_i\ H^{x,y,z,\cdots}_i$	43
3D C(CO)NH	$H^N_i\ N^H_i\ C^{x,y,z,\cdots}_{i-1}$	44
3D H(CCO)NH	$H^N_i\ N^H_i\ H^{x,y,z,\cdots}_{i-1}$	44
4D HCC(CO)NNH	$H^N_i\ N^H_i\ C^{x,y,z,\cdots}_{i-1}\ H^{x,y,z,\cdots}_{i-1}$	45

[a] The superscripts indicate the individual resonance within each residue, the subscripts refer to the residue number, e.g. C^β_i is the carbon beta resonance of residue i. x, y and z refer to nuclei separated by two or three bonds. Resonances in brackets are weaker secondary correlations.

initial build-up rate with the inverse sixth power of the internuclear distance. This relationship makes two assumptions: (i) that any two nuclei involved in a specific NOE may be considered in isolation of all others; (ii) that the molecule is conformationally rigid. Deviation from the former assumption is addressed by the refinement procedure of back-calculation (see below). Deviation from the latter approximation is considerably more difficult to address. The breakdown of the latter assumption manifests itself most obviously when a single hydrogen shares NOEs with two or more other hydrogens to which it cannot possibly be close simultaneously. The population of more than a single conformation inevitably leads to a short estimate for the internuclear distance. In the case where interconversion between conformations is faster than the overall correlation time of molecular reorientation, any apparent shortening of the estimate of internuclear distance may be opposed by a reduction in the NOE build-up rate. This results from the change in the effective correlation time associated with the internuclear distance that gives rise to the NOE.

Normally, not all NOEs can be resolved in NOESY-type spectra and care has to be taken so that this does not lead to a biased constraint set. Related to this is the common situation that a number of resonances corresponding to individual hydrogens in the same group (e.g. the three hydrogens in a methyl group) share the same frequency. In these cases the distance estimate is usually made to a pseudo-atom at the geometric average of the positions of the nuclei involved, and an appropriate correction factor applied to the upper bound of the constraint. Hence, it is important to define uniquely as many resonances as possible; for example, the stereospecific assignment of the prochiral methyl groups of valine and leucine residues, and of β-methylene groups leads to a considerable improvement in calculated structures.

The use of H–H J-coupling derived constraints is usually limited to the observation of extreme values (large or small). These values may be used to restrict the torsional angles in the structure calculation to one or more small ranges. The population of single conformations in specific ranges cannot be distinguished from averaging over a large range of torsional angles if mid-range J-coupling values are observed. Torsional constraints are commonly applied to both ϕ and $\chi 1$ angles in proteins, and lead to a considerable improvement in the success of structure calculations. There is now also considerable interest in the use of heteronuclear J-couplings to provide torsional angle constraints.

There are two main ways of converting distance and torsional constraints into atomic coordinate sets: *distance geometry* algorithms and *restrained molecular dynamics* simulations. Two classes of distance geometry calculations may be used. The metric matrix algorithms work in distance space and create a set of coordinates from initial distances selected randomly from within the bounds limits, and modified to obey triangle and tetrangle inequalities. The coordinates are obtained from the eigenvalues and eigenvectors of the corresponding metric matrix. The variable target function algorithms work in dihedral space by randomly selecting a series of dihedral angles, setting up a target function of violations of the constraints, and sequentially varying dihedral angles to minimize the target function. Both algorithm types give rise to equivalent structures but may vary in their success rate of convergence. Restrained molecular dynamics procedures include the NMR derived constraints as pseudo-energy potentials in the force field. Normally, for computational expediency, the force field is stripped down to only its bonding potentials (and the technique is termed *restrained simulated annealing*). Many of the force field potentials may be removed since it is the ability of the algorithm to sample space that is the principal quality required of the dynamics algorithm. Distance geometry algorithms commonly result in coordinate sets with poor local geometry and thus distance geometry generated structures are typically subject to a simulated annealing refinement. The structure calculation algorithms, whichever method is chosen, are repeated many times using different random starting selections to ensure a significant sampling of space. Hence NMR structures are normally represented by an ensemble of conformations as shown in *Figure 1*.

The quality of NMR structures may be judged by three criteria. First, the structures are tested for residual constraint violations, which would indicate that the coordinate set did not fully satisfy the input data. Second, root-mean-squared distance (rmsd) variations may be calculated between members of the family of generated structures. These two measurement criteria principally test the performance of the algorithms and are not ideal measures of the quality of the structures. Indeed, they are often interrelated: the more conservative the choice of constraint limits the fewer the residual violations but the worse the rmsds, and vice versa. It should also be noted that choosing too tight a series of constraint limits may fortuitously lead to a very precise series of structures that are inaccurate. The third measurement of quality of structure involves calculating the residual error between the primary experimental data (normally a series of NOESY spectra) and spectra calculated from the generated structures. This procedure, known as *back-calculation*, takes into account all the protons in the molecule simultaneously (avoiding the isolated spin pair

Figure 1. The backbone conformation of 17 solution structures of human stefin A showing regions of high definition of conformation and regions showing considerable disorder. In this case the disorder results from high-frequency conformational interconversion (as shown by nitrogen relaxation studies), although in general it could also result from lower frequency dynamics or a low constraint density in the structure calculations. Reproduced from ref. 46 with permission from Academic Press Ltd.

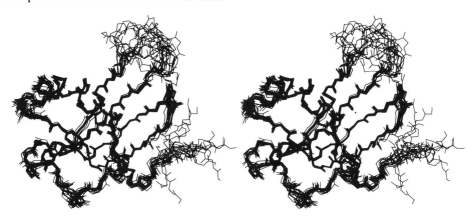

approximation). It may thus be used to create a residual error target function against which the structures are refined iteratively. However, it should be remembered that this method is still subject to errors arising from mobility effects.

In recent years there has been a considerable increase in the rate at which NMR-derived protein structures have been published. In *Table 5* is a list of over 250 protein structures, the coordinates of which have been deposited in Brookhaven protein data bank up to April 1995. The proteins included are almost all globular and cover a wide range of biological functions. In addition, since there is an upper limit on the molecular mass of proteins the structure of which can currently be solved by NMR, there are also numerous examples of the study of independently folded domains of larger proteins.

Table 5. A list of protein structures determined by NMR spectroscopy the coordinates of which have been deposited at the Brookhaven Protein Data Bank by April 1995. Not all entries are included; representative examples only are given where closely related structures have been deposited

1CTI:TRYPSIN INHIBITOR	1TAP:FACTOR XA INHIBITOR
2LET:TRYPSIN INHIBITOR II	1PCE:PEC-60
1TIN:TRYPSIN INHIBITOR V	1CCM:CRAMBIN
2BBI:TRYPSIN/CHYMOTRYPSIN BOWMAN-BIRK INHIBITOR	1PBA:PROCARBOXYPEPTIDASE B ACTIVATION DOMAIN
3CI2:CHYMOTRYPSIN INHIBITOR 2	1HWA:LYSOZYME
1BUS:PROTEINASE INHIBITOR IIA (BUSI IIA)	1AKP:APOKEDARCIDIN
	2AAS:RIBONUCLEASE A
3AIT:TENDAMISTAT	1RCL:RIBONUCLEASE F1
1EGL:EGLIN C	1BTA:BARSTAR
2HIR:HIRUDIN	1SRT:STROMELYSIN-1 CATALYTIC DOMAIN
1KST:KISTRIN	1BW3:BARWIN, BASIC BARLEY SEED PROTEIN

Table 5. Continued

1HEV:HEVEIN
1TRL:THERMOLYSIN FRAGMENT 255–316
2ECH:ECHISTATIN
1MYF:MYOGLOBIN
1FRC:CYTOCHROME C
1CCH:CYTOCHROME C551
1APC:APOCYTOCHROME B562
1FCT:FERREDOXIN CHLOROPLASTIC TRANSIT
  PEPTIDE SEQUENCE
1PIH:HIGH POTENTIAL IRON SULFUR PROTEIN
1ZRP:RUBREDOXIN
1PUT:PUTIDAREDOXIN
9PCY:PLASTOCYANIN
1TRW:THIOREDOXIN
1GRX:GLUTAREDOXIN + GLUTATHIONE
1GPS:GAMMA-1-P THIONIN
1DMC:CD-6 METALLOTHIONEIN-1 ALPHA
  DOMAIN
1DME:CD-6 METALLOTHIONEIN-1 BETA
  DOMAIN
2BCA:CALBINDIN D9K
2BBM:CALMODULIN + SKELETAL
  MLC-KINASE CAM-BD
1CTA:TROPONIN C SITE III - SITE III
  HOMODIMER
1TRF:TROPONIN C APO-TR1C
2PAS:PARVALBUMIN
1PRS:DEVELOPMENT-SPECIFIC PROTEIN S
1IRP:IL-1 RECEPTOR ANTAGONIST PROTEIN
6I1B:INTERLEUKIN-1*BETA
1BBN:INTERLEUKIN 4
1IL8:INTERLEUKIN 8
1EGF:EPIDERMAL GROWTH FACTOR
1IXA:EGF-LIKE MODULE OF HUMAN FACTOR
  IX
1APO:COAGULATION FACTOR X
  EGF-LIKE MODULE
1CCF:COAGULATION FACTOR X N-TERM EGF-
  LIKE MODULE
1HRF:HEREGULIN-ALPHA EGF-LIKE DOMAIN
2TGF:TRANSFORMING GROWTH
  FACTOR-ALPHA
1TVS:TRANSACTIVATOR PROTEIN TAT EIAVY
1HIU:HUMAN INSULIN
2GF1:INSULIN-LIKE GROWTH FACTOR
1IGL:INSULIN-LIKE GROWTH FACTOR II
1GNC:GRANULOCYTE COLONY STIMULATING
  FACTOR
1HUN:HUMAN MACROPHAGE
  INFLAMMATORY PROTEIN 1 BETA
1HFH:FACTOR H, 15TH AND 16TH C-MODULE
  PAIR
1KDU:U-T PLASMINOGEN ACTIVATOR
  KRINGLE DOMAIN

1PK2:T-T PLASMINOGEN ACTIVATOR KRINGLE
  2 DOMAIN
1TPM:T-T PLASMINOGEN ACTIVATOR F1
  DOMAIN
1TTF:FIBRONECTIN TENTH TYPE III MODULE
1BBA:BOVINE PANCREATIC POLYPEPTIDE
1MGS:HUMAN MELANOMA GROWTH
  STIMULATING ACTIVITY
1TUR:OVOMUCOID THIRD DOMAIN
1ERH:HUMAN COMPLEMENT REGULATORY
  PROTEIN CD59
1CDB:CD2
1HDN:HIS-CONTAINING PHOSPHOCARRIER
  PROTEIN
1BVH:TYROSINE PHOSPHATASE
1APS:ACYLPHOSPHATASE
1RPR:ROP
1CRR:C-H-RAS P21 CATALYTIC DOMAIN
3CYS:CYCLOPHILIN A + CYCLOSPORIN A
1FKR:FKBP12 + FK506
2PNB:PI3-KINASE P85 N-TERM SH2
1PKT:PI3-KINASE SH3
2PNI:PI3-KINASE P85 SH3
1AB2:PROTO-ONCOGENE TYROSINE KINASE
  SH2
1GFD:GRB2 C-TERM SH3
1GBR:GRB2 N-TERM SH3 + SOS-A PEPTIDE
1SRM:SRC SH3 DOMAIN
1RLP:C-SRC SH3 DOMAIN + RLP2 PEPTIDE
2HSP:PHOSPHOLIPASE C-G SH3
2PLE:PHOSPHOLIPASE C-G1 C-TERM SH2 +
  PDGFR PEPTIDE
1PCO:PORCINE PANCREATIC PROCOLIPASE B
1OLG:P53 OLIGOMERIZATION DOMAIN
1ARD:YEAST TRANSCRIPTION FACTOR ADR1
  N-TERM ZNF
1PAA:YEAST TRANSCRIPTION FACTOR ADR1
  C-TERM ZNF
1HVN:HIV-1 NUCLEOCAPSID ZNF + D(ACGCC)
1NCP:HIV-1 P7 NUCLEOCAPSID PROTEIN 2xZNF
1GDC:GLUCOCORTICOID RECEPTOR GR-DBD
1HRA:RETINOIC ACID RECEPTOR BETA DBD
1BBO:HUMAN ENHANCER-BINDING PROTEIN
  MBP-1 2xZNF
1GAT:ERYTHROID TRANSCRIPTION FACTOR
  GATA-1 ZN DBD
1CHC:EQUINE HERPES VIRUS-1 RING DOMAIN
1TFI:TRANSCRIPTIONAL ELONGATION
  FACTOR SII NA-BD
1ZTA:LEUCINE ZIPPER MONOMER
1ADN:N-ADA 10 DNA
  METHYLPHOSPHOTRIESTER REPAIR
  DOMAIN
1RET:GAMMA DELTA RESOLVASE DBD

Table 5. Continued

1TNS:MU TRANSPOSASE DBD
1HKT:HEAT SHOCK TRANSCRIPTION FACTOR
1POU:OCT-1 POU-SPECIFIC DOMAIN
1HDP:OCT-2 POU HOMEODOMAIN
1AHD:ANTENNAPEDIA PROTEIN
  HOMEODOMAIN + DNA
1FTZ:FUSHI TARAZU PROTEIN HOMEODOMAIN
1MSF:C-MYB DBD + DNA
1GHC:HISTONE H1 GLOBULAR DOMAIN
1HMA:HMG-D HMG-BOX DOMAIN
1NHM:HMG-1 DBD BOX 2
1LYP:CAP18 (106 - 137)
1LCC:LAC REPRESSOR HEADPIECE + HALF-
  OPERATOR
1ARQ:ARC REPRESSOR
1PRA:BACTERIOPHAGE 434 REPRESSOR DBD
1ADR:P22 C2 REPRESSOR N-TERM DBD
1LEB:LEXA REPRESSOR DBD
1MNT:MNT REPRESSOR DEL(77-83)
1GVA:GENE V PROTEIN MUTANT Y41H
1RIP:RIBOSOMAL PROTEIN S17
1SXL:SEX-LETHAL PROTEIN RBD-2
2GB1:PROTEIN G B1 DOMAIN
2IGG:PROTEIN G SECOND IGG-BD
2IGH:PROTEIN G THIRD IGG-BD
2PTL:PROTEIN L B1 DOMAIN
1CEY:CHEY + MG
1NTR:NTRC RECEIVER DOMAIN NTRC(1-124)
1MAJ:MURINE ANTIBODY 26-10 VL DOMAIN
1MEA:METHIONYL-TRNA SYNTHETASE ZN-BD
1ATY:F1FO ATP SYNTHASE SUBUNIT C
1BAL:DIHYDROLIPOAMIDE
  SUCCINYLTRANSFERASE E3-BD
2PDE:E2P/E3 OF PYRUVATE DEHYDROGENASE
1ACA:ACYL-CoA BINDING PROTEIN +
  PALMITOYL-CoA
1ACP:ACYL CARRIER PROTEIN
1SPF:PULMONARY SURFACTANT-ASSOCIATED
  POLYPEPTIDE C
1LPT:WHEAT LIPID TRANSFER PROTEIN
1CBH:CELLOBIOHYDROLASE I C-TERM
  DOMAIN
2PRF:PROFILIN IA
1PAJ:PILIN PEPTIDE FRAGMENT PAK
1HCE:HISACTOPHILIN
1SVQ:SEVERIN DOMAIN 2
1VIL:VILLIN DOMAIN ONE
1PDC:SEMINAL FLUID PROTEIN PDC-109
  DOMAIN B
1PCP:PORCINE SPASMOLYTIC PROTEIN
1PSM:SPAM-H1
1AMB:ALZHEIMER'S DISEASE AMYLOID BETA-
  PEPTIDE

1GNB:GUANYLIN B-FORM
1KAL:KALATA B1
1AFP:ANTIFUNGAL PROTEIN FROM A.
  GIGANTEUS
1BCT:BACTERIORHODOPSIN (FRAGMENT 163-
  231)
1BHB:BACTERIORHODOPSIN (FRAGMENT 1–71)
1EDP:ENDOTHELIN 1
1GRM:GRAMICIDIN A
1ERD:PHEROMONE ER-2
1ERP:PHEROMONE ER-10
1ABT:ALPHA-BUNGAROTOXIN + nACR
  FRAGMENT
1NTX:ALPHA-NEUROTOXIN
1AKP:APOKEDARCIDIN
1BDS:BDS-I
1BOM:BOMBYXIN-II
1NBT:BUNGAROTOXIN
2CDX:CARDIOTOXIN CTX I
2CCX:CARDIOTOXIN CTX IIB
1CXO:CARDIOTOXIN GAMMA
2CRT:CARDIOTOXIN III
1CVO:CARDIOTOXIN V
2CRD:CHARYBDOTOXIN
1CHL:CHLOROTOXIN
1COE:COBROTOXIN
1COD:COBROTOXIN II
1DEC:DECORSIN
1DRS:DENDROASPIN S5C1/SH04
1DEM:DENDROTOXIN I
1DTK:DENDROTOXIN K
2DTB:DELTA-TOXIN
  (DELTA-HAEMOLYSIN)
1ERA:ERABUTOXIN B
1KTX:KALIOTOXIN
1TCG:MU-CONOTOXIN GIIIA
1VNA:NEUROTOXIN CSE-V1
1SHI:NEUROTOXIN I
1NOR:NEUROTOXIN II
1ANS:NEUROTOXIN III
1NRB:NEUROTOXIN V
1OMA:OMEGA-AGA-IVB
1IVA:OMEGA-AGATOXIN-IVA
1CCO:OMEGA-CONOTOXIN GVIA
1OMC:OMEGA-CONOTOXIN GXIA
1SRB:SARAFOTOXIN S6B
1SIS:SCORPION INSECTOTOXIN I5A
1PNH:SCORPION TOXIN
1SCY:SCYLLATOXIN
1TER:TERTIAPIN
1NEA:TOXIN ALPHA
1TFS:TOXIN FS2

Table 6. The NMR parameters that give information on dynamic processes within proteins and the timescale of the processes that can be probed

Motional timescale	NMR probe
Nanosecond	H, C, N relaxation
Microsecond	Relaxation in a spin-lock field
Millisecond	Variation in resonance linewidths
1/10 second	Saturation transfer
Second–month	Amide NH exchange rates

Coupled to the study of the solution structure of proteins by NMR is the characterization of their dynamics. Several NMR measurable parameters give information on the range and timescale of molecular motions and conformational changes; principally these parameters are the relaxation rates of the NMR signal (including the NOE), the rate of exchange of the NMR signal between nuclei, and the rate of exchange of amide protons with those of the solvent. These NMR probes are sensitive to windows of frequencies of molecular motion on timescales ranging from picoseconds to months. A list of the timescales that can be probed by each technique is given in *Table 6*. These techniques are discussed in detail in ref. 6. Similar to the method of structure determination, the application of NMR to the study of protein dynamics is critically dependent on the identification (resonance assignment) of the nuclei involved. Indeed, it is the ability of NMR spectroscopy to provide both structural and dynamic information of proteins that makes the technique so valuable.

4. REFERENCES

1. Derome, A.E. (1987) *Modern NMR Techniques for Chemistry Research.* Pergamon Press, Oxford.

2. Evans, J.N.S. (1995) *Biomolecular NMR spectroscopy.* Oxford University Press, Oxford.

3. Wagner, G., Hyberts, S.G. and Havel, T.F. (1992) *Annu. Rev. Biophys. Biomol. Struct.*, **21**, 167.

4. Billeter, M. (1992) *Q. Rev. Biophys.*, **25**, 325.

5. Barbato, G., Ikura, M., Kay, L.E., Pastor, R.W. and Bax, A. (1992) *Biochemistry*, **31**, 5269.

6. *Methods Enzymol.* (1994) **239**.

7. Wuthrich, K. (1986) *NMR of proteins and nucleic acids.* John Wiley, New York.

8. Redfield, C. (1993) in *NMR of macromolecules* (G.C.K. Roberts, ed.). Oxford University Press, Oxford, p. 71.

9. Neuhaus, D. and Evans, P.A. (1993) *Methods Mol. Biol.*, **17**, 15.

10. Basus, V.L. (1989) *Methods Enzymol.*, **177**, 132.

11. Stockman, B.J., Euvrard, A., Kloosterman, D.A., Scahill, T.A. and Swensom, R.P. (1993) *J. Biomol. NMR*, **3**, 133.

12. Fogh, R.H., Schipper, D., Boelens, R. and Kaptein, R. (1994) *J. Biomol. NMR*, **4**, 123.

13. Anglister, J., Grzesiek, S., Wang, A. Ren, H., Klee, C.B. and Bax, A. (1994) *Biochemistry*, **33**, 3540.

14. Campbell-Burk, S.L., Domaille, P.J., Starovasnik, M.A., Boucher, W. and Laue, E.D. (1992) *J. Biomol. NMR*, **2**, 639.

15. LeMaster, D.M. (1994) *Prog. NMR Spectrosc.*, **26**, 371.

16. Edison, A.S., Abildgaard, F., Westler, W.M., Mooberry, E.S. and Markley, J.L. (1994) *Methods Enzymol.*, **239**, 3.

17. Oschkinat, H., Muller, T. and Dieckmann, T. (1994) *Angew. Chem. Int. Ed. Engl.*, **33**, 277.

18. Bodenhausen, G. and Ruben, D.J. (1980) *Chem. Phys. Lett.*, **69**, 185.

19. Marion, D., Driscoll, P.C., Kay, L.E., Wingfield, P.T., Bax, A., Gronenborn, A.M. and Clore, G.M. (1989) *Biochemistry*, **28**, 6150.

20. Carr, M.D., Birdsall, B., Frenkiel, T.A., Bauer, C.J., Jimenez-Barbero, J., Polshakov, V.I., McCormick, J.E., Roberts, G.C.K. and Feeney, J. (1991) *Biochemistry*, **30**, 6330.

21. Ikura, M., Bax, A., Clore, G.M. and Gronenborn, A.M. (1990) *J. Am. Chem. Soc.*, **112**, 9020.

22. Kay, L., Ikura, M., Tschudin, R. and Bax, A. (1990) *J. Magn. Reson.*, **89**, 496.

23. Ikura, M., Kay, L.E. and Bax, A. (1990) *Biochemistry*, **29**, 4659.

24. Bax, A. and Ikura, M. (1991) *J. Biomol. NMR*, **1**, 99.

25. Kay, L.E., Ikura, M. and Bax, A. (1991) *J. Magn. Reson.*, **91**, 84.

26. Seip, S., Balbach, J. and Kessler, H. (1992) *J. Magn. Reson.*, **100**, 406.

27. Clubb, R.T., Thanabal, V. and Wagner, G. (1992) *J. Biomol. NMR*, **2**, 203.

28. Clubb, R.T. and Wagner, G. (1992) *J. Biomol. NMR*, **2**, 389.

29. Clubb, R.T., Thanabal, V. and Wagner, G. (1992) *J. Magn. Reson.*, **97**, 213.

30. Grzesiek, S. and Bax, A. (1992) *J. Magn. Reson.*, **99**, 201.

31. Wittekind, M. and Mueller, L. (1993) *J. Magn. Reson.*, **B101**, 201.

32. Grzesiek, S. and Bax, A. (1992) *J. Am. Chem. Soc.*, **114**, 6291.

33. Wang, A.C., Lodi, P.J., Qin, J., Vuister, G.W., Gronenborn, A.M. and Clore, G.M. (1994) *J. Magn. Reson.*, **B105**, 196.

34. Grzesiek, S. and Bax, A. (1993) *J. Biomol. NMR*, **3**, 185.

35. Weisemann, R., Ruterjans, H. and Bermel, W. (1993) *J. Biomol. NMR*, **3**, 113.

36. Seip, S., Balbach, J. and Kessler, H. (1993) *J. Biomol. NMR*, **3**, 233.

37. Szyperski, T., Wider, G., Bushweller, J.H. and Wuthrich, K. (1993) *J. Biomol. NMR*, **3**, 127.

38. Boucher, W., Laue, E.D., Campbell-Burk, S.L. and Domaille, P.J. (1992) *J. Am. Chem. Soc.*, **114**, 2262.

39. Boucher, W., Laue, E.D., Campbell-Burk, S.L. and Domaille, P.J. (1992) *J. Biomol. NMR*, **2**, 631.

40. Kay, L.E., Wittekind, M., McCoy, M.A., Friedrichs, M.S. and Mueller, L. (1992) *J. Magn. Reson.*, **98**, 443.

41. Bax, A., Clore, G.M., Driscoll, P.C., Gronenborn, A.M., Ikura, M. and Kay, L.E. (1990) *J. Magn. Reson.*, **87**, 620.

42. Bax, A., Clore, G.M. and Gronenborn, A.M. (1990) *J. Magn. Reson.*, **88**, 425.

43. Weisemann, R., Lohr, F. and Ruterjans, H. (1994) *J. Biomol. NMR*, **4**, 587.

44. Grzesiek, S., Anglister, J. and Bax, A. (1993) *J. Magn. Reson.*, **B101**, 114.

45. Clowes, R.T., Boucher, W., Hardman, C.H., Domaille, P.J. and Laue, E.D. (1993) *J. Biomol. NMR*, **3**, 349.

46. Martin, J.R., Craven, C.J., Jerala, R., Kroon-Zitko, L., Zerovnik, E., Turk, V. and Waltho, J.P. (1995) *J. Mol. Biol.*, **246**, 331.

CHAPTER 20
EPR SPECTROSCOPY OF PROTEINS
R. Cammack and J.K. Shergill

1. INTRODUCTION

Electron paramagnetic resonance (EPR) spectroscopy, also known as electron spin resonance (ESR) or electronic magnetic resonance (EMR) spectroscopy is a technique for the study of paramagnetic materials, which contain unpaired electrons (1–5). In proteins, the paramagnetic species that can be observed comprise free radicals and transition metal ions. Radicals are produced in redox proteins, and may also be generated in proteins by irradiation or by active oxygen species. Transition metal ions such as iron, manganese and copper occur naturally in proteins of metal-ion transport and storage, and electron transport. EPR is often the method of choice for identifying and quantifying these paramagnetic species. In addition, EPR can examine the atoms surrounding the paramagnetic centers, their motion, and the proximity of other paramagnetic centers. These powerful methods can be applied to proteins even if the protein itself is not normally paramagnetic, by introducing a stable radical, known as a spin label, into the protein (6), or by introducing a paramagnetic metal center (7).

The types of EPR-detectable species in proteins that may be investigated by EPR are briefly reviewed in *Table 1*. There are some limitations on the measurement of these species. Transition metal ions are EPR-detectable in certain oxidation states only (such as Mn^{II}, Fe^{III}, Co^{II}, Cu^{II} and Mo^{V}), so that the addition of oxidizing or reducing agents may be needed. Also, measurement of most transition metals must be made at cryogenic temperatures. Stable radicals based on amino acids or prosthetic groups occur in some proteins, such as enzymes. Other, transient, free radicals may be generated in proteins by irradiating with ionizing or ultraviolet radiation, or by the action of oxygen radical species such as the hydroxyl radical. Such radicals may be detected by the addition of spin traps, which are diamagnetic compounds, usually nitroso compounds or nitrones, which react with the transient radicals to form stable radicals which can be identified from their EPR signals (24, 25).

2. PRINCIPLES

The principles of EPR spectroscopy are described in various texts (1, 2), some of which describe biological applications (3, 4). Further details of the operating principles, together with applications to enzymes are described in the companion volume to this book (25).

EPR is analogous to nuclear magnetic resonance (NMR) spectroscopy, in that it involves resonant absorption of electromagnetic radiation by the electron spins ($S = 1/2$), in which a splitting of energy levels is induced by an applied magnetic field (*Figure 1*). The principal differences between EPR and NMR arise from the much greater magnetic moment of the electron compared with those of nuclei (1836 times that of the proton). The spectral range is much greater than in NMR, and spin–spin interactions of electrons with other electrons (the fine-structure interaction) or with nuclei (the hyperfine interaction) are much stronger.

EPR Spectroscopy

Figure 1. Principles of EPR spectroscopy. (a) Alignment of unpaired electron spins in an applied magnetic field; (b) energy levels, absorption of a quantum of energy, $h\nu$; (c) hyperfine splitting of energy levels by a nucleus of spin $I=1/2$.

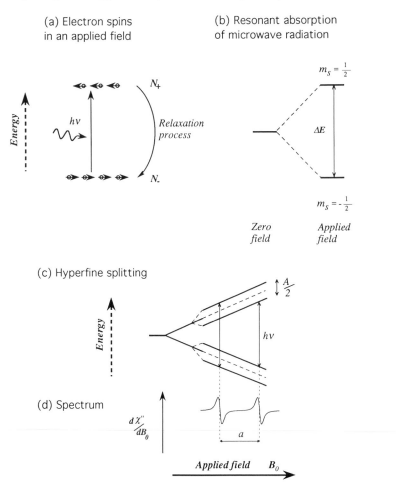

(a) Electron spins in an applied field

(b) Resonant absorption of microwave radiation

(c) Hyperfine splitting

(d) Spectrum

Statistically, there are slightly more spins with $m_s = -1/2$ (N_-) than with $m_s = +1/2$ (N_+):

$$\frac{N_+}{N_-} = \exp\left(-\frac{g\mu_B B_0}{kT}\right). \qquad \text{Equation 3}$$

This *Boltzmann distribution* is maintained by *spin-lattice relaxation*, in which the spins exchange energy with vibrational modes of the surroundings. When the spins interact with incident microwave radiation there is a net absorption, and this is detected as the EPR signal.

As can be seen from Equation 3, the difference in populations N_+ and N_- of the $m_s = \pm 1/2$ energy levels increases at lower temperatures and hence the amplitude of the EPR signal is expected to be greater. However if too much microwave power is applied the populations of the levels tend to become equalized and the signal intensity decreases; this is known as *microwave power saturation*. This is particularly a problem at low temperatures.

2.1. How the EPR spectrometer works

For a description of the operating principles and construction of spectrometers see (26). The conventional continuous-wave (cw) EPR spectrometer operates in the X-band range of microwave frequency, 9–10 GHz. The sample sits in a microwave cavity, which resonates at a fixed microwave frequency, ν. This enhances the microwave field at the sample and hence the sensitivity, but it restricts the spectrometer to a fixed frequency. Paramagnets with different g-factors are brought into resonance by sweeping the applied magnetic field, B_0; this is produced by an electromagnet. To enhance further the signal-to-noise (S/N) ratio, the magnetic field is rapidly modulated over a small range, and the signal is detected in phase with the modulation, resulting in a first-derivative display of the absorption spectrum. Thus in a typical EPR spectrum the abscissa is the magnetic field, and the ordinate is the first derivative of the microwave absorption (*Figure 1b*).

2.2. Characteristics of EPR spectra

In EPR of paramagnetic proteins the g-factor is an aid in the identification of an unknown signal. In most organic radicals the range of g is between 2.00 and 2.01. In paramagnetic metal ions it is larger (see *Table 1*). Where an ion has more than one unpaired electron, such as high-spin Fe^{III} ($S=5/2$), values may range from 10 to less than 1. For examples, see *Figure 2*. The g-factor is generally anisotropic, that is its value depends on the orientation of the molecule in the applied

Figure 2. Examples of EPR spectra of transition metal ions. (a) Transferrin, high-spin iron(III) ($S=5/2$). Note the prominent features at $g=9, 4.3$. (b) Transferrin in which the iron is substituted by copper(II) ($S=1/2$). Note the g-factor is slightly greater than 2.0, and the hyperfine splitting into four lines due to the copper nucleus ($I=3/2$). The superhyperfine splitting due to ^{14}N ($I=1$) ligands is shown in the inset (c).

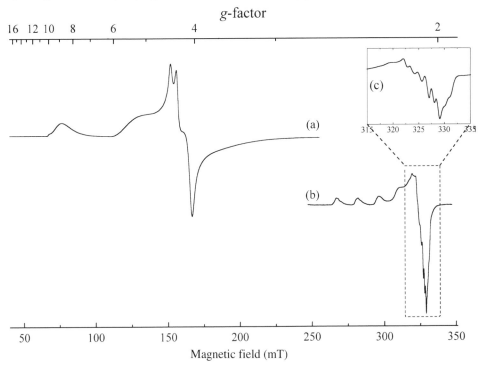

magnetic field. If the paramagnet, such as a metal ion or small radical, is rotating rapidly enough in solution, the effect of this anisotropy is averaged out, but for proteins the motion in solution is not rapid enough, and a range of g-factors is observed in the spectrum. The shape of the spectrum provides information about the coordination geometry of the complex.

2.3. Hyperfine (electron–nuclear) interactions

One of the most characteristic features of radicals and transition-metal complexes, which assists in their identification, is the hyperfine coupling of the electron spin with nearby nuclear spins. The values of relevant nuclear spins are given in *Table 2*. Nuclei with spins greater than $I = 1/2$, such as ^{2}H and ^{14}N, also possess a nuclear electric quadrupole moment, which may cause further splitting of the electron energy levels.

The hyperfine interaction with a nuclear spin I is defined by the number of lines and the magnitude of the *hyperfine coupling, A*. If the EPR spectrum is a narrow line, the hyperfine interaction is seen as a splitting into $(2I + 1)$ lines; for example a proton, ^{1}H, $I = 1/2$, splits the line into two. If the electron also interacts with a second nucleus, each of the two lines will be further split; for example ^{14}N, $I = 1$, will cause each line to be further split into three, and so on (see *Figure 2b, c*). For cases such as radicals where the EPR spectra show so many splittings as to be uninterpretable, or for transition metal spectra where the smaller hyperfine splittings are concealed by the linewidth of the spectrum, greater resolution is offered by electron-nuclear double resonance (ENDOR) spectroscopy (*Table 3*). This method may be described essentially as an NMR experiment with EPR detection (27).

In spectroscopic measurements the hyperfine splitting is often expressed in magnetic field units, in which case it is given the symbol a. For molecular orbital calculations the hyperfine coupling is expressed in energy units, such as joules or MHz, and given the symbol A. The relationship between these parameters is:

$$A = \frac{g\mu_B}{h \times 10^9} . a \qquad \qquad \text{Equation 4}$$

where A is in MHz and a in mT. The magnitude of a or A depends on the magnetic moment μ_N of the nucleus, and the extent to which the electron spin interacts with the nucleus. It decreases proportionately as the electron density is distributed over more atoms. The distribution of unpaired electron density over the molecule may thus be determined.

2.4. Electron–electron interactions

The interaction between electrons observed in EPR is stronger than the interactions between nuclei in NMR, and occurs over longer distances. The interactions occur both through electron orbitals (the exchange interaction), and through space (the dipolar interaction).

The strongest interactions occur when two or more electrons occur in the same spin system. In transition-metal ions with $S > 1/2$, the electrons become coupled to give $(2S + 1)$ spin states. For example for high-spin Fe^{III} or Mn^{II} (3d^5) there are three pairs of states, with $m_s = \pm 1/2, \pm 3/2, \pm 5/2$. EPR transitions can occur between these states. Depending on the coordination geometry of the protein metal site, the energy levels may be shifted by electrostatic interactions, giving rise to the so-called *zero-field splittings*. It is these splittings that give rise to large variations in the effective g-factor of these metal ions. The zero-field splittings can provide information about the

Table 2. Some nuclei with magnetic moments

Isotope	% Natural abundance	Nuclear spin I	μ_N
^{1}H	99.985	1/2	+2.793
^{2}H	0.015	1	+0.857
^{13}C	1.11	1/2	+0.702
^{14}N	99.63	1	+0.404
^{15}N	0.37	1/2	−0.283
^{17}O	0.037	5/2	−1.89
^{19}F	100	1/2	+2.629
^{31}P	100	1/2	+1.132
^{33}S	0.76	3/2	+0.643
^{55}Mn	100	5/2	+3.444
^{57}Fe	2.19	1/2	+0.090
^{59}Co	100	7/2	+4.649
^{63}Cu	69.09	3/2	+2.226
^{65}Cu	30.91	3/2	+2.385
^{95}Mo	15.72	5/2	−0.914
^{97}Mo	9.46	5/2	−0.934

coordination geometry (28, 29); for example in the asymmetric Fe^{III} center of transferrin (*Figure* 2), the signal around $g = 4.3$ arises from the $\Delta m_s = \pm 3/2$ transition, and the signals around $g = 9$ from the $\Delta m_s = \pm 1/2$ and $\pm 5/2$ transitions (29).

Interactions between discrete paramagnets may give rise to splitting or broadening of the EPR signals, and/or enhancement of the electron-spin relaxation rates. The magnitude of the effects depends on various factors, including the distance between them and their respective electron-spin relaxation rates. This may be used to estimate, for example, the average distance between two spin-labeled molecules moving in solution, or the fixed distance between two redox centers in a protein.

Triplet states are systems with two unpaired electrons, $S = 1$, or two $S = 1/2$ centers <1 nm apart, coupled to give a net spin of $S = 1$. They are observed, for example, in photosynthetic systems at low temperatures (22). If induced by photochemical processes, they may be spin-polarized (i.e. the populations N_- and N_+ deviate substantially from the equilibrium values), so that intensities are high, and parts of the spectrum are in emission (appearing upside-down).

2.5. Specialist EPR techniques

Extensions of the EPR method are being made possible by the development of new types of spectrometers (30, 31). These include measurements at different microwave frequencies. Higher frequencies are used for the resolution of g-factors from hyperfine splittings. Lower frequencies are used for measurements involving large aqueous samples, such as EPR imaging of radicals in whole-body samples (32).

Pulsed EPR spectrometers, in contrast to the continuous wave (cw) spectrometers described above, use short, high-power pulses of microwaves, and detect the EPR signal as an electron spin-echo. Because of the broad bandwidth of EPR spectra these instruments are not as widely used as in NMR spectroscopy, but they have a number of special applications, some of which are outlined in *Table 3*. Electron spin-echo envelope modulation (ESEEM) spectroscopy is a

Table 3. Types of EPR spectroscopy

Mode of spectroscopy	Types of information available	Instrumental requirements	Sample requirements[a]
EPR—Liquid state	Radicals Kinetics Motion	cw spectrometer. Flat cell or capillary for aqueous samples	Volume ≈0.25 ml Concentration ≈0.1–100 μM
EPR—Frozen state	Transition ions Reaction intermediates	Cryostat-liquid nitrogen or helium	3 mm quartz tubes Volume ≈0.15 ml Concentration ≈1–100 μM
Multifrequency cw EPR	Resolution of g-factors and hyperfine interactions in complex spectra	Microwave bridges operating at different frequencies	
ENDOR	Hyperfine coupling, e.g. ^{13}C, ^{1}H, giving an indication of the electron–nuclear distance and electron delocalization	ENDOR spectrometer	Concentration ≈0.5 mM
ESEEM	Hyperfine couplings, e.g. ^{14}N, ^{2}H	Pulsed EPR spectrometer	Concentration ≈0.5 mM
Pulsed ENDOR	Hyperfine couplings, e.g. ^{1}H, ^{2}H, ^{13}C, ^{14}N	Pulsed EPR spectrometer, interfaced to an ENDOR unit	Concentration ≈0.5 mM
Flash photolysis	Transient radicals Photosynthesis	Rapid scan cw or pulsed EPR spectrometer	Liquid or frozen state Concentration depends on radical concentration
Relaxation rates	Spin–spin interactions Distances between spins, determined from either: (a) power saturation or (b) saturation recovery profiles	(a) cw EPR spectrometer or (b) pulsed EPR spectrometer	Liquid or frozen state Concentration 0.1–0.5 mM
EPR imaging	Location of paramagnets in living body	Low-frequency (0.25 GHz) spectrometer, gradient coils	Narrow-line radical

[a] The concentration referred to in this table indicates the concentration of the paramagnetic species, or active protein, and *not* the total protein concentration.

technique for measuring anisotropic hyperfine interactions (33). The intensity of the electron spin-echo is measured as a function of the temporal spacing between the microwave pulses. The echo intensity is modulated as a result of interactions with the nearby nuclear spins. The Fourier-transformed frequency-domain spectrum detected corresponds to hyperfine transition frequencies. Pulsed ENDOR spectrometers also exist; they avoid some of the limitations of cw ENDOR, for example the requirement for microwave power saturation.

2.6. Sample requirements

Samples are held in quartz tubes, since glass contains paramagnetic impurities. It is usually an advantage to have as large a sample as possible, to enhance the signal intensity, since paramagnetic centers tend to be dilute in biochemical systems. However aqueous samples are 'lossy', due to dielectric microwave absorption by water, and the volume is typically limited to $<50 \mu l$. Frozen solutions are less lossy and the size of sample is limited by the dimensions of the cryostat and cavity; the optimum volume required for signal detection is of the order of $100 \mu l$. The concentration required is dependent upon the properties of the EPR signal detected. The EPR signal amplitude is inversely proportional to the square of the signal linewidth, so that free radical signals which typically display narrow linewidths are detectable at concentrations down to $0.1 \mu M$. Higher concentrations of $10-50 \mu M$ may be required for a species with broad spectra, such as Fe^{III}.

3. APPLICATIONS IN THE STUDY OF PROTEINS

3.1. Measurements in whole-cell and tissue systems

EPR spectra of proteins that contain naturally occurring paramagnetic centers may be recorded in animal and plant tissues, or in bacterial cell preparations. Since the majority of molecules contain only paired electrons, EPR is a selective technique. A number of proteins, such as iron–sulfur proteins, which are difficult to detect by other means such as spectrophotometry, have been shown by EPR spectroscopy to be major components of respiratory chains. EPR can therefore be applied to complex mixtures such as cell extracts, or to monitor protein purification (see *Table 4*). This may be valuable where the center is labile, so that measurement of protein concentration, for example by Western blotting, does not give a measure of the *active form* of the protein. EPR spectra of cell extracts may be complex, but overlap between EPR spectra may be resolved in a number of ways:

(i)   temperature dependence of the signals;
(ii) redox state of the proteins, for example by adding oxidizing or reducing agents;
(iii) g-factor selection; if some features of the spectrum are narrow and readily distinguished they may be detected in the presence of overlapping signals;
(iv) sequestering of paramagnetic species, for example chelation of Mn^{2+} by ethylene-diaminetetraacetic acid (EDTA) broadens out the Mn^{2+} spectrum, rendering it almost undetectable.

3.2. Information on paramagnetic centers in proteins

For transition-metal complexes, the redox state is readily determined from the EPR spectrum, since oxidation or reduction of a center with an odd number of electrons changes it to a state with either zero spin (and thus undetectable), or an even number of electrons (e.g. $S = 1, 2 \ldots$) which is usually difficult to detect. By measuring the spectrum as a function of the applied redox potential it is possible to measure the midpoint redox potential, even in complex or membrane-bound proteins (34).

Table 4. Types of information obtained by EPR spectroscopy

Information	Method	Notes
Identity of paramagnetic species in proteins	Characteristics of spectrum: temperature dependence, g-factors and hyperfine interactions	Often by comparison with known spectra
Spin quantification	Double integration of the EPR signal	Spectrum is compared with a standard of known concentration. Signals measured under nonsaturating conditions
Monitoring of protein purification	Conventional EPR	EPR can be used in relatively impure systems, such as whole tissue samples or bacterial cells
Modifications by site-directed mutagenesis	Conventional EPR	Spectra are often sensitive to subtle changes in protein structure, providing information on the role of individual amino acids
Distances between centers	Spin–spin interactions Relaxation rates Spectral simulation	cw EPR: power saturation experiments Pulsed EPR: using the saturation-recovery technique
Types and number of metal ions	Hyperfine interactions	For nuclei, see *Table 2*
Types and number of ligands	Superhyperfine interactions, detected by EPR, ENDOR or the ESEEM technique	Isotopes may be substituted by growth of cells on labeled media
Oxidation states and redox potentials	Samples prepared under controlled redox conditions	Alternate oxidation states may be EPR-silent or even-spin
Spin states	Conventional EPR	Higher spin states show zero-field splittings
Radicals–determination of electron density distribution	Spectral simulation ENDOR Isotope substitution	Detailed description requires molecular-orbital calculations
Mobility of protein or side-chain	Spin labels–lineshape analysis	Determined by computer simulation of spectra

Ligands to transition metal centers may be identified by hyperfine couplings to ligand nuclei, sometimes called superhyperfine couplings (for an example, see *Figure 2c*). The spin state of the metal center is also readily determined from the EPR spectrum; for example low-spin Fe^{III} ($S = 1/2$) typically has spectra with g-factors up to 4, whereas high-spin Fe^{III} ($S = 5/2$) may have g-factors up to 10.

Multinuclear and mixed-valent metal complexes are also detected by their unusual g-factors, and by hyperfine interactions with metal nuclei; for example, the copper spectrum of cytochrome *c* oxidase shows a seven-line hyperfine spectrum which has been interpreted as due to a mixed-valent Cu^I–Cu^{II} dinuclear center. This is not well resolved at X-band (9 GHz), but is more readily observed at S-band frequency (4 GHz) (35).

For organic radicals, the electron density distribution over the radical may be determined by hyperfine interactions with nuclei such as ^1H or ^{13}C in isotopically labeled molecules. It may also be estimated by molecular-orbital calculations; for example in the tyrosyl radical of Photosystem II (*Figure 3c*) the hyperfine splitting is principally due to the two protons on the β-carbon of tyrosine (23).

The distance between a paramagnetic center in a protein and the aqueous environment may be estimated from the effects of relaxation probes. These are small paramagnetic molecules, for example chromium oxalate for RT measurements or dysprosium complexes for cryogenic temperatures. The probes cause broadening of the spectrum of the paramagnet, or enhancement of the electron-spin relaxation rate, which, with suitable calibration, can be used to estimate distances (36). This method may be used on membrane proteins and has been applied to determine the position of paramagnetic centers relative to the inner and outer surfaces.

3.3. Spin labels and spin probes

The types of information about proteins that may be obtained from EPR spectra of spin labels are summarized in *Table 5*. Many types of spin labels are made commercially though some require organic synthesis (37).

Spin probes

Spin probes are paramagnets that are introduced into the environment of a protein, to act as a reporter of the parameters such as polarity and oxygen concentration. An example is the nitroxide 2,2,6,6,-tetra+methylpiperidino-oxyl (TEMPOL) (*Figure 3*). They may be designed to partition into specific environments such as hydrophobic or soluble regions of the cell.

Spin labels

Spin labels are available commercially with functional groups to attach to different types of amino acid residue, such as *N*-ethyl maleimide for cysteines. If there are several residues of this type in the protein, the binding may be unpredictable, usually nonspecifically to multiple binding sites. However a recent development is to use genetic engineering to introduce cysteines into unique sites in a protein such as bacteriorhodopsin, to which a spin label is attached (38). Such labels can provide specific structural information, for example, an increase in the linewidth of the spin-label signal indicates the presence of O_2 in the hydrophobic medium, and the distance of the label from the aqueous medium can be determined using water-soluble relaxation probes.

Molecular dipstick

This approach has been used (39) to determine the depth of a binding site from the surface of a protein, such as an antibody. A series of spin labels was synthesized, with a nitroxide at one end

Figure 3. Examples of EPR spectra of free radicals. (a) The nitroxide spin label (TEMPOL) in ethanol, and (b) in glycerol, at 25°C. Note the splitting of the nitroxide radical into three lines by hyperfine interaction with the ^{14}N nucleus ($I = 1$), and the broadening of (b) due to slower motion in the more viscous solvent. (c) Tyrosine cation radical in Photosystem II of spinach chloroplasts. The nuclei mainly responsible for the hyperfine couplings (three lines due to ^{14}N, two lines due to ^{1}H) are circled.

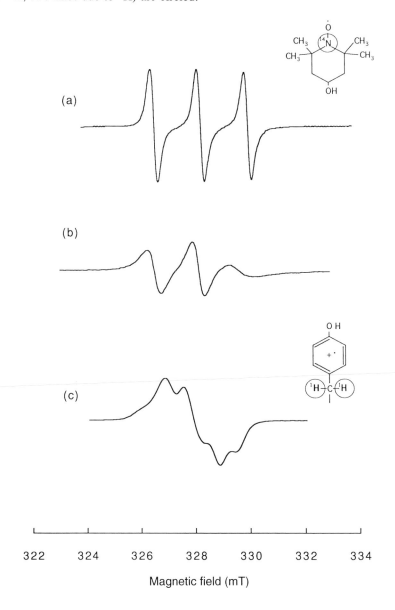

Magnetic field (mT)

and a suitable binding group such as an antigen at the other, connected through a hydrocarbon chain of variable length. When the length is sufficient to reach the surface of the protein, the nitroxide is more mobile in solution and the EPR signal is narrow (39).

Table 5. Applications of spin labels to studies of proteins

Application	Example	Notes	Ref.
Distance of paramagnet from surface of protein or membrane	Cytochrome bc_1 complex of the mitochondrial system	Uses paramagnetic relaxation probes such as Cr^{3+} or Dy^{3+}	9
Motion of spin-labeled protein in solution	Hemoglobin	Assumes label is rigidly attached to protein	40
Librational motion of protein side-chains	Myosin headgroups	Anisotropy of motion can be calculated	41
Saturation transfer EPR— slow motion of protein	Protein motion in membranes	Spectra are interpreted by computer simulation	42
Spin–spin interaction with other spin labels	Rhodopsin in retinal membranes	Interaction of spin labels attached to protein and lipid molecules	43
Structure of membrane proteins	Bacteriorhodopsin	Depth of protein in membrane; interactions with spin-labeled lipids	44
Polarity of environment around spin label	Membrane-bound proteins	The hyperfine splitting a by the ^{14}N nucleus is 1.72 mT in water, and decreases in an apolar environment, for example 1.39 mT in decane	6
Molecular dipstick— determination of distance of binding site from solution	Antibody binding sites	Requires custom synthesis of spin labels	39

3.4. Measurement of rates of motion

Whole protein motion

Molecular motion is usually described by a *correlation time*, given the symbol τ_c. Roughly speaking, τ_c may be considered as the time (in seconds) required for a molecule to rotate through an angle of one radian, or move laterally by one molecular radius. EPR spectra of spin labels are sensitive to τ_c in the range $\approx 10^{-7}–10^{-11}$ s. The lines are narrowest for a nitroxide in free solution ($\approx 10^{-11}$ s, the rapid motion limit), and broaden as the motion becomes slower. At $\tau_c > 10^{-7}$ s the spectrum is indistinguishable from a solid sample (rigid limit). By means of an experimental technique known as *saturation-transfer EPR*, the spectra may be made to be sensitive to correlation times up to 10^{-3} s, as found, for example in membrane-bound proteins.

Often the motion of a spin label attached to an amino acid residue is not characteristic of the protein as a whole, but of the local region or side-chain. Such librational (or oscillatory) motion gives rise to narrower EPR linewidths. The rate and anisotropy of motion of spin labels may be calculated by computer simulations of the spectra (45).

3.5. Interactions between paramagnetic centers and between macromolecules

As previously noted, the spin–spin interaction between paramagnets causes splitting, broadening or relaxation effects on EPR spectra of proteins which may be detected either in solution or the frozen state. Broadening effects may be observed over distances of 0.5–2.0 nm, depending on the shape of the spectra. The effects on relaxation may be detected at even greater distances, up to 4 nm in favorable cases (46). The relaxation enhancement may be observed in power saturation measurements on a conventional EPR spectrometer. Estimates of the distances involved are semi-quantitative. More precise estimates may be obtained from pulsed EPR spectroscopy, which can give values of the spin-lattice relaxation rate T_1 and the spin–spin relaxation rate T_2. In the case of solid-state spectra, precise distance estimates have been obtained from computer simulation of the spectral lineshapes (47). For membrane proteins, lipid–protein interactions may be observed by the effects of spin–spin interactions between spin-labeled lipids and spin labels attached to proteins.

4. REFERENCES

1. Weil, J.A., Bolton, J.R. and Wertz, J.E. (1994) *Electron Paramagnetic Resonance: Elementary Theory and Practical Applications.* John Wiley & Sons, New York.

2. Atherton, N.M. (1993) *Principles of Electron Spin Resonance.* Ellis Horwood, Chichester.

3. Knowles, P.F., Marsh, D. and Rattle, H.W.E (1976) *Magnetic Resonance of Biomolecules.* John Wiley & Sons, London.

4. Swartz, H.M., Bolton, J.R. and Borg, D.C. (1972) *Biological Applications of Electron Spin Resonance.* Wiley-Interscience, New York.

5. Cammack, R. (1993) in *Methods in Molecular Biology, Vol. 17: Spectroscopic Methods and Analyses* (C. Jones, B. Mulloy and A. Thomas, eds). Humana Press, Totowa, NJ, p. 327.

6. Berliner, L.J. (1978) *Methods Enzymol.*, **69**, 418.

7. Maret, W. and Zeppezauer, M. (1988) *Methods Enzymol.*, **158**, 79.

8. Feuerstein, J., Kalbitzer, H.R., John, J., Goody, R.S. and Wittinghofer, A. (1987) *Eur. J. Biochem.*, **162**, 49.

9. Ohnishi, T., Schagger, H., Meinhardt, S.W., LoBrutto, R., Link, T. and von Jagow, G. (1989) *J. Biol. Chem.*, **264**, 735.

10. Blumberg, W.E. and Peisach, J. (1971) in *Bioinorganic Chemistry* (R. Dessy, J. Willard and L. Taylor, eds), Advances in Chemistry Series, Vol. 100. American Chemical Society, Washington, DC, p. 271.

11. Orme-Johnson, W.H. and Sands, R.H. (1973) in *Iron–Sulfur Proteins*, Vol. 2 (W. Lovenberg, ed.). Academic Press, London, p. 195.

12. Henry, Y. and Banerjee, R. (1973) *J. Mol. Biol.*, **73**, 469.

13. Hori, H., Ikeda-Saito, M., Froncisz, W. and Yonetani, T. (1983) *J. Biol. Chem.*, **258**, 12368.

14. Solomon, E.I., Baldwin, M.J. and Lowery, M.D. (1992) *Chem. Rev.*, **92**, 521.

15. Hille, R. and Massey, V. (1985) in *Molybdenum Enzymes* (T. Spiro, ed.). John Wiley & Sons, Chichester, p. 443.

16. Massey, V. and Palmer, G. (1966) *Biochemistry*, **5**, 3181.

17. Prince, R.C. and George, G.N. (1990) *Trends Biochem. Sci.*, **15**, 170.

18. Gordy, W. and Kurita, Y. (1960) *J. Chem. Phys.*, **34**, 282.

19. Henrikson, T. (1969) in *Solid State Biophysics* (S. J. Wyard, ed.). McGraw-Hill, New York, p. 203.

20. Davies, M.J., Gilbert, B.C. and Haywood, R.M. (1991) *Free Radical Research Communications*, **15**, 111.

21. Kelman, D.J., DeGray, J.A. and Mason, R.P. (1994) *J. Biol. Chem.*, **269**, 7458.

22. van Meighem, F.J.E., Satoh, K. and Rutherford, A.W. (1991) *Biochim. Biophys. Acta*, **1058**, 379.

23. Barry, B.A. and Babcock, G.T. (1987) *Proc. Natl Acad. Sci. USA*, **84**, 7099.

24. Li, A.S.W., Cummings, K.B., Roethling, H.P., Buettner, G.R. and Chignell, C.F. (1988) *J. Magn. Reson.*, **79**, 140.

25. Cammack, R. and Shergill, J.K. (1995) in *Enzymology Labfax* (P. Engel, ed.). BIOS Scientific Publishers, Oxford/Academic Press, San Diego.

26. Poole, C.P. Jr (1983) *Electron spin resonance (2nd Edn)*. John Wiley & Sons, New York.

27. DeRose, V.J. and Hoffman, B.M. (1995) *Methods Enzymol.*, **246**, 554.

28. Pilbrow, J.R. (1990) *Transition Ion Electron Paramagnetic Resonance*. Oxford University Press, Oxford.

29. Cammack, R. and Cooper, C.E. (1993) *Methods Enzymol.*, **227**. 353.

30. Hoff, A.J. (ed.) (1989) *Advanced EPR: Applications in Biology and Biochemistry*. Elsevier, Amsterdam.

31. Berliner, L.J., Reuben, J. (eds) (1993) in *EMR of Paramagnetic Molecules, Biological Magnetic Resonance*, Vol. 13. Plenum Press, New York.

32. Eaton, G.R., Eaton, S.S. and Ohno, K. (1991) *EPR Imaging and in Vivo EPR*. CRC Press, Boca Raton, FL.

33. Dikanov, S.A. and Tsvetkov, Y.D. (1992) *Electron Spin Echo Envelope Modulation (ESEEM) Spectroscopy*. CRC Press, Boca Raton, FL.

34. Cammack, R. (1995) in *Bioenergetics: a Practical Approach* (G.C. Brown and C.E. Cooper, eds). IRL Press, Oxford, p. 85.

35. Kastrau, D.H.W., Heiss, B., Kroneck, P.M.H. and Zumft, W.G. (1994) *Eur. J. Biochem.*, **222**, 293.

36. Hirsh, D.J., Beck, W.F., Lynch, J.B., Que, L. and Brudvig, G.W. (1992) *J. Am. Chem. Soc.*, **114**, 7475.

37. Esmann, M., Sar, P.C., Hideg, K. and Marsh, D. (1993) *Anal. Biochem.*, **213**, 336.

38. Farahbakhsh, Z.T., Hideg, K. and Hubbell, W.L. (1993) *Science*, **262**, 1416.

39. Sutton, B.J., Gettins, P., Givol, D., Marsh, D., Wain-Hobson, S., Willan, K.J. and Dwek, R.A. (1977) *Biochem J.*, **165**, 177.

40. Thomas, D.D., Dalton, L.R. and Hyde, J.S. (1976) *J. Chem. Phys.*, **65**, 3006.

41. Barnett, V.A. and Thomas, D.D. (1989) *Biophys J.*, **56**, 517.

42. Birmachu, W., Voss, J.C., Louis, C.F. and Thomas, D.D. (1993) *Biochemistry*, **32**, 9445.

43. Marsh, D. and Horváth, L.I. (1989) in *Advanced EPR: Applications in Biology and Biochemistry* (A.J. Hoff, ed.). Elsevier, Amsterdam, p. 707.

44. Millhauser, G.L. (1992) *Trends Biochem. Sci.*, **17**, 448.

45. Schneider, D.J. and Freed, J. (1989) in *Biological Magnetic Resonance*, Vol. 8 (L.J. Berliner and J. Reuben, eds). Plenum Press, New York.

46. Eaton, G.R. and Eaton, S.S. (1988) *Accts Chem. Res.*, **21**, 107.

47. Guigliarelli, B., Guillaussier, J., More, C., Setif, P., Bottin, H. and Bertrand, P. (1993) *J. Biol. Chem.*, **268**, 900.

CHAPTER 21
PROTEIN STABILITY
A. Giletto and C.N. Pace

1. INTRODUCTION

Most small globular proteins closely approach a two state unfolding mechanism F⇌U, where only the folded (F) and unfolded (U) states are substantially populated at equilibrium. Protein conformational stability is then defined as the difference in free energy (ΔG) between these two states:

$$\Delta G = -RT \ln K, \qquad \text{Equation 1}$$

where K is the equilibrium constant of unfolding or denaturation, R is the gas constant (1.98 cal (8.31 J) mol^{-1} deg^{-1}) and T is the absolute temperature.

2. MEASURING PROTEIN STABILITY

The main purpose of this chapter is provide practical information for measuring protein stability.

2.1. Selecting a technique to monitor unfolding
Table 1 provides a list of the most common techniques used to monitor unfolding, with comments and a few references where they were used. Selection of a particular technique depends on a variety of considerations some of which are listed in *Table 2*. Other techniques to monitor unfolding, such as nuclear magnetic resonance (NMR), viscosity and enzyme activity measurements have been omitted because they are rarely used.

2.2. Determining an unfolding curve
Once a method to monitor unfolding has been chosen, the observable, y, is measured as a function of concentration of chemical denaturant or temperature.

Chemical denaturation
The most common chemical denaturants are urea and guanidine hydrochloride (GdnHCl). *Table 3* lists some physical properties useful for preparing stock urea and GdnHCl solutions. Frequently, about 30 samples are prepared, each with a different concentration of denaturant and the same amount of protein. Once the samples have reached equilibrium with respect to unfolding, the observable, y, is measured for each. (See refs 14 and 20 for a description of the chemical denaturation of proteins.)

Thermal denaturation
Many spectrophotometers can be fitted with a thermostatic cell holder. This allows the temperature of the holder to be varied and the observable measured as a function of temperature. (See ref. 20 for a technical description of the thermal denaturation of proteins.)

Table 1. Techniques used to monitor protein unfolding

Technique	Comments	Refs
Fluorescence	Reflects the difference in the environment of Trp and Tyr residues in the folded and unfolded states. Best for chemical denaturation studies. Particularly useful if the protein contains a buried Trp or Tyr which becomes exposed during unfolding. Generally requires microgram quantities of protein	RNase T1 (1), Barnase (2), Barstar (3), *trp* repressor (4), chymotrypsin inhibitor 2 (5)
UV Absorbance	Similar to fluorescence in that it reflects the difference in the environment of the Trp and Tyr residues in the folded and unfolded states. Best for thermal denaturation studies. Generally requires milligram quantities of protein	Myoglobin (6), RNase T1 (1)
Circular dichroism (CD)	Reflects the difference in secondary structure (α helix and/or β sheet) content in the folded and unfolded states. Good for both chemical and thermal denaturation studies. Generally requires milligram quantities of protein	Turkey ovomucoid third domain (7), actin (8), interleukin 4 (9), RNase T1 (1), HPr (10)
Differential scanning calorimetry (DSC)	Measures the excess heat capacity of a protein solution as a function of temperature. The melting temperature, T_m, the calorimetric and van't Hoff enthalpies, ΔH_{cal} and ΔH_{vH}, and the change in heat capacity, ΔC_p, can be obtained from a single experiment. Thus, information about the stability of the protein and about the applicability of a two-state folding mechanism can be obtained. Generally requires milligram quantities of protein	RNase T1 (11), transferrin (12), *Staphylococcal* nuclease (13), chymotrypsin inhibitor 2 (5)

Table 2. Criteria for selecting a technique to monitor unfolding

(i)	Amount of protein available
(ii)	Minimum amount of protein required to generate a usable spectroscopic signal
(iii)	Magnitude of the difference in signal between the folded and unfolded states
(iv)	Signal-to-noise (S/N) ratio

Table 3. Urea and GdnHCl solutions (14)

Property	Urea	GdnHCl
Molecular mass	60.056 Da	95.533 Da
Solubility (25°C)	10.49 M	8.54 M
d/d_0[a]	$1+0.2658\ W+0.0330\ W^2$	$1+0.2710\ W+0.0330\ W^2$
Molarity[b]	$117.66(\Delta N)+29.753(\Delta N)^2$ $+185.56(\Delta N)^3$	$57.147(\Delta N)+38.68(\Delta N)^2$ $-91.60(\Delta N)^3$
Activity[c]	$0.9815(M)-0.02978(M)^2$ $+0.00308(M)^3$	$0.6761(M)-0.1468(M)^2$ $+0.02475(M)^3+0.00132(M)^4$
Grams of denaturant per gram of water to prepare		
6 M	0.495	1.009
8 M	0.755	1.816
10 M	1.103	–

[a] W is the weight fraction denaturant in the solution, d is the density of the solution, and d_0 is the density of water (15).
[b] ΔN is the difference between the refractive index of the denaturant solution and water at the sodium D line. The equation for urea solutions is based on the data in reference (16); and the equation for GdnHCl solutions is from reference (17).
[c] The equation for urea is based on the data from reference (18). The equation for GdnHCl gives the mean ion activity and fits the unpublished data of E.P.K. Hade above 0.5 M with an average deviation of 0.007 and the data from reference (19) at 0.5 M and below with an average deviation of 0.006.

2.3. Equilibrium and reversibility

Since thermodynamic parameters are being measured, it is essential that equilibrium is reached at each point in the unfolding curve. In general, the nearer the midpoint of the transition, the longer it takes to reach equilibrium. Also, unfolding must be reversible. (See ref. 20 for a complete discussion of equilibrium and reversibility.)

2.4. Data analysis

Figures 1 and *2* are typical unfolding curves of RNase T1. The curves can be divided into three regions as described in *Table 4*. The fraction of unfolded protein (f_u) can be determined at any point in the transition region using

$$f_u = \frac{(y_f - y)}{(y_f - y_u)},$$

Equation 2

Figure 1. Urea denaturation of RNase Tl monitored by fluorescence. The excitation wavelength was 278 nm and the emission wavelength was 320 nm. The data were fitted to Equation 5 in the text.

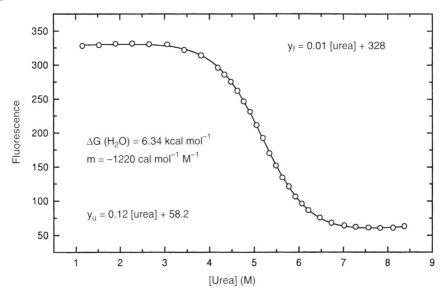

and the equilibrium constant, K, can be calculated using

$$K = \frac{f_u}{(1 - f_u)} = \frac{(y_f - y)}{(y - y_u)},$$

Equation 3

which allows the calculation of ΔG using Equation 1.

Specific analysis of chemical denaturation data

The free energy of unfolding in the absence of urea, $\Delta G(H_2O)$, can then be determined by assuming a linear extrapolation model (21–23) and fitting the data to

$$\Delta G = \Delta G(H_2O) - m[\text{denaturant}],$$

Equation 4

where m is the dependence of ΔG on the denaturant concentration and $\Delta G(H_2O)$ is the free energy of unfolding at 0 M denaturant. Other extrapolation methods, such as the Tanford model (22,24) and the binding model (22,25,26) can be used, but they will not be discussed here.

If a nonlinear least squares fitting software package is available, such as Origin, Kaleidoscope or Enzfitter, then Equations 2, 3, and 4 can be combined and the raw data fit directly to the following six parameter equation:

$$y = b_f + m_f[D] + (b_u + m_u[D]) \frac{e^{\left(\frac{-\Delta G(H_2O)}{RT} - \frac{m[D]}{RT}\right)}}{1 + e^{\left(\frac{-\Delta G(H_2O)}{RT} - \frac{m[D]}{RT}\right)}},$$

Equation 5

where b_f and m_f are the intercept and slope of the pre-transition baseline (folded protein), b_u and m_u are the intercept and slope of the post-transition baseline (unfolded protein), and $\Delta G(H_2O)$ and m are defined by Equation 4.

Figure 2. Thermal denaturation of RNase T1 monitored by CD at 244 nm. The data were fitted to Equation 6 in the text.

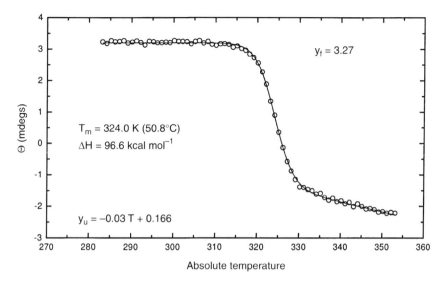

Table 4. Regions of an unfolding curve

Region	Comments
Pre-transition	where $y = y_f$. y_f is the y value for folded protein
Transition	where $y = f_f(y_f) + f_u(y_u)$. y_u is the y value for unfolded protein and f_f and f_u are the fractions of folded and unfolded protein respectively ($f_f + f_u = 1$)
Post-transition	where $y = y_u$

Specific analysis of thermal denaturation data
The raw data from *Figure 2* can be fit to a six parameter equation analogous to Equation 5:

$$y = b_f + m_f T + (b_u + m_u T)\frac{e^{\left(\frac{-\Delta H_m \left(1 - \frac{T}{T_m}\right)}{RT}\right)}}{1 + e^{\left(\frac{-\Delta H_m \left(1 - \frac{T}{T_m}\right)}{RT}\right)}} \qquad \text{Equation 6}$$

where T_m and ΔH_m are the melting temperature and the enthalpy at T_m, respectively. Unlike chemical denaturation, ΔG is not linear with respect to temperature because there exists a difference in heat capacity, ΔC_p, between folded and unfolded protein. Some methods to measure ΔC_p are listed in *Table 5*. Once ΔC_p, T_m and ΔH_m are known, ΔG can be calculated at any temperature using the Gibbs–Helmholtz equation:

$$\Delta G(T) = \Delta H_m \left(1 - \frac{T}{T_m}\right) - \Delta C_p \left[(T_m - T) + T \ln\left(\frac{T}{T_m}\right)\right]. \qquad \text{Equation 7}$$

Protein Stability

Table 5. Methods used to measure ΔC_p

Method	Comments
van't Hoff analysis of thermal unfolding curve	Requires a very accurate thermal denaturation curve because ΔC_p is the second derivative with respect to temperature (27–29). This method is only useful with certain proteins
Measure ΔH_m and T_m with thermal unfolding curves versus pH	Thermal denaturation curves are measured at different pHs and ΔH_m values are plotted versus T_m values. The slope of the line is ΔC_p (23,30–33). This method may be difficult if the thermal denaturation is not reversible at all pHs, particularly near the pI of the protein. This method assumes that ΔH_m and ΔC_p are independent of pH (23,30,31)
Measure ΔH_m and T_m from thermal unfolding curves versus denaturant concentration	Thermal denaturation curves are measured in the presence of different amounts of denaturant and ΔH_m values are plotted versus T_m values. The slope of the line is ΔC_p. Corrections can be made for the enthalpy of binding of denaturant (10,34)
Measure ΔH_m and T_m from a thermal unfolding curve and $\Delta G(H_2O)$ values from a few urea denaturation curves at different temperatures	ΔH_m and T_m are determined from a single thermal denaturation curve and chemical denaturations are performed at different temperatures outside of the thermal denaturation range. ΔG is then plotted versus temperature and the value of ΔC_p is determined that gives the best fit of the data to the Gibbs–Helmholtz equation (10,35)
Differential scanning calorimetry (DSC)	ΔC_p is obtained directly by measuring the difference between the pre- and post-transition baselines extrapolated to T_m (5,11–13,36)

3. THE CONFORMATIONAL STABILITY OF GLOBULAR PROTEINS

Analysis of the denaturation data has allowed $\Delta G(H_2O)$ values for a number of proteins, such as RNase T1, bovine pancreatic trypsin inhibitor, dihydrofolate reductase and T4 lysozyme, to be measured. The values obtained are surprisingly low (between 5 and 15 kcal mol^{-1} (21 and 63 kJ mol^{-1})) (37). It has been suggested that the relatively low stability of the folded conformation of most globular proteins could be an important factor contributing to their turnover in the cell, or could provide the basis of flexibility in their function.

4. REFERENCES

1. Thomson, J.A., Shirley, B.A., Grimsley, G.R. and Pace, C.N. (1988) *J. Biol. Chem.*, **264**, 11614.

2. Pace, C.N., Laurents, D.V. and Erikson, R.E. (1992) *Biochemistry*, **31**, 2728.

3. Khurana, R. and Udgaonkar, J.B. (1994) *Biochemistry*, **33**, 106.

4. Fernando, T. and Royer, C.A. (1992) *Biochemistry*, **31**, 6683.

5. Jackson, S.E., Moracci, M., elMasry, N., Johnson, C.M. and Fersht, A.R. (1993) *Biochemistry*, **32**, 11259.

6. Pace, C.N. and Vanderburg, K.E. (1979) *Biochemistry*, **18**, 288.

7. Swint, L. and Robertson, A.D. (1993) *Protein Science*, **2**, 2037.

8. Bertazzon, A., Tian, G.H., Lamblin, A. and Tsong, T.Y. (1990) *Biochemistry*, **29**, 291.

9. Windsor, W.T., Syto, R., Le, H.V. and Trotta, P.P. (1991) *Biochemistry*, **30**, 1259.

10. Scholtz, J.M. (1995) *Protein Science*, **4** 35.

11. Hu, C.Q., Sturtevant, J.M., Thomson, J.A., Erikson, R.E. and Pace, C.N. (1992) *Biochemistry*, **31**, 4876.

12. Lin, L.N., Mason, A.B., Woodworth, R.C. and Brandts, J.F. (1994) *Biochemistry*, **33**, 1881.

13. Carra, J.H., Anderson, E.A. and Privalov, P.L. (1994) *Protein Science*, **3**, 944.

14. Pace, C.N. (1986) *Methods Enzymol.*, **131**, 266.

15. Kawahara, K. and Tanford, C. (1966) *J. Biol. Chem.*, **241**, 3228.

16. Warren, J.R. and Gordon, J.A. (1966) *J. Phys. Chem.*, **70**, 297.

17. Nozaki, Y. (1972) *Methods Enzymol.*, **26**, 43.

18. Bower, V.E. and Robinson, R.A. (1963) *J. Phys. Chem.*, **67**, 1524.

19. Bonner, O.D. (1976) *J. Chem. Thermodyn.*, **8**, 1167.

20. Pace, C.N., Shirley, B.A. and Thomson, J.A. (1989) in *Protein Structure: A Practical Approach* (T.E. Creighton, ed.). IRL Press, Oxford, p. 311.

21. Greene, F.R. and Pace, C.N. (1974) *J. Biol. Chem.*, **249**, 5388.

22. Ahmad, F. and Bigelow, C.C. (1982) *J. Biol. Chem.*, **257**, 12935.

23. Becktel, W.J. and Schellman, J.A. (1987) *Biopolymers*, **26**, 1859.

24. Tanford, C. (1970) *Adv. Prot. Chem.*, **24**, 1.

25. Prakash, V., Loucheux, C., Scheutele, S., Gorbunoff, M.J. and Timasheff, S.N. (1981) *Arch. Biochem. Biophys.*, **210**, 455.

26. Lee, J.C. and Timasheff, S.N. (1974) *Biochemistry*, **13**, 257.

27. Pace, C.N. and Tanford, C. (1968) *Biochemistry*, **7**, 198.

28. Nojima, H., Ikai, A., Oshima, T. and Noda, H. (1977) *J. Mol. Biol.*, **116**, 429.

29. Alexander, S.S. and Pace, C.N. (1971) *Biochemistry*, **10**, 2738.

30. Shiao, D.F., Lumry, R. and Fahey, J. (1971) *J. Am. Chem. Soc.*, **93**, 2024.

31. Privalov, P.L. and Khechinashvili, N.N. (1974) *J. Mol. Biol.*, **86**, 665.

32. Shortle, D., Meeker, A.K. and Freire, E. (1988) *Biochemistry*, **27**, 4761.

33. Privalov, P.L. (1979) *Adv. Prot. Chem.*, **33**, 167.

34. Chen, B. and Schellman, J.A. (1989) *Biochemistry*, **28**, 685.

35. Pace, C.N. and Laurents, D.V. (1989) *Biochemistry*, **28**, 2520.

36. Sturtevant, J.M. (1987) *Annu. Rev. Phys. Chem.*, **38**, 463.

37. Pace, C.N. (1990) *Trends Biochem. Sci.*, **15**, 14.

Protein Stability

CHAPTER 22
COMPUTER ANALYSIS OF PROTEIN STRUCTURE

E.J. Milner-White

1. GENERAL BIOCOMPUTING

1.1. Hardware and software requirements

The aim of this chapter is to describe how protein structures and sequences can be examined by computer. Mostly, all that is needed is a reasonably powerful PC or Macintosh connected to the outside world. Recent improvements in hardware mean that protein three-dimensional (3D) structures can be examined effectively on such machines. However, UNIX-based workstations will also be covered.

A minimal set of software might only consist of a molecular graphics package and a World Wide Web (WWW) browser. Other programs can be accessed on remote computers via WWW. There are now many molecular graphics packages such as 'O', Molscript, Setor, PROMOD, Interchem, DTMM, Imdad, Kinemage, MolView and RASMOL that are low cost or free of charge for academic users at least. One of these, RASMOL, is described in Section 2.2 because it is especially useful and easy to use.

1.2. World wide web

The WWW browsers I use are called Mosaic and Netscape. They are available free of charge via anonymous ftp from *ftp.ncsa.uiuc.edu* (ncsa is the National Center for Supercomputer Applications). Versions can be obtained for UNIX X-windows, PCs or Macintoshes. It is useful to keep a hotlist or bookmark list of sites that are especially valuable for molecular biology. Below are some of the key items in our hotlist, together with their URLs (uniform resource locator, the name used for WWW addresses). Several, for example, Pedro's Biomolecular Research Tools, the EBI (European Biomolecular Institute at Cambridge), BNL (Brookhaven National Laboratory), ExPASy (University of Geneva), NCBI (US National Center for Biotechnology Information), GenBank or SEQNET provide compilations of available databases, which are good places to start.

1.3. Gopher

WWW uses a hypertext graphical interface which can be somewhat slow. Gopher provides comparable facilities for searching and downloading files. It works much the same way except for the graphical interface. If the letters 'http' and 'www' are replaced by 'gopher' in *Table 1* the WWW browsers Mosaic and Netscape access gopher sites. Further information about both WWW and Gopher may be obtained from ref. 1.

1.4. ftp

File transfer protocol (ftp) is the standard means of transferring files between computers. To access another computer a user identification and password are normally needed, but certain centrally placed computers on the network are set up so that they can be accessed by anyone via

Table 1. Some key URLs

Pedro's Biomolecular Research Tools	*http://www.fmi.ch/biology/research_tools.html*
As above, USA site	*http://www.public.iastate.edu/~pedro/research_tools.html*
As above, Japanese site	*http://www.peri.co.jp/Pedro/research_tools.html*
EBI at Cambridge, UK	*http://www.ebi.ac.uk/*
PDB at Brookhaven (BNL), USA	*http://www.pdb.bnl.gov/*
ExPASy at University of Geneva	*http://expasy.hcuge.ch*
NCBI, GenBank, USA	*http://www.ncbi.nlm.nih.gov/*
SEQNET, Daresbury, UK	*http://www.dl.ac.uk/SEQNET/home.html*

what is called anonymous ftp. As user identification the word anonymous is typed and for the password the user's home e-mail address is given. On the Macintosh there is an application called fetch that makes ftp straightforward to use, but Mosaic and Netscape also allow anonymous ftp access to appropriate sites by replacing the letters 'http' and 'www' in *Table 1* with 'ftp'.

1.5. Telnet

An emulator program allows users to log in, as opposed to simply looking at files or transferring them from remote computers. Telnet is the one most widely used. It is often easier to run programs in this way on distant computers than to set the program to work on the local computer. To access a computer remotely as a normal user, an account with a user identification and password is needed. However, some WWW servers are set up so that users accessing them via WWW browsers can run certain selected programs; for example, the sequence analysis program BLAST can be run over the WWW at GenBank.

1.6. Network news

To keep up-to-date in the field it can be helpful to belong to a user group and exchange information with those working in the same area. To do this the computer has to have a newsreader. The one we use is called Network News. A news group relevant to the area under discussion is *bionet.molbio.proteins*. This topic is reviewed in refs 2 and 3. Netscape has a Network News reader built in so a separate one is not needed.

2. HOW TO ACCESS, VIEW AND ANALYZE PROTEIN 3D STRUCTURES

2.1. Finding and transferring protein structures

Protein and nucleic acid 3D structures are stored as text files. The database containing them is called the protein databank (PDB), or, sometimes, the Brookhaven protein databank. Each entry in this databank has the data for a biological macromolecule or its complex. *Table 2* shows a small section of the protein databank entry for serine 41 of the A chain of an adenylate kinase from *Escherichia coli*. Each line corresponds to a nonhydrogen atom; there are four main chain atoms (N, CA (α-carbon), C and O) and two side-chain atoms (CB (β-carbon) and OG (γ-oxygen)). The x,y,z coordinates give each atom's position in space. To examine coordinates, or to alter them, the PDB files can be opened in a word processing application, or other text editor. Coordinate data for proteins or nucleic acids are indicated by the word ATOM at the left hand side. It is necessary to scroll some way before it is reached. Ligand data are listed after this, distinguished by the word HETATM at the left hand side. Fairly soon the format of PDB files is to change to a new format called CIF (Crystallographic Information File); however the essentials are likely to be fairly similar.

Table 2. A small section of a PDB file

	aa		resno	x(Å)	y(Å)	z(Å)		B	code no.	
ATOM 295	N	SER	A 41	37.272	52.721	11.806	1.00	30.04	1AKE	449
ATOM 296	CA	SER	A 41	38.507	53.491	11.905	1.00	33.63	1AKE	450
ATOM 297	C	SER	A 41	39.777	52.736	11.541	1.00	34.75	1AKE	451
ATOM 298	O	SER	A 41	40.837	53.367	11.412	1.00	43.02	1AKE	452
ATOM 299	CB	SER	A 41	38.784	53.990	13.294	1.00	31.94	1AKE	453
ATOM 300	OG	SER	A 41	37.700	54.780	13.703	1.00	53.72	1AKE	454

The residue number from the PDB file (41 in *Table 2*) is stored and used as the residue label by most molecular graphics programs. In many proteins there are gaps, so certain residues are not recognized by the program. Sometimes there are also residues labeled, say 41A, to allow for consistent labeling of aligned residues. Some programs are fooled by this.

It is sometimes useful to edit the PDB file before viewing it. Although most molecular graphics programs allow the user to display only part of a file, there are several reasons for doing this before entering the molecular graphics program, for example when displaying a subunit, an active site or a domain. Editing is done simply by editing the PDB file (on a Macintosh or a PC, WORD might be used) and deleting ATOM and HETATM lines not required.

For proteins with multiple subunits related by symmetry axes, often only one set of coordinates is given and the others have to be generated by rotation about the axis. Often, though not always, the coordinate axes are coincident with the symmetry axes, which makes generation of other subunits easier. Examination of the information in the first part of the PDB file generally reveals which axes are which. To rotate about a dyad (this is the word for a twofold axis of symmetry), say the z-axis, the z-coordinates remain the same and the signs of the x and y coordinates are reversed.

PDB files can be retrieved via the WWW, using Netscape or Mosaic WWW browsers, either from EBI, Cambridge, UK or from Brookhaven, USA, as outlined below.

Fetching protein coordinates from EBI Cambridge, UK: The URL at Cambridge is *http:// www.ebi.ac.uk/htbin/pdbfetch/* Simply type in the PDB ID in the box (for example *3dfr* for dihydrofolate reductase). When it appears save it as a file on your own filestore.

Fetching protein coordinates from Brookhaven, USA: The URL at Brookhaven is *http:// www.nih.gov/* Go to Gopher server. Choose 'Search by entry ID Only'. Type the PDB ID in the box (for example *3dfr*). Double click on the full entry.

PDB codes mostly consist of a sequence such as *3dfr*, which means version 3 of dihydrofolate reductase. To help find the code number for a given protein database servers may have a file giving full names of proteins which can be listed and queried. The easiest way to do this is to use the Xwindows-based facility PDB Browser (4), accessible at *http://www.pdb.bnl.gov/cgi-bin/ browse*; alternatively the browser itself may be installed by the user via anonymous ftp to *ftp.pdb.bnl.gov*. PDB Browser uses RASMOL (see the next section) if displaying molecules. Proteins are accessed by categories, by authors and so on.

2.2. Viewing proteins

To obtain information about molecular graphics software the URL *http://www.nih.gov/*

Table 3. RASMOL instructions

To start up
    (a) Double click the icon RasMac 2.5 [a]
    (b) Arrange the two windows that appear so you can use both easily. One is the graphics, showing your molecule, the other is the RasMol command line
    (c) In the FILE menu select OPEN, go to the correct folder and double click on the file desired. It must have the suffix *.pdb*, meaning it is a Protein Databank File
    (d) To replace a protein select CLOSE in the FILE menu and redo step (c)

Graphics display
    Molecules are rotated about any axis within the x,y plane by dragging the cursor in an appropriate direction
    To rotate by large angles in the x,y plane, use the scroll bars
    Zoom molecules by dragging the cursor vertically with the shift key pressed
    Translate molecules by moving the cursor with the command (or option) key pressed
    Moving the cursor with both shift and command (or option) keys pressed down causes rotation about the z-axis
    Selecting GROUP in the COLOR menu is useful in BACKBONE mode as it colors the N-terminus blue and the C-terminus red, with the other residues in between

Useful command line commands
    Pressing the key with the arrow pointing up retrieves the last command typed. If repeated, the one before is retrieved, and so on. Type (on the Command line):

zap	(deletes the current molecule so RasMol is ready for another)
center	(centers the rotation on the midpoint of the whole molecule)
center 26	(centers the rotation point on residue 26)

Displaying hydrogen bonds
    Type (on the Command line):

hbonds on	(mainchain–mainchain or basepair hydrogen bonds only displayed)
hbonds 20	(the same hydrogen bonds made thicker)
hbonds off	(all hydrogen bonds deleted)
set hbonds backbone	(makes hydrogen bonds join α-carbon atoms, useful for BACKBONE displays; for DNA the basepairing is shown in abbreviated form)
set hbonds sidechain	(makes hydrogen bonds join the usual atoms, the default)

Displaying ligand and protein together
    It is often useful, say, to display the protein as BACKBONE (the program author really meant α-carbon), and the ligand as STICKS
    In the DISPLAY menu select BACKBONE, then type (on the Command line):

select ligand	(display the ligand in the mode required)

Selecting and restricting displays
    Select means that subsequent commands only apply to what is selected. Restrict is the same except that only what is chosen is displayed. Select and restrict are reversed by *select all* or *restrict all*. Type (on the Command line):

restrict ligand	(only ligand displayed; best in WIREFRAME or STICKS mode)
restrict protein	(only the protein parts displayed)

Table 3. Continued

restrict nucleic	(only nucleic acid parts displayed)
*restrict *a*	(only chain a displayed)
*restrict **	(everything displayed)
restrict 0	(removes everything displayed, which is often useful)
restrict 20–30	(only residues 20–30 displayed)
select 3–10:a,5:b	(residues 3–10 of subunit a plus residue 5 of subunit b selected)
select his,cys	(then STICKS, handy for adding detail to BACKBONE displays)
restrict within	(displays ligand plus atoms within 0.4 nm (4.0 Å) of any ligand atom;
(4.0,ligand)	useful for displaying active sites)

Labeling atoms
Clicking on an atom produces a list of its attributes in the Command line window. Or:

label on	(selected atoms are labeled as: *residue type and number:chain type.atom*)
label off	(selected atoms are unlabeled)
*select *.ca*	(selects α-carbon atoms only)
label %r	(labels residue numbers of selected atoms)
label %c	(labels chain identifiers of selected atoms)

Displaying metal ions and water molecules

select mg	(selects magnesium ions)
or:	
select water	(not all PDB files include water molecules)

Ions and waters are made visible using the BALL and STICK representation

Coloring
Use the COLOR menu, or:

color blue	(or red, yellow, green, etc.; colors whatever is selected)

Expressions
The terms center, select, restrict and label act on 'expressions'. Examples of expressions are:
protein, ligand, water, 30 (means residue 30), *20-30,32* (means residues 20 to 30 and residue 32),
a .ca (means α-carbon atoms of chain a)

[a] Note added in proof: version 2.6 was released in August 1995.

molecular_modelling/net_services is useful. I shall concentrate on one public domain molecular graphics program called RASMOL that is available for PCs, Macintoshes and UNIX-based computers. It is fast, easy to use and versatile but first time users may not discover all the facilities initially. The instructions in *Table 3* which are for the Macintosh version, are intended to rectify this. RASMOL works best on PowerMacs, but is adequate on LCIIIs. The instructions for PCs or Xwindows/UNIX are the same, but differ in regard to use of the mouse. This is explained in *Table 4*. RASMOL can be obtained by anonymous ftp from *ftp.dcs.ed.ac.uk(129.215.160.5)* in the directory */pub/rasmol*.

2.3. Structural analysis

Having gained an impression of the protein, it is often useful to know the exact positions of α-helices and β-strands. This is derived, usually, from the the hydrogen bond arrangement, but

Table 4. Mouse button and key combinations for RASMOL on PCs and UNIX

	Left button	Right (or middle) button
Normal	Rotation about x and y	Translation about x and y
Shift key	Zoom	Rotation about z

there are a number of uncertainties, so a standard procedure has been created by Kabsch and Sander (5) and the results stored in their Dictionary of Secondary Structure Information in proteins (DSSP) database (a sample of a DSSP File is shown in *Table 5*). The DSSP database contains much further useful information.

Other databases
Another potentially more convenient and general way of finding out information about proteins is to employ a query system such as IDITIS (6). This makes use of a special query language called sql which allows users to make enquiries about protein structure, the details being stored in a special database. IDITIS is available as part of the Oxford Molecular package. Other, even more sophisticated, means of interrogating and storing data about proteins are being developed (7).

2.4. Making evolutionary comparisons

The HSSP database
A question often asked about a protein is whether a given residue or set of residues is conserved in the family of related proteins. This can be answered by reference to the HSSP database (8) (a sample of an HSSP File is shown in *Table 6*). HSSP stands for homology-derived secondary structure of proteins. The sequence of each target protein (usually of known 3D structure) is optimally aligned with all protein sequences that exhibit 30% or more sequence identity (after optimal alignment). The database goes through the residues, one by one, in the target proteins and finds the aligned residues in the partially homologous proteins. From this a parameter, called the variability, is calculated to measure the degree of conservation of each residue.

One possible disadvantage of this database is the way it deals with gaps. Gaps in the partially homologous proteins are shown as dots, which is satisfactory. However, insertions in the partially homologous proteins are not shown at all, which can be misleading.

3D superimposition and the 3D-ALI database
Because core regions of 3D structures are so much better conserved than their sequences, comparisons based on 3D structure are inherently better than those based on sequence; for example, two sequences with 50% or more identical residues after optimal alignment are observed to have virtually identical 3D structures with regard to the main chain part of the protein. Sequences with 25% to 50% identical residues usually have very similar core regions with perhaps a few non-superimposable loops. On the other hand, pairs of sequences with less than 25% identity are in the region where alignment by sequence alone is uncertain. Sequences of related proteins in this low identity region can only be partially aligned, provided the 3D structures are known, by performing 3D superposition of the core region(s).

Superposition, rather than sequence alignment, is the best way to compare 3D structures of distantly related proteins. Algorithms (9) are available for optimal 3D superposition. The effectiveness is measured by a single parameter, the RMS fit (root mean squared fit). After

Table 5. A sample of a DSSP file (the P-loop of the G-protein p21^{ras})

Res	AA	Structure	Acc	Hydrogen bonding					κ	α	φ	ψ			
9	V	E +b 0 80A	0	70,−2.5	72,−2.9	−2,−0.6	2,−0.2	−0.874	13.4	106.1	−126.9	162.5	2.8	25.2	16.2
10	G	–	0	−2,−0.3	72,−0.1	49,−0.3	3,−0.1	−0.776	64.3	−46.2	147.0	169.9	5.2	24.3	19.0
11	A	S> S−	8	70,−0.5	3,−1.5	78,−0.3	5,−0.3	−0.040	72.1	−75.4	−65.4	161.3	6.1	25.1	22.6
12	G	T 3 S+	47	48,−0.8	−1,−0.2	1,−0.2	47,−0.1	−0.257	112.5	11.8	−59.7	136.7	6.4	28.6	24.1
13	G	T 3 S+	59	−3,−0.1	−1,−0.2	1,−0.1	−2,−0.1	0.502	83.6	120.4	80.0	5.3	9.5	30.6	23.1
14	V	S< S−	0	−3,−1.5	70,−0.1	67,−0.1	−2,−0.1	0.672	89.7	−97.6	−76.7	−14.0	10.8	28.3	20.3
15	G	S> S+	15	−4,−0.2	4,−2.3	66,−0.1	5,−0.2	0.659	74.3	142.5	108.5	22.6	10.7	31.2	17.7
16	K	H> S+	13	−5,−0.3	4,−1.9	2,−0.2	5,−0.2	0.946	81.5	40.9	−57.2	−47.9	7.3	30.5	15.9

Each row corresponds to one residue and mostly relates to the main chain atoms only. The following parameters are particularly useful.

Res and AA: the residue number and residue type (single letter code).

Structure: the type of secondary structure is given according to a code specified by the left-hand letter: E: β-strand; H: α-helix; T: turn; S: bend. A dash means none. For the meanings of the symbols to the right, see ref. 5.

Acc: is a measure of the accessibility of the residue to water (10 times the number of water molecules estimated to be in contact with the residue). 0 means entirely buried.

Hydrogen bonding: the top row reveals that the NH of residue 9 forms a hydrogen bond with the CO of a residue 70 ahead (residue 79) in sequence with an energy of −2.5 kcal mol^{-1}, whereas the CO of residue 9 forms a hydrogen bond with the NH of a residue 72 ahead in sequence with an energy of −2.9 kcal mol^{-1}. Negative energies are favorable. Two other hydrogen bonds are listed but they are hardly significant because their energies (−0.6 and −0.2 kcal mol^{-1}) are near to zero. κ: is a measure of the degree to which the main chain bends. It is equal to 180° minus the angle between the α-carbon atoms of residue i−1, i and i+1. For an extended chain the value approaches 0°. A value above 90° indicates a significant bend.

α: is the torsion (or dihedral) angle between the four α-carbon atoms of residues i−1, i, i+1 and i+2.

Computer Analysis

Table 6. A sample of an HSSP file (the P-loop of the G-protein p21^{ras})

Res	AA	Structure				Acc		Var		Homology
9	V	E + b	0	80A	0	175	14			VVVVVVVVVVVVVVVVVVVVVVVVVVVVVVVVVVVMVVVVVVVVVVVMVVVVVVVVVLLLLLLVVVVVVVVVLLLIIIVIMIIIIIVVII
10	G	–		0	0	0	175	0		GGGGGGGGGGGGGGGGGGGGGGGGGGGGGGGGGGGGGGGGGGGGGGGGGGGGGGGGGGGGGGGGGGGGGGGGGGGGG
11	A	S > S–		0	8	0	177	26		AAAAAAAAAAAAAAAAAAADGGGGGGDGGGGGGDGGGPGGSSSSSSSGASSDNDDDDNDD
12	G	T 3 S+		0	47	0	179	28		GGRKGRRCGGGGGVGGGSGGGGGGKGGGGGGGGGGGGSSSSGGSPSSSSS
13	G	T 3 S+		0	59	0	179	19		GGGGGGGGGGGGGGGGGGGGGDGGGGGGGGLGGGSGGAAGHGGGGG
14	V	S < S–		0	0	0	179	11		VVVVVVVVVVVVVVVVVVVVVVVVVVVVVVCVVVVVCVV
15	G	S > S+		0	15	0	181	0		GGGGGGGGGGGGGGGGGGGGGGGGGGGGGGGGGGGGGGGGGG
16	K	H > S+		0	13	0	181	0		KKKKKKKKKKKKKKKKKKKKKKKKKKKKKKKKKKKKKKK

Each row gives data relating to one residue or residue position. The following parameters are particularly useful.
Res and AA: the residue number and amino acid type (single letter code).
Structure: the lefthand letter gives the type of secondary structure: E, β-sheet; H, α-helix, S, bend; T, turn, etc. For the meaning of the symbols to the right, see ref. 5.
Acc: is a measure of the accessibility of the whole residue to water (10 times the number of water molecules estimated to be in contact with the residue). 0 means entirely buried.
Var: the measure of variability for that amino acid position, independent of the number of sequences. Zero means no variation and high values are above 50. Note the values for G, G and K within the consensus GxxxxGK sequence of the well conserved P-loop are all zero.
Homology: a list of the homologous residues. For each of the 70 most homologous protein sequences, the corresponding residue is listed.

optimal superposition of two structures, the pairs of corresponding coordinates are observed to be at certain distances apart (x_1, x_2, x_3, etc.). The RMS fit is calculated by summing the squares of the distances ($x_1^2 + x_2^2 + x_3^2 +$, etc.), taking the average of this sum and calculating the square root of the average. A value of zero for the RMS fit means structures are identical while values for a medium sized protein larger than 0.5 nm might indicate that structures are unrelated.

A database that provides alignments of protein sequences based on 3D superposition is 3D-ALI (10). The number of such sequences is not so large as that for those based on sequence similarity alone at present because there are fewer data, but it is nevertheless a useful resource. Some of the 3D-ALI files have only one protein in them, which initially can be confusing.

2.5. Assigning a protein to categories

The high 3D structure conservation observed in distantly related proteins also means that it is worthwhile to be able to categorize protein domains according to their 3D structure rather than their sequence. The SCOP (structural classification of proteins) database (11, 12) provides a way of telling which superfamily protein domains belong to, according to its known 3D structure. Domains are grouped into categories such as all-α, all-β, α/β and $\alpha + \beta$; these in turn are grouped into successively smaller subcategories. SCOP can be accessed by a WWW browser; the major European URL is *http://scop.mrc-lmb.cam.ac.uk/scop*; that in the USA is *http://ncbi.nlm.nih.gov/repository/scop/index.html* Viewing of the 3D structures themselves using RASMOL, or another viewer, is built into the package.

An alternative database for categorizing proteins in a similar sort of way is PRODOM (13). This database can be searched at the URL *http://www.sanger.ac.uk/~esr/prodom.html*.

2.6. Conformational changes

An important aspect of proteins is the way their conformations can change. Although X-ray crystallography gives a static picture of proteins, the use of crystals produced in different ways has generated much data on proteins with different conformations. This has been assembled into the Protein Motions Database (14) which may be accessed at the URL *http://hyper.stanford.edu/~mbg/ProtMotDB/*.

3. HOW TO RETRIEVE AND INTERPRET PROTEIN SEQUENCES

3.1. Finding sequences

We are concerned with protein, rather than nucleic acid, databases. The three main standard protein sequence databases are listed at the top of *Table 7*. They are standard in the sense that they include the full-length protein products of virtually all genes that have been sequenced. Two other databases are listed in *Table 7*: NRL3D, which contains the sequences of the proteins in the protein databank, allowing them alone to be searched, and PROSITE, a dictionary of sites and patterns in proteins which will be described later.

It is worth being aware that sequence files exist in different formats. Two types are mentioned in *Table 7* and Genetics Computer Group (GCG) files (see later) also have their own formats. A program called READSEQ, which enables sequence files to be interconverted between different formats, is useful. It is available via anonymous ftp at the site *ftp.bio.indiana.edu* in the directory *IUBio-Software+Data/Molbio/readseq*. GCG also has a file format interchange facility.

Table 7. Protein databases

Name	Ref.	Format
PIR or NIH-Protein	15	PIR/NBRF
SWISS-PROT	16	SWISS-PROT
OWL	17	PIR/NBRF
NRL3D	18	PIR/NBRF
PROSITE	19	SWISS-PROT

I shall describe three ways to obtain protein sequence data. The first and easiest is from the Web. Sequences may be retrieved from the URL *http://www.ebi.ac.uk/srs/srsc*, or, indirectly, via the URLs in *Table 1*.

The suite of GCG (20) programs is widely used for sequence analysis. If not present locally, it might be possible to access it on a remote computer. A way to retrieve sequences in GCG with a nongraphical interface is outlined as follows:

To find the sequence of a named protein, type:

stringsearch

You are then asked for a database to search. For the three commonly used *protein* databases, Owl, SWISS-PROT and NIH-protein, type either:

Owl:* or *Sw:** or *Pir:**

You are asked for the text pattern. Type, for example: *'lactate dehydrogenase'*.
You then have to confirm a name for the output file. The output file contains the names of the proteins sequences that match, plus descriptions of them. The names are of two kinds, one of the form Owl:A3b_Bovin, and the other of the form Owl:A32952. The latter is the accession number. The sequence files themselves are placed in the user's current directory or folder by the command:

fetch Owl:A3b_Bovin or *fetch Owl: A32952*.

A simpler alternative to GCG is called GDE, which stands for Genetic Data Environment. This Xwindows package may be obtained from the ftp site *ftp://golgi.harvard.edu/pub/* without charge and allows users to run a comprehensive set of programs for sequence manipulation.

Another useful way to retrieve sequences is via Entrez (produced by NCBI, the US National Center for Biotechnology Information) (21, 22). Entrez, called Nentrez on Macintoshes, conveniently combines a protein sequence database with the bibliographical database Medline. The software is obtained by anonymous ftp to *ncbi.mlm.nih.gov*. With Entrez it is possible to find the relevant references for a particular protein and then to go directly on to find the sequence and also all available sequences that are related. This employs the program BLAST.

3.2. Comparing sequences

Given a novel sequence the first question is often whether similar sequences exist. There are a number of different ways of doing this, all useful for different purposes. In *Table 8* they are arranged roughly in order of increasing sophistication.

Table 8. Some alternative ways of finding matching sequences

FIND	A GCG program. For simple matching
STATSEARCH	A more complex GCG matching program that includes statistical analysis of results
FASTA	The most widely used complex matching program; available within GCG
BLAST	A rapid yet sophisticated search program (23)
PROFILESEARCH	The most sophisticated means of finding matches within GCG; uses data from multiple alignments
BLITZ (or MPsearch)	The most thorough method, makes use of parallel processors (24)

The PROSITE database

Although, in general, sequence is poorly conserved compared to 3D structure, certain short sequences of up to say 30 residues that are often functionally important *are* found to be conserved in distantly related proteins. These consensus sequence motifs are characteristic for a given family and can reveal to which one a newly sequenced protein belongs. PROSITE (19) is a widely used database that stores these motifs. An example of its use is given below. Another database that stores such motifs is PRINTS (25), derived from the OWL database. It complements PROSITE in that it employs multiple sequence motifs to characterize protein families.

Searching the PROSITE database on WWW for the P-loop:
  Open URL: *http://expasy.hcuge.ch/sprot/prosite.html*
  Access to PROSITE: *by description*
  Enter search keywords: *ATP*
    A list of consensus motifs appear with ATP in the name. The one wanted is: *PDOC00017 ATP/GTP binding site motif A (P-loop).* Select it.
    The full entry appears, revealing that the Consensus sequence is: *[AG]-X(4)-G-K-[ST].*
  The entry can be saved to the user's filestore.

3.3. Multiple alignment

The simplest way to optimally align two related sequences is to use the program GAP, within GCG. This can only be done for sequences alone reliably where there is a better than 25% identity (after optimal alignment). Below this value the uncertainty of positioning gaps correctly means that alignment is unreliable.

For multiple alignments of more than two sequences, CLUSTAL (26) and PILEUP, available within GCG, are two widely used programs which are available by anonymous ftp to the URL *ftp://s-ind2.dl.ac.uk/*. The program MSA (27) can be run via the WWW at the URL *http://ibc.wustl.edu/msa.html*. Multiple alignment (28) is a tricky operation as it is necessary first to work out an evolutionary tree to measure the relatedness of the proteins and then to perform the alignment itself leaving the right gaps in the right places.

Databases of multiple alignments are starting to be widely used. One, called ALIGN, that is linked to the PROSITE database may be accessed via the WWW by the URL *ftp://s-ind2.dl.ac.uk/pub/database/prints/align*. At the same site a program called XALIGN, written by D.N. Perkins, is available that allows multiple alignments to be viewed in a number of convenient formats, in an Xwindows environment. Another database of protein sequence alignments called PIRALN is available at the URL *http://www.embl-heidelberg.de/srs/srsc?-info//PIRALN*. A further database called BLOCKS (29) based on PROSITE stores multiple alignments with no gaps that correspond to

Computer Analysis

the most highly conserved regions of proteins; this database is available at the URL *http:/www.blocks.fhcrc.org/*. Finally, a program called AMSA (Analyze Multiple Sequence Alignments; ref. 30) provides a facility for making best use of multiple alignments.

4. ACKNOWLEDGMENT

I wish to thank Wai-Yan Wan for help in the preparation of this chapter.

5. REFERENCES

1. Schatz, B.R. and Hardin, J.B. (1994) *Science,* **265**, 895.

2. Bleasby, A.J., Griffiths, P., Hines, D., Marshall, S., Staniford, L., Hoover, K. and Kristofferson, D. (1993) *Trends Biochem. Sci.,* **18**, 310.

3. Smith, U. (1993) *Trends Biochem. Sci.,* **18**, 398.

4. Peitsch, M.C., Wells, T.N.C., Stampf, D.R. and Sussmann, J.L. (1995) *Trends Biochem. Sci.,* **20**, 82.

5. Kabsch, W. and Sander, C. (1983) *Biopolymers,* **22**, 2577.

6. Thornton, J.M., Gardner, S.P. and Hutchinson, E.G. (1992) in *Computer Modelling of Biological Processes* (J.M. Goodfellow and D.S. Moss, eds). Ellis-Horwood, Chichester, p. 211.

7. Gray, P.M.D., Paton, N.W., Kemp, G.J.L. and Fothergill, J.E. (1990) *Protein Engineering,* **3**, 235.

8. Sander, C. and Schneider, R. (1991) *Proteins,* **9**, 56.

9. Holm, L. and Sander, C. (1994) *Proteins,* **19**, 165.

10. Pascarella, S. and Argos, P. (1992) *Protein Engineering,* **5**, 121.

11. Barton, G.J. (1994) *Trends Biochem. Sci.,* **19**, 554.

12. Murzin, A.G., Brenner, S.E., Hubbard, T. and Chothia, C. (1995) *J. Mol. Biol.,* **247**, 536.

13. Sonnhammer, P. and Kahn, R. (1994) *Protein Science,* **3**, 482.

14. Gerstein, M., Lesk, A.M. and Chothia, C. (1994) *Biochemistry,* **33**, 6739.

15. Sidman, K.F., George, D.G., Barker, W.T. and Hunt, L.T. (1988) *Nucleic Acids Res.,* **16**, 1869.

16. Bairoch, A. and Boekmann, B. (1994) *Nucleic Acids Res.,* **22**, 3578.

17. Bleasby, A.J., Akrigg, D.A. and Attwood, T.K. (1994) *Nucleic Acids Res.,* **22**, 3574.

18. Pattabiraman, N., Lowrey, G.A., Gaber, B. and George, D.G. (1990) *Prot. Seq. Data Anal.,* **3**, 387.

19. Bairoch, A. and Bucher, P. (1994) *Nucleic Acids Res.,* **22**, 3583.

20. Devereux, J., Haeberle, P. and Smithies, O. (1984) *Nucleic Acids Res.,* **12**, 387.

21. Henikoff, S. (1993) *Trends Biochem. Sci.,* **18**, 267.

22. Cockerill, M. (1994) *Trends Biochem. Sci.,* **19**, 94.

23. Altshul, S.F., Gish, W., Miller, W., Myers, E.W. and Lippmann, D.J. (1990) *J. Mol. Biol.,* **215**, 403.

24. Collins, J.F. and Coulson, A.F.W. (1990) *Methods Enzymol.,* **183**, 474.

25. Attwood, T.K., Beck, M.E., Bleasby, A.J. and Parry-Smith, D.J. (1994) *Nucleic Acids Res.,* **22**, 3590.

26. Higgins, D.M. and Sharp, P.M. (1988) *Gene,* **73**, 237.

27. Lipman, D., Altschul, J. and Kececioglu, P. (1989) *Proc. Natl Acad. Sci. USA,* **86**, 4412.

28. Bell, L.H., Coggins, J.R. and Milner-White, E.J. (1993) *Protein Engineering,* **6**, 683.

29. Henikoff, S. and Henikoff, J.G. (1991) *Nucleic Acids Res.,* **19**, 6565.

30. Livingstone, C.D. and Barton, G.J. (1993) *Comput. Appl. Biosci.,* **9**, 745.

CHAPTER 23
PROTEIN CHEMISTRY METHODS, POST-TRANSLATIONAL MODIFICATION, CONSENSUS SEQUENCES

A. Aitken

1. INTRODUCTION

This chapter is intended to provide a guide to basic protein chemistry methods, techniques for identifying consensus sequences and sites of post-translational modification in proteins. Some modifications are considered to be co-translational, while many modifications occur after the protein translation is complete. The latter processes may be reversible, for example if their role is primarily one of regulation. Many modifications, including post-translational proteolysis of a proprotein, are clearly not reversible.

It is clearly not sufficient to determine the primary structure of a protein (its amino acid sequence) or deduce it from DNA sequence and expect that this will explain all the properties of a protein. Knowledge of the complete covalent structure of a protein entails the primary structure, plus the chemical nature and positions of all the modifications to the protein that take place in the cell and are necessary for the correct function, regulation and antigenicity of that protein.

Post-transcriptional editing of mRNA sequences involves the modification, removal or addition of nucleotides (for example, to change cytosine to uracil by deamination at the 6-position). This phenomenon, 'RNA editing', which occurs in mammals, plants, RNA viruses, and in mitochondria of trypanosomes (1), may be defined as any process that changes the nucleotide sequence of an RNA molecule from that of the DNA template encoding it. In this way the coding potential of a gene may be altered. Protein splicing is another mechanism by which an internal sequence can be removed from a precursor protein followed by ligation of the flanking sequences to form the mature protein (in a post-translation event analogous to removal of an intron from RNA). The mechanism is encoded within the internal protein sequence (or 'intein', 2). These and many other examples confirm that we cannot predict the exact protein sequence from the gene sequence. It is clear that direct protein sequence information is required. There is, for example, an enduring misconception that sites of phosphorylation can be easily deduced from the motif of basic amino acid followed by a serine or threonine residue. There are many different types of protein kinase, whose activity is regulated in entirely different ways with completely different linear consensus sequences (see *Table 1*). Careful analysis of the possible phosphorylation site motif and direct structural analysis of the peptide(s) containing the modification(s) are essential. The consensus sequence for a phosphorylation site may also include a phosphoamino acid as an essential recognition feature that is due to a prior phosphorylation by the same or another protein kinase. This would not be immediately apparent by examination of the amino acid sequence deduced from the gene. Many of the co- and post-translational modifications described in *Table 2* do not occur when recombinant proteins are expressed in bacterial cells or even in a eukaryotic expression system (3). The enzymes for the proper processing may not be present to produce, for example, correct glycosylation structures.

Protein Chemistry

Table 1. Consensus sequences and functional motifs

1. Phosphorylation site consensus sequences (phosphorylated residue underlined)

(i) *Cyclic-AMP dependent protein kinase* -Arg-Xaa$_{1-2}$-S/T- or-Arg-Arg-Xaa-S/T-(Hyd)-

(ii) *Cyclic-GMP dependent protein kinase* -R/K$_{1-3}$-Xaa$_{1-3}$-S/T-Arg-

(iii) *Protein kinase C* -(1-5 basic residues)-S/T-Xaa-R/K- where Xaa is uncharged (C-terminal basics are the strongest determinants)

(iv) *Ca^{2+}-calmodulin-dependent protein kinase II* -Arg-Xaa-Xaa-S/T-Hyd-

(v) *Ribosomal protein kinase S6 (including MAPKAP kinase-1)* -Hyd-Xaa-Arg-Xaa-Xaa-Ser-Xaa-

(vi) *Glycogen phosphorylase kinase* -R/K-Xaa-Xaa-Ser-Hyd-(may also phosphorylate Tyr, in the presence of Mn^{2+})

(vii) *Myosin light chain kinase* -K/R$_{-3}$-Xaa$_2$-Arg-Xaa$_2$-Ser-

(viii) *AMP-activated kinase* -Hyd-(Baa, Xaa$_2$)-Xaa-S/T-Xaa$_3$-Hyd- (Hyd is Met, Val, Leu or Ile; Baa is Lys, Arg or His, this may be at -2, -3 or -4)

(ix) *Proline-directed kinases (cell-cycle/MAP kinases)* -S/T-Pro-Xaa-K/R- or -(K/R)-S/T-Pro-(K/R)-Xaa-

(x) *Glycogen synthase kinase-3* -S/T-Xaa^1-Xaa-Xaa- Ser*- (Ser* is already phosphorylated by the same or another kinase. The residue at +1 is frequently Pro)

(xi) *Casein kinase I* -S/T*-Xaa-Xaa-Ser- (the Ser or Thr that is already phosphorylated, can be substituted by Asp or Glu but this becomes a poor substrate)

(xii) *Casein kinase II* -S/T-Aaa-Aaa-Aaa-Aaa- where Aaa are acidic residues, Glu, Asp, phospho-serine or -tyrosine, particularly at +3

(xiii) *2-oxoacid dehydrogenase kinases* -Ser-Xaa$_{1-2}$-D/E- (In pyruvate dehydrogenase:- Tyr-Xaa-Gly-His-Ser-Met-Ser-Asp-Pro-Gly-T/V-S/T-Tyr-Arg-)

(xiv) *Light harvesting complex II kinase* -Arg-Lys-Aaa-Aaa-Thr-Aaa/Lys-Xaa-(Lys)- (Aaa is Ser, Thr or Ala)

(xv) *NIMA kinase* -Phe-Arg-Xaa-S/T-

(xvi) *Autophosphorylation-dependent kinase* -Arg-Xaa$_{1-2}$-S/T-Xaa$_3$-S/T-

(xvii) *MAP kinase kinase (Mek) (in the intact protein)* -Thr-Glu-Tyr-

(xviii) *Tyrosine-protein kinases* -Aaa$_{1-3}$-Xaa-Tyr-Xaa$_2$-Hyd- where Aaa is hydrophilic. The consensus includes one or more acidic residues, Glu > Asp at -2 to -4 (the best substrate is also the optimal phosphorylated sequence for binding to the appropriate SH2 domain, see section 11 in this table)

(xix) *Receptor tyrosine kinases* (preferentially phosphorylate peptides recognized by group III SH2 domains) -Aaa$_{1-3}$-Glu-Tyr-Hyd-Hyd-Hyd- (e.g. insulin receptor -Aaa$_{1-3}$-Glu-Tyr-Met-Hyd-Met-)

(xx) *Cytosolic/intracellular tyrosine kinases* (preferentially phosphorylate peptides recognized by their own or group I SH2 domains) -Aaa$_{1-3}$-I/V-Tyr-E/G-Glu-Hyd- (in c-Abl: -Hyd-Tyr-Ala-Ala-Pro-)

2. N- and C-terminal consensus sequences

(i) *N-terminal myristoylation* myr- Gly-Xaa^2-Xaa$_2$-Aaa-Xaa^6 (the initiator Met is removed). Positions 2 and 6 are not Pro; Aaa is Ser,Thr, Ala, Gly, Cys, Asn; i.e. mainly hydrophilic

(ii) *N-terminal palmitoylation.* Proteins that are myristoylated in the above consensus are also palmitoylated if Xaa^2 is Cys (i.e. in the initial transcript sequence: Met- Gly-Cys- - -)

Protein Chemistry

(iii) *N-terminal methylation*   (Me)$_3$-Ala-Pro-Lys- and (Me)$_2$-Pro-Pro-Lys- *In bacterial pilins* -Gly-M/F*-S/T-L/T-Hyd-Glu-; Precursor cleaved after Gly and Met/Phe methylated (* - generally M and F but L, I, V and Y can also occur)

(iv) *Bacterial lipoprotein glyceride cysteine thioethers*   -Leu-Ala-Gly-Cys-Ser-Ser-Asn -   Other neutral residues occur at 2, 3, 5 or 6, mostly Ser, Ala or Asn

lipid

1   5   7

cleavage site

(v) *Amidation (generally)*:   -Xaa-Gly-R/K-R/K- The precursor is cleaved between Gly and Arg/Lys then Gly processed to an amide   *In thyrotropin releasing hormone and related peptide precursors* :   - Gln-Xaa-Pro-Gly- (Gln is cyclized to pyroglutamate)

(vi) 'C-A-A-X boxes'   -Cys-Aaa^1-Aaa^2-Xaa-CO$_2$H where Aaa^2 is Leu, Ile, Met or Val; Aaa^1 is generally Asp, Glu, Asn or Gln. If Xaa is Ser, Cys, Met, Ala or Gln, C-terminus is farnesylated; if Xaa is Leu or Phe, protein is geranylated   Prenylation may be followed by carboxymethylation of *Cys-*   Cys-Xaa$_3$ → C(isoprenyl)-carboxyMe;   Cys-Xaa-Cys → Cys-Xaa-Cys(isoprenyl)- carboxyMe;   Cys-Cys → Cys-Cys(isoprenyl$_2$) (not carboxymethylated)

3. Esterase and serine active sites   -Gly-Xaa-Ser-Xaa-Gly-

(i) *Serine proteinases*   -T/S-Ala-Ala-His-Cys$_{\sim40}$-Asn-Asn-Asp-Ile-T/M/A-Leu-Leu-Lys-Leu$_{\sim85}$- *Gly-Xaa-Ser-Xaa-Gly-*   In the *subtilisin subclass*, the linear arrangement of the active site residues is different: -Asp-....-His-....-Ser-

(ii) *Cysteine proteinases*   -Gly-Xaa-*Cys*-Hyd-....-Asp- *His*-Ala-Val-T/A-L/A-....- *Asn*-...

(iii) *Aspartate proteinases* -Asp-Thr-Gly- (occurs twice). The aspartates are generally found in the following consensus sequence: -Phe - *Asp*-Thr/Ser- Gly-Ser-....-I/L/V-Val- *Asp*-Thr-Gly-

(iv) Active site aspartates in retroviruses show slight variation (rarely, serine is found in place of threonine):   -Leu-Val - *Asp*-T(S)-Gly-Ala-....-I/L/V-Gly- Arg-Asp-

(v) *Thioesterase active site serine*   -Gly-Xaa- *Ser*-Xaa-Gly-....-Gly-Asx- *His*-Xaa-Xaa-Leu-

(vi) *Cysteine switch*   Pro-Arg-*Cys*-Gly-Xaa-Pro-Asp- (Cys is zinc ligand)

4. Proteolytic processing motifs

(i) *Signal peptidase cleavage sites*   -Aaa-Xaa-Baa-   Mainly small neutral residues, alanine, glycine and serine at the Baa position - also common at position Aaa. In addition, Leu, Ile and Val occur at Aaa

(ii) *Mitochondrial processing peptidase*   -Arg-Xaa-Yaa-

(iii) *Prohormone processing*   -Lys-Arg-Xaa- or Xaa-Arg-Lys- or -Arg-Arg-

Table 1. Continued

(iv) Viral processing by endogenous proteinases

(a) In picornaviruses the recognition sequence is frequently: Thr-Xaa-$\overset{\rightarrow}{\text{Gly}}$- where Xaa is frequently Tyr or Phe, or Gln

(b) -Gly-$\overset{\rightarrow}{\text{Gly-A}}$/G- where A/G is alanine or glycine. (The Gly-A/G-G/A- motif is found in alphaviruses)

(c) -Gly-$\overset{\rightarrow}{\text{Gly-Hyd}}$- where Hyd is a hydrophobic residue, Ile, Val, Phe, Met or Ala.

(d) Aspartate proteinases from retroviruses -Hyd^1-$\overset{\rightarrow}{\text{Hyd}^2\text{-Pro}}$-$\text{Hyd}^3$ where the hydrophobic residue Hyd^2 is leucine or aromatic, Hyd^3 is a small hydrophobic residue (Ala, Val, Ile, Leu) and Hyd^1 is more variable

(v) Thiol ester -P/G-Xaa-G/S-Cys-G/A-Glu-Glx-Xaa-M/L/I/V- Glx is the γ-thio-glutamyl residue, specified in the gene by glutamine codons

5. Calcium binding

(i) EF hand consensus sequence Ca^{2+} coordinating residues:

$$\begin{array}{cccccc} X & Y & Z & -Y & -X & -Z \\ \text{- Asp -Xaa - Asx} & \text{-Xaa-} & \text{Asx -Gly- Oaa} & \text{-I/V- Oaa} & \text{-Xaa-Neg-} & \text{Glu-F/L-} \\ & \text{(Lys)} & \text{(Gly)} & & & \end{array}$$

(most common residues)

General pattern of the helix-loop-helix:
-Glu-Hyd-Xaa_2-Hyd_2-Xaa_2-Hyd-Oaa-Xaa-Oaa_2-Gly-Xaa-Hyd-Oaa-Xaa_2-Neg-Hyd-Xaa_2-Hyd_2-Xaa_2-Hyd-

← helix → ← loop → ← helix →

(ii) Calcium-lipid binding proteins (annexins)
-Lys-Gly-Hyd- Gly-Thr-Asp-Glu-Asp-S/A/T-L/I-Ile-Oaa-I/L/V-I/L-C/A/T-Xaa-Arg-
     (Leu)   (Glu)

← loop → ← α-helix →

The most common residue at two of the positions is shown underneath. Some alternative residues shown in single letter code. Oaa is oxygen-containing (Glu, Asp, Ser, Thr or Asn); Xaa is normally polar

(iii) Gla-containing proteins -Glu-Xaa_3-Glu-Xaa-Cys- Leu-*Gla*-(*Gla*)-Hyd- and -Arg-*Gla*-(*Gla*)-Hyd- where Hyd = hydrophobic amino acid, frequently cysteine. Two consecutive Gla residues occur frequently

Possible recognition site in the γ-glutamyl-precursor proteins:

signal peptide | site of cleavage   secreted protein →
-A/G-S/N/H-Xaa-V/L/I/-V/L-Q/N-Arg-L/A/   P/I-R/K-Arg↓
The cleavage after the dibasic sequence takes place prior to secretion

6. Zinc-binding

(i) *Zinc fingers*

    C_2H_2 *class*   -Phe/Tyr-Xaa-Cys-Xaa$_{2-5}$-Cys-Xaa$_3$-Phe(Hyd)-Xaa$_5$-Leu-Xaa$_2$-His-Xaa$_{2-5}$-His-

    C_3HC_4 *type*   -Cys-Xaa$_2$-Cys-Xaa$_{11-27}$-Cys-Xaa-His-Xaa$_2$-Cys-Xaa$_2$-Cys-Xaa$_{6-17}$-Cys-Xaa$_2$-Cys-

    GATA type   -Cys-Xaa$_2$-Cys-Xaa$_{17}$-Cys-Xaa$_2$-Cys-

    C_X *classes* C_4 -Cys-Xaa$_2$-Cys-Xaa$_{13}$-Cys-Xaa$_2$-Cys-

                     C_5 -Cys-Xaa$_5$-Cys-Xaa$_9$-Cys-Xaa$_2$-Cys-Xaa$_4$-Cys-

                     C_6 -Cys-Xaa$_2$-Cys-Xaa$_6$-Cys-Xaa$_{5-9}$-Cys-Xaa$_2$-Cys-Xaa$_{6-8}$-Cys-

(ii) *LIM domains* (in homeobox and other gene products)

    -Cys-Xaa-Cys-Xaa$_{16-23}$-Hyd-His-Xaa$_2$-Cys-Xaa$_2$-Cys-Xaa$_{16-21}$-Cys-Xaa$_{2-3}$-C/H/D-

(iii) *Ring finger*

    Cys-Xaa$_2$-Hyd-Cys-Xaa$_{9-27}$-Cys-Xaa-His-Xaa-Phe-Cys-Xaa$_2$-Cys-Xaa$_{6-17}$-Cys-Pro-Xaa-Cys-

(iv) *B box*   -Cys-Xaa$_2$-His-Xaa$_7$-Cys-Xaa$_7$-Cys-Xaa$_2$-Cys-Xaa$_5$-His-Xaa$_2$-His-

(v) *Diacylglycerol/phorbol ester binding*

    -His-Xaa-Hyd-Xaa$_{10-11}$-Cys-Xaa$_2$-Cys-Xaa$_3$-Hyd-Xaa$_{(2-7)}$-A/G-Hyd-Xaa-Cys-Xaa$_2$-Cys-Xaa$_4$-*His*-Xaa$_2$-Cys-Xaa$_{(6-7)}$-Cys-

    (the residues underlined and italicized form the first and second zinc coordination sites respectively)

(vi) *Thermolysin('metzincin'-type)* -His-Glu-Xaa$_2$-His-Xaa$_2$-Gly-Xaa$_2$-His-

(vii) *HIT family (PKCI-1).*   (Residues that are known to occur in unique cases are shown in parentheses)

    -Asn-Xaa-G(E)-Xaa$_2$-G/A-Xaa-Gln-T/S/E-V(I)-Xaa-His-L/V(S/T)-His-Hyd-His-L/V(I)-L/I(F)-

7. Miscellaneous metal binding motifs

(i) *Ferredoxins and related iron-sulfur proteins [2Fe-2S] cluster:* -CYS-Xaa$_4$-CYS-Xaa$_2$-CYS-

                                         [4Fe-4S] cluster:-CYS-Xaa$_2$-CYS-Xaa$_2$-CYS-Xaa$_3$-CYS-Pro*-

                              (* there are two exceptions, with Gly or Glu)

(ii) *Reiske iron-sulfur proteins* -Cys-Thr-His-Leu-Gly-Cys-L/I/V- and -Cys-Pro-Cys-His-Gly-Ser

(iii) *Rubredoxin turn (knuckle)* -Trp-Xaa-Cys-Pro-Xaa-Cys-G/A/D-

(iv) *Covalent heme-binding site (in c-type cytochromes)* -Cys-Xaa-Xaa-Cys-His- Xaa$_n$-Met-(Pro)-

    The heme (covalently attached to the cysteines) supplies four ligands to the iron. His and Met are the 5th and 6th. The 6th iron ligand is Tyr in cytochrome f; 5th and 6th are His in cytochrome b$_5$; 5th and 6th are Met in bacterioferritin

(v) *Type I copper binding* -His-Asn-Xaa$_{4-39}$-Tyr-Xaa-Y/F-Y/F-Cys-Xaa-Pro-*His*-Xaa$_{2-6}$-Met-

    (copper ligands are in italics)

(vi) *Metal binding site in dehydrogenases* *(Fe^{2+} or other M^{2+} with two oxidation states)*

    -His-Xaa-L/I/M-Xaa-His-Xaa$_9$-His-Gly-

Table 1. Continued

(vii) *Nickel-dependent hydrogenases* -Cys-Xaa$_2$-Cys-

(viii) *Metallothionein* -Cys-Xaa-Cys-Xaa$_3$-Cys-Xaa-Cys- or -Cys-Xaa$_3$-Cys-Xaa-Cys-Xaa$_2$-Cys-Xaa-Cys-Xaa$_2$-Cys-

(ix) *Phospholipase A$_2$* -Cys-Cys-Xaa$_2$-His-Xaa$_2$-Cys- (the His, along with a conserved Asp, is an active site residue)

(x) *Aminopeptidase* -Asn-Thr-Asp-Ala-Glu-Gly-Arg-Leu- (Asp and Glu are Zn or Mn ligands)

(xi) *Pyruvate kinase* -Hyd$_2$-Xaa-Lys-Hyd-Glu-N/R-Xaa-E/Q-G/A- (Hyd is Leu, Ile, Val or Met; Lys is active site residue and <u>Glu</u> is a Mg^{2+} ligand)

8. Redox proteins

(i) *Thioredoxins* -Trp-<u>Cys</u>-Gly-Pro-<u>Cys</u>- (other residues than W, G and P occur)

(ii) *Glutaredoxin* -<u>Cys</u>-Pro-Aro-<u>Cys</u>-Xaa$_2$-T/A- Aro is Phe,Tyr or Trp

9. Covalent modifications and prosthetic group attachment sites (residues involved are underlined)

(i) *N-Glycosylation* -Asn-Xaa-Ser/Thr- Xaa is any amino acid but is rarely proline or aspartate (Pro, C-terminal to S/T is also rare).
The motif -Asn-Xaa-Cys- is also known but is rare

(ii) *O-Glycosylation* Not as clearcut as *N*-glycosylation; frequent occurrence of Ser, Thr, Ala and Pro at positions -4 to +3, e.g. Xaa-Pro-Xaa-Xaa-(glycosylated Thr within 1 or 2 residues of Pro); Ser-Xaa-Xaa-Pro-, i.e. Pro at +3
Within EGF modules:-C-X-S-X-P-C- and -C-X-X-G-G-T/S-C- *glucosaminoglycans:* -(two acidics at -2 to -4)-Ser-Gly-Xaa-Gly-

(iii) *Tyrosine sulfation* -Neg$_{(1-5)}$-<u>Tyr</u>-Neg$_{(1-5)}$- At least three negatively charged acidics, Asp, Glu or Tyr SO$_4$, at positions ±1 to 5. The residue at position -1 (mostly Asp) is the strongest determinant. Glu at -2 to -5 or Tyr SO$_4$ at ±2 is frequently found

(iv) *Phosphopantetheine binding site* -Leu-Gly-Xaa-Asp-<u>Ser</u>-L/I/T- (or -Hyd-Gly-Hyd-D/K-<u>Ser</u>-Hyd-)

(v) *Biotinyl-lysine* -A/I/V-Met-<u>Lys</u>-Met^2- (one exception is Met^2 to Ala)

(vi) *Lipoyl-lysine binding sites* -Hyd^1-Glu-T/S-Asp-<u>Lys</u>-Ala-Xaa- Hyd^1-Neg- where Hyd^1 is a nonaromatic hydrophobic residue (Ile, Leu, Met or Val); Hyd^2 is any hydrophobic; Neg is aspartate, glutamate or occasionally glycine

(vii) *Bilin attachment* -Ala-Xaa-<u>Cys</u>-Hyd-Arg-Asp-

(viii) *Dipyrromethane cofactor* -Hyd-Xaa-Gly-Ali-<u>Cys</u>-Xaa-Val-Pro-Hyd-Ali- (Hyd is L,I,V or M and Ali is G, A or S)

(ix) *FAD binding site* -Arg-Ser-His-S/T-Xaa$_2$-Ala-Xaa-Gly-Gly-

(x) *Hypusine attachment (in eIF-5A)* -Thr-Gly-<u>Lys</u>-His-Gly-Xaa-Ala-Lys-

(xi) *Retinal binding* -Leu-Asp-Hyd-Xaa-Ala-<u>Lys</u>-Xaa$_2$-Aro- (Hyd is Leu,Ile,Val or Met and Aro is aromatic, Trp, Tyr or Phe)

(xii) *Carbamoyl–phosphate binding* -Phe-Xaa-E/K-Ser-G/T-Arg-Thr- (E in ATCase; K in OTCase)

(xiii) *Phosphohistidine active site* (ATP-citrate lyase and succinyl CoA ligases) -Hyd-Gly-<u>His</u>-Ala-Gly-Ala- (phosphoglycerate mutase) -Hyd-Xaa-Arg-His-Gly-Xaa$_2$-Asn- (Hyd is L/I/V/M)

10. Nucleotide-binding proteins

General feature (the P-loop) -Gly-Xaa-Xaa^3-Xaa^4-Xaa-Gly-(*Lys/Hyd)-   There is normally a third glycine residue at position 3 or 4, depending on the individual type of nucleotide binding protein. *lysine or a hydrophobic residue depending on subtype

(i) *Guanine nucleotide binding*   -Gly-Xaa_4-Gly-Lys-Xaa_{40-80}-Asp-Xaa_2-Gly-Xaa_{40-80}-Asn-Lys-Xaa-Asp- The last four residues comprise the
     nucleotide specificity region. If -Asn-Lys-Xaa-Trp- is present, ITP as well as GTP may be utilized

(ii) *G-protein consensus sequence*   -Lys-Hyd_4-Gly-(Ala)-Gly-G/E-V/S-Gly-Lys-Ser-....-Asp-(Thr)-Xaa-Gly-(Gln)-....-Asn-Lys-Xaa-Asp-
     Hyd is leucine, isoleucine or valine. The alanine in the first motif may be replaced by glycine or asparagine. Similarly, the other residues in
     parentheses are not totally invariant

(iii) *Protein synthesis initiation and elongation factors*   -Gly-His-I/V-Asp-Xaa-Gly-Lys-T/S-....-Asp-C/A/S/T-Pro-Gly-His-....-Asn-Lys-(Met)-Asp-
     Xaa is frequently His, Ser or Ala. Met may be replaced by Cys, Val or Glu

(iv) *Dinucleotide binding proteins*   -Gly-Xaa-Gly-Xaa-Xaa-Gly-Hyd-

(v) *Mononucleotide binding proteins*   -Gly-Xaa-Xaa-Gly-Xaa-Gly-Lys-S/T-

(vi) *Protein kinase catalytic domain consensus sequences*

$$\xleftarrow{\text{nucleotide binding}}\qquad\qquad\xleftarrow{\hspace{2cm}\text{catalytic domain}\hspace{2cm}}$$

-L/V/I-Gly-Xaa-Gly-Xaa-Y/F-Gly-Xaa-Val-Xaa $_{9-26}$-Ala-Xaa-Lys-Xaa- Hyd-Asp-Phe-Gly-$Xaa_{\sim20}$ -Ala-Pro-Glu-
                    (catalytic Lys)   (ATP-binding)

Consensus for serine/threonine kinase specificity   -Asp-Leu-Lys-Pro-Glu-Asn- and -Gly-T/S-Xaa-Xaa-T/F-Xaa-*Ala-Pro-Glu-*
Consensus for protein-tyrosine kinase specificity   -Asp-Leu-Ala-Ala-Arg-Asn- and -Pro-I/V-L/R-Trp-T/M-*Ala-Pro-Glu-*
Variation for Src viral tyrosine kinases first subdomain: -Asp-Leu-Arg-Ala-Ala-Asn-

(vii) *DEAD- and DEAH-box ATP helicases*   -Hyd_2-Asp-Glu-Ala-Asp-K/R/E/N- and -Hyd_3-Asp-Glu-A/L/I/V-His-C/R/E/N-

(viii) *Tyrosine specific phosphatase active site cysteine* -Hyd-His-Cys-Xaa_2-Gly-Xaa_2-Arg-S/T-S/T/A/G-Xaa-Hyd-

11. Protein binding to DNA, protein and carbohydrate

(i) *DNA binding (AT-rich sequences)*   tandem repeats of - Ser-Pro-Lys-(variations: SPRR, TPSR; SPRK; GRP, KPK or RPR; AK; AKP)
     (e.g. A+T hook in HMG proteins) -T/A-Xaa_{1-2}-K/R-R/K-P/G-Arg-Gly-Arg-Pro-R/K-

(ii) *Cell adhesion motif*   -Arg-Gly-Asp- (-Lys-Gly-Asp- in platelets and megakaryocytes, specific for the integrin GPIIb-IIIa)

(iii) *Cell division motif*   -Asx-Xaa_2-Cys-Xaa-T/E/S-Xaa_{1-8}-D/E-E/D/T/S-D/E-

$$\xleftarrow{\beta\text{-turn}}\qquad\xleftarrow{\alpha\text{ helix}\longrightarrow}$$

(iv) *Actin-binding*   Leu-Xaa_2-Ile-Gly-Xaa_2-I/L-Val-Asp-Xaa_6-Leu-Gly-Leu-Ile-Trp-(T/N/Q)-(Ile)-Ile-Leu-   (some variation in hydrophobic residues)

(v) *P-domain (trefoil motif), mucin-associated.*   -Cys-Xaa_{3-4}-Pro-Xaa_2-Arg-Xaa-N/D-Cys-Gly-Y/F-Pro-Gly-Ile-Thr-Xaa_2-Q/E-Cys-Xaa_2-K/R-Gly-
     Cys-Cys-Phe-Asp-Xaa-T/S-V/I-Xaa_2-V/T-P/K-Trp-Cys-Phe-Xaa-Pro-

(vi) *Methionine bristles*   The amphipathic helix repeating unit is: -MET-Xaa-K(R)-Xaa_{4-5}-MET- Xaa_{4-5}-MET-Xaa_{1-2}-MET-
     Xaa are polar or hydrophobic residues as appropriate, to maintain the amphipathicity

Table 1. Continued

(vii) *Collagen* -Gly-P/A-Hyx- where Hyx is hydroxyproline or hydroxylysine

(viii) *SH2 domain* -W-Y-F-G-X-I/L-G/S-X$_{0-5}$-R-K-D/E-A-E-X-L-L-X$_{3-11}$-G-S/T-F-L-V-R-E-S-X$_{5-7}$-Y-S-L-S-V-X$_{4-22}$-V-K-H-Y-K-I-X$_{3-25}$-Y-Y-I-X$_{4-6}$-F-X-S-L-Q-E-L-V-X-H-Y- (this is the consensus sequence plus some common alternatives only)

(ix) *SH2 domain binding* (examples, Y* is phosphotyrosine, see also Section 1 in this table) *PI-3-kinase binding*, Y*-X-X-M- ; *Src*, Y*-E-E-I; *Grb*, Y*-X-N-X ; *Crk*, Y*-X-X-P; *T cell receptors*, Y*-X-X-L/I-(repeated); *Syp/PTP2*, -Y*-Hyd$_5$-

(x) *SH3 domains* A-L-Y-D-Y-X-A-X$_{5-10}$-D/E-L-T-F-X$_{1-3}$-K/R-G-D/E-X-Hyd-Hyd-X-Hyd-Hyd-X$_{3-11}$-G-X-W-W-X-A-X$_{3-9}$-G-X$_2$-G-Hyd-Hyd-P-S-N-Y-N-Y-V- (consensus sequence only, some members contain an RGD, KGE or KGD motif, underlined, that may be involved in interactions at the cell membrane, see Section 11(ii) in this table)

(xi) *SH3 domain binding* class I Arg-X/P-Leu-Pro-(Pro)-Xaa-Pro (more generally -Pro-(Pro)-Xaa-Pro) e.g. *PI3K*, Arg-Xaa-Leu-Pro-Pro-Arg-Pro ; *Src*, R-(A)-L-P-P-L-P-. class II -L/P-Pro-Pro-Leu-Pro-Xaa-Arg- (bind in opposite orientation) *Abl*-SH3 domain-binding -Pro-Xaa-Xaa-Pro-Pro-Hyd-Xaa-Pro-; *Src*, -P-P-L-P-X-R

(xii) *WD-40 repeats* (β-transducin repeats) Xaa$_{\sim 8}$-L/F-Xaa-Gly-His-Xaa$_3$-I/L/V-Xaa$_2$-Hyd-Xaa-Naa-Xaa$_{\sim 6}$-Hyd-Hyd-S/T-G/A-G/A-Xaa-D/N-Xaa$_2$-Hyd-Xaa-I/L/V-W(F/Y)-D(N)- (approx. 42–43 residue repeat) Naa is uncharged

(xiii) *PH (pleckstrin) homology domain* -V-I/V-K-E-G-Y-L-K-K-G-S-X$_N$-K-S-W-K-R-R-Y/W-F-V-L-R/T-D/E-X$_N$-L-S-Y-Y-K-D-S-X$_N$-P-K-G-L/S-I-D/P-L-E-N/G-I/C-Q-I/V-V-E-V-E-D-X$_N$-K-H-C-F-E-I-V-T-K/P-D-G-X$_N$-L-I/L-L-Q-A-E/S-S-E-E-E-R-E/Q-E-W-V/I-A/K-A-L/I-R/Q-R-A-I- (this is the consensus sequence plus some common alternatives only — many differences are known)

(xiv) *Phosphotyrosine-binding (PTB) domain recognition* -Asn-Pro-Xaa-Tyr*- (where Tyr* is phosphotyrosine)

(xv) *Ankyrin repeat* -Gly-Xaa-Thr-P/A-Leu-His-Ala-Ala-Xaa$_7$-V/A- Xaa$_2$ -Leu-Leu-Xaa$_2$-Gly-Ala-Xaa$_{2-6}$-D/N-

(xvi) *Cadherin (cell–cell adhesion) motif* -Leu-Asp-Arg-Glu-Xaa$_4$-Tyr-Xaa-Leu-

(xvii) *Sugar-binding in lectins* WGEI or WGER/K (motif in galectins)

12. Nucleic acid-binding and eukaryotic transcriptional regulatory motifs

(i) *Leucine zippers* KRXRHX*XAAXKXRXRK-Xaa$_6$-Leu-Xaa$_6$-Leu-Xaa$_6$-Leu-Xaa$_6$-Leu-Xaa$_6$-Leu- (* frequently Lys or Arg)

← basic motif → ← leucine repeat →

13. Cell targeting motifs

(i) *C-terminal endoplasmic reticulum-retention signal* In vertebrates it is KDEL (RDEL maintains efficient retention; KNEL, DKEL, KEEL maintain at least partial retention). KDEL also in *Drosophila* and *C. elegans*.
In viruses, KTEL; budding yeast, HDEL; fission yeast, ADEL ; plants use KDEL and HDEL

(ii) *Nuclear localization signal* -Baa-Baa-Xaa$_{10}$-Baa$_{(3/4\ \text{out of}\ 5)}$- (where Baa is Lys or Arg)

(iii) *Nuclear export signal* LALKLAGLDI

(iv) *C-terminal microbody targeting signal* -Ser-Lys-Leu-

(v) *Nucleoporin anchoring* -Xaa-Phe-Xaa-Phe-Gly-

(vi) *Ca^{2+}-binding parallel β-roll motif* (in proteins secreted by Gram-negative bacteria) -Gly-Gly-Xaa-Gly-Xaa-Asp-

14. Protein splicing motif

-Hyd-Hyd-Hyd-V/A/T-His-Asn-C/S/T-

15. Disulfide bond patterns

(i) *EGF domain* -Cys-Xaa$_{4 \text{ or } 7}$-Cys-Xaa$_{2 \text{or} 3}$-Gly-Xaa-Cys-Xaa$_{1 \text{or} 3}$-D/N-Xaa$_4$-F/Y-Xaa-Cys-Xaa-Cys-Xaa$_2$-Gly-Aro-Xaa-Gly-Xaa$_2$-Cys- (D/N may be β-hydroxy Asp or Asn)

(ii) *Kunitz domain (protease inhibitor superfamily)* -Cys-Xaa$_8$-Cys-Xaa$_6$-F/Y/W-F/Y-Y/F- Xaa$_6$-Cys-Xaa$_2$-Phe-Xaa- Y/W-Xaa-Gly-Cys-Xaa$_4$-Asn- Xaa-Phe-Xaa-S/T-Xaa$_3$-Cys- Xaa$_3$-Cys- (more generally -Phe-Xaa$_3$-Gly-Cys-Xaa$_6$-F/Y-Xaa$_5$-Cys-)

Protein Chemistry

Table 2. Co- and post-translational modifications in proteins

Amino acid	Protein sequence	Residue mass of parent amino acid	Mass change
Alanine		71	
	D-Ala		0
	N-acetyl-; N-butyl-; N hexanoyl-;		42, 70, 98, 126,
	N-octanoyl-; N-decanoyl		154
Arginine	NH ‖ H$_2$N-C-NH-CH$_2$-CH$_2$-CH$_2$- ω	156	
	Ornithine		−42
	Citrulline		1
	N-methyl-		14
	N$^\omega$-dimethyl- and N $^\omega$, N $^{\omega\prime}$-dimethyl-		28
	N$^\omega$-phospho-		80
	N-(ADP-ribosyl)-		541
	Arginyl-protein		
Asparagine	O ‖ H$_2$N- C - CH$_2$- β	114	
	N$^\varepsilon$-(β-aspartyl)-lysine		−17
	Aspartate		1
	N-methyl-		14
	N-(ADP-ribosyl)-		541
	Glycosyl-		∼892–2770
Aspartic acid	O ‖ HO - C - CH$_2$- β	115	
	D-Asp (may be methylated)		0 (14)
	β-hydroxy-		16
	β-carboxy-		44
	O-phospho-		80
Cysteine		103	
	Lanthionine		−34
	Dehydroalanine		−34
	S-γ-glutamyl-		−18
	Internal thiol ester (with Gln)		−18
	S-(3-tyrosyl)		−2
	Lysinoalanine		−2
	S-(2-histidyl)-		−2
	Cystine (disulfide)		−2

Table 2. Continued

Amino acid	Protein sequence	Residue mass of parent amino acid	Mass change
	Cystine (trisulfide)		30
	S-(sn-1-glyceryl)-		74
	S-phospho-		80
	S-sulfo-		80
	p-coumaric acid (thiol ester-linked)		146
	S-farnesyl-		204
	S-farnesyl(-OH)-		220
	Biotinyl-		226
	S-palmitoyl-		238
	S-stearoyl- (or myristoyl-, oleoyl-)		266 (210, 182)
	S-geranylgeranyl-		272
	Retinoic acid		282
	Glutathionyl-		305
	S-(hexosyl)$_{2-3}$- (e.g. galactose, glucose)		324–486
	4′-phosphopantetheinyl-		339
	Pyrromethane cofactor		418
	S-(ADP-ribosyl)-		541
	S-(sn-1-dipalmitoyl-glyceryl), i.e. S-2,3 bis(palmitoyloxypropyl)-		548
	S-phycocyanobilin (and others, phycobiliviolin; phycoerythrobilins; phycourobilins; phytochromobilin, mono and di-linked to Cys)		587
	S-heme (plus others, e.g. photoporyphyrin IX)		617
	S-coenzyme A		765
	S-(6-flavin[FAD])-		784
	S-(6-riboflavin-5′-phosphate-, FMN)-		455

C-terminus

	−glycine → amide		−58
	−amide		−1
	O-methyl- (Cys, Leu)		14
	−(N$^\alpha$-Tyr) (reversible)		163
	O-(ADP-ribosyl)-		541
	−N-ethanolamine-glycan-phosphoinositides)		~1900

Glutamine   128

$$\underset{\gamma}{H_2N\text{-}\overset{\displaystyle O}{\overset{\displaystyle \|}{C}}\text{-}CH_2\text{-}CH_2\text{-}}$$

	Internal thiol ester (with Cys)		−18
	N-pyrrolidone carboxyl-		−17
	N$^\varepsilon$-(γ-glutamyl)-lysine		−17
	Glutamate		1

Protein Chemistry

Table 2. Continued

Amino acid	Protein sequence	Residue mass of parent amino acid	Mass change
	N^{α}-(γ-glutamyl)-Glu; N^{α}-(γ-glutamyl)-Glu$_2$; N^{α}-(γ-glutamyl)-Glu$_3$		130, 259, 388
Glutamic acid		129	
	HO - C - CH$_2$-CH$_2$- (O double bond on C), γ		
	S-γ-glutamyl-; O-γ-glutamyl (to Cys and Ser)		−18
	O-methyl		14
	γ-carboxyl-		44
	γ-spermidine		127
	O-(ADP-ribosyl)-		541
	γ-carboxyl-(poly-ADP-ribosyl)-		
Glycine		57	
	−amide (C-terminal)		-58
	α-hydroxy- (intermediate in α-amidation)		16
Histidine		137	
	HC====C—CH$_2$— (positions 5, 4); HN$_\tau$, N$_\pi$, 2 C H		
	Alanino-histidyl		−2
	2-S-cysteinyl-		−2
	N^{π}-methyl-		14
	N^{τ}- and N^{π}-phosphoryl-		80
	Diphthamide		143
	4-iodo-		126
	N^{τ}(ADP-ribosyl) diphthamide-		684
	N^{τ}- and N^{π}-(8α-flavin)		784
	Lysinealdol- (see lysine crosslinks)		
Leucine		113	
	Hydroxy-		16
	Carboxymethyl (C-terminal)		14
Lysine	H$_2$N − CH$_2$ − CH$_2$ − CH$_2$ − CH$_2$− (ε, δ)	128	
	N^{ε}-methyl, N^{ε}-dimethyl-, N^{ε}-trimethyl-		14, 28, 43

Table 2. Continued

Amino acid	Protein sequence	Residue mass of parent amino acid	Mass change
	δ-hydroxy-		16
	N^ε-acetyl-		42
	N^ε-trimethyl-δ-hydroxy-		59
	Carboxyethyl- (through the C2 of propionic acid)		71
	N^ε-phosphoryl-		80
	N^ε-($N\alpha$-methylalanyl)-		85
	N^ε-(4-NH_2-2-OH-butyl)- (hypusine)		87
	N^ε-malonyl-		88
	N^ε-nicotinyl-		233
	N^ε-glucosyl- (aminoketose)		162
	δ-glycosyloxy- (i.e. O-glycosyl-δ-hydroxy-; e.g. D-galactosyl)		177
	N^ε-octanoyl-		126
	N^ε-lipoyl-		188
	N^ε-biotinyl-		226
	N^ε-phosphopyridoxyl-		231
	N^ε-11-*cis*-retinal- , N^ε-(all *trans*)-retinal-		266
	3-hydroxyretinal-		282
	N^ε-ubiquitinyl-		~8565
	GMP(5′ →N^ε)-lysine		362
	AMP(5′ →N^ε)-lysine		346
	N^ε-(ADP-ribosyl)-		541
	αCO_2-(poly-ADP-ribosyl)-		
	Lysine cross-links:		
	These are formed through allysine (α-amino adipic acid semialdehyde) and δ-hydroxy-allysine		
	Lysinoalanine (from cysteine)		−2
	N^ε-(γ-glutamyl)-		−17
	N^ε-(β-aspartyl)-lysine (from asparagine)		−17
	Desmosine, allo-desmosine and isodesmosine		−5
	Allysine aldol; dehydroallysine aldol; allysine aldol-His;		
	Cyclopentenosine (formed from condensation of three allysine);		−1
	Dehydrolysinonorleucine; dehydromerodesmosine; merodesmosine; lysinonorleucine; lysylpyridinoline (deoxypyridinoline); dihydrodesmosines; tetrahydrodesmosines		
	The same compounds also occur as derivatives of hydroxylysine, e.g. syndesines (hydroxyallysinealdols)		
	Peptidoglycan- (also known as murein), linked through N^ε-diaminopimelyl-Lys		

Table 2. Continued

Amino acid	Protein sequence	Residue mass of parent amino acid	Mass change
Methionine		131	
	D-methionine		0
	Sulfoxide		16
	Selenomethionine		48
***N*-Terminus**			
	N-pyrrolindone carboxyl-		−17
	N-methyl, *N*-dimethyl-, *N*-trimethyl-		14, 28, 43
	N-formyl-		28
	N-acetyl-		42
	Carbamoyl-		44
	N-α-ketoacyl- (from Ser via dehydroAla or from Thr via dehydroamino butyrate)		
	N-α-pyruvoyl-		70
	N-α-ketobutyryl-		84
	N-α-succinyl-(Trp)		100
	Short-chain alkyl (attached through Ala), C4 (butyl-); C6 (hexanoyl-); C8; C10		70, 98, 126,154
	N-glucosyl (aminoketose)		162
	N-glucuronyl-		176
	Lauryl-(C12:0);tetradecadienoyl(C14:2); (attached through Gly)		182, 206
	N-myristoeyl- (i.e. tetradecaenoyl, C14:1)		208
	N-myristoyl-(Gly)		210
	Peptidoglycan- (murein-)		
Phenylalanine		147	

$$\langle \rangle\!\!-\!CH_2\text{-}$$
$$\beta$$

	o-, *m*- and *p*-bromo- (^{79}Br or ^{81}Br)		78, 80
	β-glycosyloxy (*O*-glycosyl-β-hydroxy-, from β-hydroxy-Phe)		177
Proline		97	

$$\overset{4}{H_2C}\!\!-\!\!\overset{3}{CH_2}\!\!-$$
$$|\qquad|$$
$$\overset{5}{H_2C}\qquad\overset{2}{CH}\!\!-$$
$$\backslash\ /$$
$$N$$
$$H$$

	Dehydro		−2
	3-hydroxy-, 4-hydroxy-		16
	3,4-dihydroxy-		32
	4-glycosyloxy-(pentosyl, C5)		147

Table 2. Continued

Amino acid	Protein sequence	Residue mass of parent amino acid	Mass change
	4-glycosyloxy-(hexosyl, C6) (e.g. hydroxyproline O-glycosylated with xylose, arabinose, galactose)		177
Serine		87	
	Lanthionine		−34
	Dehydroalanine		−18
	O-γ-glutamyl-		−18
	Pyruvoyl-		−17
	Alanine		−16
	Alanino (τ or π-histidine)		−2
	O-methyl-		14
	O-acetyl-		42
	O-phosphoryl-		80
	Phosphate diester i.e. O-(O-phosphoserine) serine		63
	O-fucosyl-		146
	O-glucosyl-		162
	α-glycerophosphoryl-		168
	O-palmitoyl-		238
	O-stearoyl-		266
	O-(N-acetylglucosaminyl-O-phospho)-		283
	O-pantetheinephosphoryl-		324
	(ADP-ribosyl)-phosphoserine		468
	O-linked carbohydrate		1000s
Threonine		101	
	Methyllanthionine		−34
	Dehydrobutyrine		−18
	α-ketobutyrate		−17
	O-methyl-		14
	O-phosphoryl-		80
	O-fucosyl-		146
	O-glycosyl-		162
	α-glycerophosphoryl-		168
	O-palmitoyl-		238
Tryptophan		186	

	β-hydroxy-		16
	2,4-bisTrp-6,7-dione		28
	Formyl		28
	6,7-dione		30
	N-α-succinyl-		100

Table 2. Continued

Amino acid	Protein sequence	Residue mass of parent amino acid	Mass change
Tyrosine		163	

HO–⟨◯⟩– CH$_2$ –
β

Protein sequence	Mass change
Dehydroalanine (formed by β-elimination of iodophenols from 3-iodo-and 3,5-diiodo-tyrosyl residues- intermediates in the formation of the thyronines)	−94
β-hydroxy-	16
3,4-dihydroxy-phenylalanyl- (DOPA)	16
(DOPA can be oxidized to o-quinone	14)
3,4,6-trihydroxy-phenylalanyl-(TOPA, 6-hydroxy-DOPA)	32
3-chloro- (^{35}Cl or ^{37}Cl)	34, 36
3,5-dichloro- (^{35}Cl, ^{37}Cl or a mixture of isotopes)	68, 70, 72
3-bromo- (^{79}Br or ^{81}Br)	78, 80
O-phosphoryl	80
O-sulfo-	80
5-bromo, 3-chloro-	112, 114, 116
3-iodo-	126
3,5-dibromo- (^{79}Br, ^{81}Br or a mixture of isotopes)	156, 158, 160
O-(glucose-1)-	162
β-glycosyloxy- (O-glycosyl-β-hydroxy-)	177
3,5-diiodo-	252
O^4-uridylyl-	306
O^{4-} adenylyl-	329
3,5,3′-triiodothyronine	470
3,5,3′,5′-tetraiodothyronine (thyroxin)	596
O-(8 α-flavin [FAD])-	783
Protein-tyrosine (reversible O-tyrosinylation of tubulin)	
Poly-ADP-ribosyl-	
RNA and DNA (attached through 5′phosphate) i.e. 5′-(O^4-tyrosylphospho)-DNA and -RNA	
Cross-links:- 3,3′; 5,5′-terTyr	−4
3,3′-bityr; isodiTyr and 3-(S-cysteinyl)	−2
Pulcherosine (trivalent cross-link, 5-[4″-(2-carboxy-2-aminoethyl)phenoxy]-3,3′-dityrosine)	

The search for functional motifs in proteins and for sites of covalent modification will attain increasing importance as the plan to undertake the sequence analysis of the entire human genome comes to fruition. The continued development of protein microsequencing methodology, combined with more powerful techniques for the identification of the nature and location of post-translational modification, is required for the study of a purified protein of unknown structure and function.

2. POST-TRANSLATIONAL MODIFICATIONS

There are more than 150 known (at least 130 chemically distinct) post-translational modifications in proteins. These modifications to the linear amino acid sequence vary widely in chemical structure and amino acid to which they are attached. These have been well documented by Wold and co-workers (4–6). *Table 2* is an updated list of known co- and post-translational modifications (see ref. 7 and references therein).

Access to a 'list of useful biological masses' termed Delta_Mass is possible via world wide web (WWW) via the uniform resource locator (URL): *http://www.medstv.unimelb.edu.au/ WWWDOCS/SVIMRDocs/MassSpec/Delta_Mass_V1.01.html*. This useful list includes masses of protecting groups that may be incompletely removed from synthetic peptides during work-up.

Selenocysteine is now considered to be a 21st amino acid since it has its own tRNA (UGA), although it is first aminoacylated with serine and converted to the final product via dehydroalanine (8).

3. CONSENSUS SEQUENCES

Table 1 lists the consensus sequences for motifs that are shared between distinct proteins. The one and three letter abbreviations for amino acids (including nonstandard ones) that are used in this table are listed in *Table 3*. Predictions of higher order structure are not considered when they are mainly secondary structure motifs such as β-sheet nucleotide binding domains or protein 'signatures' (i.e. sequence profiles that are confined to one particular family). Cell wall peptides and peptide antibiotics that are not synthesized on ribosomes are also outside the scope of *Table 1*. As with *Table 2*, an extensive reference list is impracticable due to space constraints. (References up to 1990 are found in 7.)

3.1. Computer analysis on the internet/WWW
Computer analysis of protein structure and the relevant programs have been covered extensively in Chapter 22. With regard to consensus sequences, it is, however, worth mentioning the most useful of these; PROSITE is accessed from the user-friendly ExPASy WWW server (from the University of Geneva) which also contains the SWISS-PROT, ENZYME, SWISS-2DPAGE and SWISS-3DIMAGE databases.

PROSITE lists consensus sequences as well as an exhaustive treatment of protein 'signatures' (9). PROSITE can enable one to recognize quickly that a sequence from a new protein belongs to a particular family through its database of biologically significant sites, patterns and profiles. Other servers which will enable similar searches include BLOCKS (from the group of Henikoff, Seattle, USA); PRINTS (from T. Attwood, University College, London) and ProDom (D. Kahn at INRA, Toulouse, France).

Protein Chemistry

Table 3. One and three letter abbreviations for the amino acids

Amino acid	Abbreviations		Amino acid	Abbreviations	
Alanine	Ala	A	Valine	Val	V
Arginine	Arg	R	Aspartic acid or Asparagine	Asx	B
Asparagine	Asn	N	Glutamic acid or Glutamine	Glx	Z
Aspartic acid	Asp	D	Any amino acid[a]	Xaa	X
Cysteine	Cys	C	Aromatic residues	Aro	
Glutamine	Gln	Q	Aliphatic	Ali	
Glutamic acid	Glu	E	γ carboxy glutamic acid	Gla	
Glycine	Gly	G	Pyrollidone carboxylic acid	Glp	
Histidine	His	H	Oxygen-containing amino acid	Oaa	
Isoleucine	Ile	I			
Leucine	Leu	L	Hydrophobic amino acid	Hyd	
Lysine	Lys	K	Negatively charged amino acids (Asp or Glu)	Neg	
Methionine	Met	M			
Phenylalanine	Phe	F	Hydroxyproline	Hyp	
Proline	Pro	P	Hydroxylysine	Hyl	
Serine	Ser	S	(Either of these)	Hyx	
Threonine	Thr	T	Homoserine	Hse	
Tryptophan	Trp	W	Homoserine lactone	Hsl	
Tyrosine	Tyr	Y	Cysteic acid	Cya	

[a] Occasionally Aaa for aliphatic; Baa or Yaa are also used, to distinguish specific sites.

4. CLEAVAGE AND HYDROLYSIS OF PROTEINS AND PEPTIDES

4.1. Hydrolysis of proteins for amino acid analysis

Methods for hydrolysis of proteins to free amino acids are listed in *Table 4*. These include conditions that will optimize recovery of labile amino acids (including phosphoamino acids and amino sugars). The elution order of amino acids and derivatives (including many modified amino acids and compounds containing amino groups) from commercial ion-exchange analyzers is shown in *Table 5*. This is not a definitive elution order and will vary according to the exact conditions. The table is only intended as a guide to preliminary information on nonprotein or modified amino acids in a mixture. Ion-exchange amino acid analyzers (with post-column detection by ninhydrin or OPA) normally employ a sulfonated styrene-divinylbenzene resin and three buffers which are usually within the range pH 3.2–3.4, 0.2 M sodium citrate, to separate acidic and hydrophilic compounds (Nos 1–34); pH 4.2–4.25, 0.2–0.4 M sodium citrate to separate hydrophobic compounds (Nos 35–58); pH 5.25–6.25, 1.2–1.7 M sodium citrate (or 0.2 M sodium borate, pH 10.0) to separate basic and some very hydrophobic compounds (Nos 59–77). The precise buffer compositions (containing other additives such as thioglycol) depend on the instrument and the particular range of amino acids whose separation is to be optimized. A lithium buffer system is used for separation of amino acids in physiological fluids (where, for example, Lys, His and their derived compounds elute, in that order, after ammonia). Depending on the exact details of buffer pH and ionic strength the separation will show some variation. A change in temperature from approximately 45°C to 55°C and/or additives such as isopropanol are frequently incorporated in the program to effect an improved separation of amino sugars or to separate Asn, Gln and Hse from closely eluting amino acids.

At the ABRF workshop at the Protein Society Meeting in Boston (1995) a booklet entitled 'Tables of Amino Acids' was distributed by Lowell H. Ericsson (Dept. of Biochem., University

Table 4. Protein hydrolysis methods

Method and conditions	Ref.
Standard conditions[a]	
(1) 6 M HCl (+ 2 mM phenol), 110°C for 24–72 h	
(2) 150°C for 1–4 h, *in vacuo*	10
Rapid hydrolysis	
(3) HCl/propionic acid (1:1)150–160°C for 15–30 min or 130°C for 2 h, *in vacuo*	11
(4) HCl/TFA (2:1) 166°C for 25 min, *in vacuo*	12
Improved CYS, MET and TYR recoveries	
(5) 6 M HCl plus $Na_2 SO_3$, 110°C for 24 h, *in vacuo*	13
Improved TRP yields	
(6) 6 M HCl, 0.5-6% (v/v) thioglycolic acid, 110°C for 24–64 h,	14
(7) 3 M *p*-toluenesulfonic acid[b] + 0.2% tryptamine (3-(2-aminoethyl) indole), 24–72 h, 110°C, *in vacuo*	15
(8) 4 M methanesulfonic acid[b] + 0.2% tryptamine, 110°C for 24–72 h, *in vacuo*	16
(9) 3 M mercaptoethanesulfonic acid,110°C for 24–72 h, *in vacuo*	17
Phospho amino acids	
O-Phospho-Ser	
(10) 6 M HCl, 110°C, 1–2 h	18
O-Phospho-Thr	
(11) 6 M HCl, 110°C, 2–4 h	
O-Phospho-Tyr	
(12) 6 M HCl, 110°C, 1 h	
(13) 5 N KOH,155°C for 30 min	19
Amino sugars	
(14) Optimal recovery (approx 80%) best achieved after hydrolysis for 14 h at 110°C in 3.3 M HCl	20

[a] Longer times are necsssary for complete hydrolysis of bonds containing Val and Ile.
[b] Not recommended if protein contains more than ~20% carbohydrate.

of Washington, Seattle, WA 98195 e-mail: *ERICSSONLH@u.washington.edu*). The booklet contains a very full list of retention times of amino acids and dipeptides for a postcolumn ninhydrin system (from the Beckmann Instruments Application Data Bulletin DS-656), precolumn derivatization (PTC amino acids on Picotag and FMOC-), Dabsyl amino acids and AQC amino acid analysis systems.

Asn and Gln are converted to Asp and Glu during normal acid hydrolysis conditions but may be seen if analyzing physiological fluids or products of carboxypeptidase digestion. Tryptophan, in particular, requires special hydrolysis conditions to be recovered in good yield (see *Table 4*). Serine and threonine are usually destroyed by 6–10 and 3–5% respectively per 24 h, employing 'standard' acid hydrolysis conditions and a correction is usually made. Tyr recovery may be low and variable. Cysteine residues are recovered (if not oxidized despite precautions to exclude oxygen) as cystine during normal HCl hydrolysis (often referred to as '1/2 cystine' on analysis report sheets). Met is also partially oxidized.

Norleucine is often used as an internal standard. Strongly acidic amino acids (Nos 1–5) are not actually retained by this type of ion-exchange column and usually elute together in the 'breakthrough'.

Table 5. Elution order of amino acids and derivatives from ion-exchange analyzers

1. *o*-Phospho-serine, -threonine and-tyrosine; tyrosine sulfate
2. γ-carboxyglutamic acid
3. Cysteic acid
4. Taurine
5. Glycero-phosphoethanolamine
6. Phosphoethanolamine
7. Urea
8. Glucosaminic acid
9. Methionine sulfoxides
10. Hydroxyproline
11. Carboxymethylcysteine
12. β-hydroxy aspartic acid
13. Aspartic acid
14. Methionine sulfone
15. Asparagine
16. Glutamine
17. Threonine
18. Serine
19. α-methyl aspartic acid
20. α-methyl serine
21. Muramic acid
22. Homoserine
23. Ianthionine
24. Glutamic acid
25. α-methyl glutamic acid
26. Sarcosine
27. α-amino adipic acid
28. α-amino, ε-hydroxy pimelic acid
29. Citrulline
30. Proline
31. Cysteine
32. S-methylcysteine
33. β-hydroxy leucine
34. Glycine
35. Alanine
36. S-ethylcysteine
37. Cystine
38. α-aminobutyric acid
39. Valine
40. Selenocysteine
41. Methionine
42. Dihydroxyphenylalanine (DOPA)
43. Cystathionine/allo-cystathionine
44. Diaminopimelic acid
45. Selenomethionine
46. α-methylmethionine
47. Alloisoleucine
48. Isoleucine
49. Leucine
50. Norleucine
51. Tyrosine
52. 3-nitro-tyrosine
53. Phenylalanine
54. Fluorophenylalanine
55. Chlorotyrosine
56. β-amino-isobutyric acid
57. β-alanine
58. Glucosamine
59. Mannosamine
60. Galactosamine
61. Talosamine
62. γ-aminobutyric acid
63. π-methylhistidine (1-MeHis)
64. Histidine
65. γ-methylhistidine (3-Me-His)
66. Homoserine lactone
67. Hydroxylysine/allohydroxylysine
68. Aminoethylcysteine
69. Ornithine
70. Lysine
71. *N*-mono, di- and tri-methyllysines
72. Ethanolamine
73. Ammonia
74. Aminoguanidino-propionic acid
75. Tryptophan
76. Pyridylethylcysteine
77. Homocysteine thiolactone
78. Dimethylarginine
79. Arginine
80. 3,5-diiodotyrosine

4.2. Enzymic and chemical cleavage of proteins

Table 6 includes a review of (a) methods for cleavage of proteins and peptides using all the commonly used proteinases and proteolysis conditions. The chemical cleavage of proteins and peptides using all the commonly used reagents is also described, in (c).

Table 6. Digestion and cleavage of proteins and peptides

(a) Endoproteinases and proteolysis conditions

Trypsin is specific for cleavage at the C-terminal side of arginine, lysine and S-aminoethylcysteine bonds. Some chymotryptic-like anomalous cleavages may be seen, especially cleavage at tyrosine. This may be due to inherent trypsin specificity. The cleavage may be limited to arginyl bonds by citraconylation, succinylation or maleylation of the ε-amino group of lysine. Specificity may be restricted to lysine bonds by reaction of arginine with 1,2-cyclohexanedione or *p*-hydroxyphenylglyoxal. The use of trypsin as an arginine-specific enzyme is widely and successfully practiced but the converse is not satisfactory. Both arginine-specific and lysine-specific proteinases are now available commercially

Chymotrypsin cleaves at the C-terminal side of Phe, Tyr, Trp, Leu, Met and, more slowly, His, Asn, Gln. Digestion conditions are as for trypsin

Staphylococcus aureus (*endoproteinase Glu-C*), or V-8 proteinase is, as the name suggests, specific for the C-terminus of Glu-Xaa bonds. With careful choice of buffer conditions (ammonium bicarbonate, pH 8.0 or ammonium acetate, pH 4.0) these bonds will be almost exclusively susceptible. Incubation in sodium phosphate, pH 7.8, will enhance cleavage of Asp-Xaa bonds, although this will not be as efficient as Glu-Xaa cleavage. This proteinase is also stable in 4 M urea or 0.2% SDS. Extended reaction times may also be employed (12–48 h at 25°C) to achieve optimal cleavage

Thermolysin specifically cleaves peptide bonds at the N-terminal side of hydrophobic residues, especially leucine, isoleucine, phenylalanine, valine, alanine and methionine (in descending rate order). The presence of proline at the C-terminal side of the hydrophobic residue inhibits cleavage, i.e. -Xaa-Hyd-Pro- where Xaa is any amino acid and Hyd is hydrophobic. Although the enzyme is active at 50–60°C the buffer should contain 2–3 mM calcium to enhance thermostability (especially above 60°C). This proteinase is also active in 6 M urea and 0.5% SDS. The pH optimum of 7.0–8.0 is approximately 1 unit lower than the serine proteinases (trypsin, chymotrypsin, subtilisin and elastase)

Pepsin will cleave at the N- and C-terminal side of hydrophobic and neutral amino acids. It would appear to have a particular preference for leucine and phenylalanine (and to a lesser extent, methionine and tryptophan). Isoleucine-containing and Tyr-Xaa bonds are relatively resistant. Although it has this broad specificity, pepsin is a very useful proteinase in the study of disulfide bonds. At the pH optimum there is little disulfide exchange. Normal digestion conditions are incubation at 25°C for 1–2 h at pH 2.0–3.0 (for example dilute HCl or 50% formic acid)

Subtilisin, papain and bromelain are also broad-specificity proteinases that find particular application for subdigestion of large fragments

Subtilisin appears to have a particular specificity for peptide bonds adjacent to Asp, Glu, Ala, Gly and Val as well as other Hyd-Xaa bonds

Papain is a thiol proteinase and requires 1 mM 2-mercaptoethanol or dithiothreitol (DTT) to retain activity. Thiol reagents such as iodoacetate must be rigorously excluded. Papain will cleave bonds containing Arg, Lys, Gln, His, Gly and Tyr most readily

Elastase has a preference for cleavage at the C-terminus of small nonpolar residues, especially alanine; also glycine, serine, valine and leucine and isoleucine-containing bonds are cleaved

Pronase is a crude mixture of different proteinases (from *Streptomyces griseus*) which will, in theory, digest a polypeptide to free amino acids. This is useful for very extensive digestion and for release of a covalently modified amino acid from a peptide if the chemical bond is acid or alkali labile

Lysine proteinase from *Armillaria mellea* cleaves at the N-terminal side of lysine residues, Xaa-Lys (not at present commercially available). The pH optimum is 7.5 to 8.0 in *N*-ethylmorpholine acetate. Certain Arg-Xaa bonds may be cleaved in good yield, particularly Arg-Ile and Arg-Arg bonds. This enzyme will cleave 2-aminoethylcysteine bonds. Combined with trifluoroacetylation to block lysine side-chains, this can be a very useful method of obtaining specific cleavage at modified cysteine residues (see also Asp-N proteinase)

Protein Chemistry

Table 6. Continued

Endoproteinase Lys-C is another useful proteinase (from *Lysobacter enzymogenes*) that is specific for the C-terminal side of lysine bonds (available from Boehringer, or Promega). This enzyme may also give minor cleavage at some Asn-Xaa bonds. There are other sources, from other organisms, some of which may be stable in urea or up to 2% SDS (e.g. from Waco, isolated from *Chromobacter lyticus*, pH optimum around 9.0)

Arginine-specific proteinases that cleave Arg-Xaa bonds include *clostripain, thrombin* and an enzyme from mouse submaxillary gland (called *endoproteinase Arg-C*). Certain Lys-Xaa bonds may also be cleaved with the latter

Clostripain, like papain, is a thiol proteinase that requires activation with DTT (2 mM). Calcium acetate is also included in the digestion buffer (buffer may be ammonium bicarbonate, pH 7.8 or 25 mM phosphate, pH 7.5)

Thrombin is also a very specific enzyme, cleavage is restricted to certain Arg-Xaa bonds and is useful for generating large protein fragments. Both thrombin and endoproteinase Arg-C are used in bicarbonate buffer. *Thrombin* specificity: Leu-Val-Pro-Arg-Xaa

Factor Xa is (like thrombin) used to cleave fusion proteins. Its specificity is: Ile-Glu-Gly-Arg-Xaa

Endoproteinase Asp-N from Boehringer is specific for cleavage at Xaa-Asp and Xaa-cysteic acid bonds. The latter specificity can be used as a means of producing limited cleavage of a protein. If all the carboxyl groups, including the side-chains of the aspartate residues, are blocked by glycinamide-carbodiimide modification, digestion of the oxidized protein will result in specific cleavage at the positions of Xaa-Cya bonds. Cya = cysteic acid

Proline-specific endopeptidase is now available commercially (Miles laboratories). This will cleave Pro-Xaa bonds (and some Ala-Xaa bonds slowly) in phosphate buffer, pH 7.0 at 30°C

Glycine-specific endopeptidase is also available commercially (Calbiochem) but I have no experience of its use. Claimed to be stable in SDS; employed at pH 6.8, 40°C

(b) Exopeptidases

These include *aminopeptidases* which remove amino terminal residues, e.g. *aminopeptidase M* and *leucine aminopeptidase* (they cleave all except Xaa-Pro bonds)

Carboxypeptidases remove carboxy-terminal residues

Carboxypeptidase A (removes most amino acids except Arg and Pro; Asp, Glu (and oxidized MetSO$_2$ and Cya, CMCys) and Gly are removed very slowly). Asn, Ser and Lys release is slow

Carboxypeptidase B is specific for Lys and Arg

Carboxypeptidase P and *Carboxypeptidase Y* are now widely used. They remove all L-amino acids including proline (although CpP may release Ser and Gly slowly; CpY may remove Ser, Gly, Lys and Arg relatively slowly). The actual rates of release depend, to a certain extent, on the particular penultimate residue; specificity of CpA and B may be enhanced towards release of charged residues by adjusting pH

Pyroglutamate aminopeptidase cleaves N-terminal Glp-Xaa (except where Xaa is Pro or Lys). Glp is pyrollidone carboxylic acid, i.e., cyclized glutamine

Acylamino acid releasing enzyme removes preferentially *N*-acetyl-Ser, -Thr, -Ala and -Met from *peptides*. These are the most commonly found at the N-terminus in any case

(c) Chemical cleavage

One of the advantages of chemical cleavage of proteins is the specificity for residues that may be relatively rare in proteins, especially methionine and tryptophan, for which specific proteinases are not available. Relatively few large fragments may result, an advantage for automated microsequencing

Cyanogen bromide cleaves Met-Xaa bonds in a wide variety of acid conditions (70% formic acid is commonly used). When Met-Ser and Met-Thr bonds are cleaved, $N \rightarrow O$ acyl shifts may occur that result in an isopeptide bond. Incubation in 70% trifluoracetic acid (TFA) may help reduce this. Amino acid or size analysis may suggest that the Met-Ser/Thr bond is intact. However, when this bond is reached during Edman degradation, no further sequence is seen. The protein is normally preincubated with 0.1 M 2-mercaptoethanol for 2 h at 37°C to ensure that methionine residues are fully reduced and 100–1000-fold molar excess of cyanogen bromide is added. The reaction is carried out for 18–24 h, under nitrogen, in the dark, at room temperature. Some cleavage of acid-labile peptide bonds (particularly Asp-Pro) may occur and incubation at 4°C will minimize this. Occasionally some tryptophan bonds are cleaved. This may be enhanced, if desired, by incubation in heptafluorobutyric acid. Since it is carried out in acid conditions, cyanogen bromide digestion is a useful method when *retention of disulfide bond* specificity is required

Met-Xaa bonds can also be cleaved by carboxymethylation or carboxyamidomethylation at low pH (Na formate, pH 3.5) followed by heating. This converts CM-Met to homoserine and S-carboxymethylhomoserine. Can be used to radiolabel Met-containing peptides

Specific cleavage of *tryptophan bonds* is achieved by incubation with *o-iodosobenzoic acid*. The protein is dissolved in 80% acetic acid with 4 M guanidine-HCl. *O*-iodosobenzoic acid (3 μg μg^{-1} protein) is added and incubated for 24 h at room temperature in the dark, under nitrogen. High yields (70–90%) of cleavage at most tryptophan bonds are achieved

Other tryptophanyl bond cleaving reagents include:

N-bromosuccinimide or chlorosuccinimide, that may give variable yields. Some cleavage at Tyr-Xaa bonds and, to a lesser extent, His-Xaa will result from *N*-bromosuccinimide treatment (*N*-chloro-succinimide may be more specific for Trp cleavage)

BNPS-skatole (probably the reagent of choice) gives up to 70% cleavage (conditions are 10-fold molar excess of reagent in 50–70% acetic acid, 37°C, 24 h)

Partial acid hydrolysis: Asp-Xaa peptide bonds (particularly Asp-Pro bonds) are 2 to 3 orders of magnitude more labile to acid hydrolysis than any other peptide bond. A substantial degree of specific cleavage of Asp-Pro may be achieved in dilute formic or acetic acid. In a particularly favorable case, this bond was cleaved in 70% yield when incubated for 48 h at 40°C in 1 M acetic acid buffered to pH 2.9 with pyridine. Partial acid hydrolysis in dilute hydrochloric acid (25 mM, pH 2.0) has also been used. Incubation times are found by carrying out pilot experiments, but 4 h to overnight at 37°C are frequently optimal. These conditions do not lead to extensive deamidation of amide side-chains of asparagine or glutamine

If high concentrations of organic acids are used in any of the above chemical cleavage methods, the incubation mixture must be diluted with high purity water to at least 5% formic or acetic acid before freeze drying. Azeotropic mixtures will otherwise form, water will preferentially evaporate and the eventual result will be an oily residue of organic acid

Table 6. Continued

Hydroxylamine will specifically cleave *Asn-Gly bonds*. The protein is dissolved in 6 M guanidine-HCl and incubated with 2 M hydroxylamine-HCl (adjusted to pH 9.0 with 4 M lithium hydroxide) for 4 h at 25°C. The yield is approximately 50%. The reaction proceeds through a cyclic intermediate. This five-membered ring succinimide derivative reopens to form α and β-aspartylhydroxamate and another fragment with N-terminal glycine. Some of the products are uncleaved peptides containing rearranged β-Asn-Gly peptide bonds that will sequence normally until the β-peptide is reached, when the Edman degradation cannot continue

(d) Specific cleavage at cysteine residues

Because of its relative rarity, cysteine represents an attractive candidate for cleavage. However, the development of efficient specific methods has so far met with only limited success. The best of these utilizes conversion of the -SH moiety to the thiocyano group. This can cyclize to an acyliminothiazolidine followed by rapid hydrolysis to give fragments where the N-terminus is blocked by the 2-iminothiazolidinyl moiety. Without generation of a free amino group, therefore, the Edman degradation procedure will not proceed. Successful unblocking has been reported using Raney nickel. The blocked amino terminus would not pose a problem during sequencing by mass spectrometric techniques. The protein is incubated in 6 M guanidine-HCl, 0.2 M Tris-acetate (pH 8.0) containing 1 mM DTT (if disulfides are present 10 mM DTT is added and incubated at 37°C for 1 h or at room temperature for 2 h). A five-fold molar excess (over *total* thiol) of NTCB is added, the pH readjusted to 8.0 and incubated at 37°C for 30 min to a few h. For cleavage, the protein is dissolved in 6 M guanidine-HCl, sodium borate or Tris acetate (0.2 M) pH 9.0 and incubated for 18–72 h at 37°C

Several proteases will cleave at cysteine residues if these are first converted to *S*-aminoethylcysteine with ethylenimine. C-terminal aminoethylcysteine residues are, like lysine and arginine, susceptible to cleavage with *carboxypeptidase B* and *Lysine proteinase* from *Armillaria mellea*

276

PROTEINS LABFAX

The reaction conditions described in *Table 6* may be varied depending on whether or not complete digestion of the vulnerable peptide bonds is required. Limited digestion may be preferred. This will produce relatively large fragments for automated microsequencing. Complete digestion is desirable if it is wished to identify the position of a particular covalent modification by mass spectrometry or to reduce ambiguity if it is known that a particular covalent modification is on a certain amino acid; for example, if a protein is known to contain phosphoserine, it will be advantageous to digest to a peptide that contains only one serine. Modification of a residue in a particular peptide bond will frequently affect its ability to be cleaved by a protease; for example, Lys-Xaa bonds, where lysine is acetylated, are not susceptible to trypsin cleavage, in contrast to those containing methylated lysines (the latter residues are still basic). In addition, trypsin will not cleave Arg-Xaa-Ser(P) where (P) indicates a phosphorylated residue. In contrast, Arg-Ser(P) and Lys-Ser(P) and Lys/Arg-Thr(P) bonds may be cleaved in good yield. Details of individual proteinases that require substantially different conditions are noted in the appropriate section of *Table 6*. Further details of enzymes and conditions are found in (7 and 21). The commercially available enzymes (particularly those from Boehringer) are supplied with full instructions for use and conditions for storage. The buffer and pH conditions used are generally around pH 8.0. Volatile buffers are desirable since they may be removed by lyophilization. Ammonium bicarbonate, which gives a pH of around 7.8 when dissolved (without adjustment), and *N*-ethylmorpholine acetate are the buffers of choice. Detergents such as sodium dodecyl sulfate (SDS) may be used if solubility of the substrate protein is a problem, since most proteinases are stable for some time in 0.05–1.0% SDS. Proteolytic digestion will, in any case, proceed more readily when the substrate is denatured. It is also prudent to add the enzyme in aliquots over the total digestion period of 2–4 h at 30–37°C, without exceeding a final enzyme/substrate ratio of 1/20. Suspension of an insoluble protein by sonication is also useful before initiating digestion.

The enzymes listed are in the order of use in sequence studies, acccording to an analysis (22) of 1911 publications in the literature. Trypsin has been used in 31% of digests, chymotrypsin in 22%, endoproteinase Glu-C (*S. aureus* V8) in 13%, thermolysin in 12%, pepsin in 6%, subtilisin, papain, clostripain, lysobacter Lys-specific proteinase, approximately 2% each; thrombin, elastase, pronase, submaxillary Arg-specific and *A. mellea* proteinases approximately 1% each.

Although there are exceptions, most of the proteinases will not cleave the designated peptide bond if Xaa is proline.

5. IDENTIFICATION AND PURIFICATION OF MODIFIED PEPTIDES BY HPLC

High performance liquid chromatography (HPLC) is the most widely used method for purification of polypeptides. A very large range of packing materials is now available, including reverse-phase, ion-exchange, hydrophobic interaction, gel exclusion and chiral packings. Details of the methodology for purification of peptides and proteins can be found in Chapter 6.

The most widely applicable technique is reverse-phase (RP)HPLC. Water/acetonitrile gradients at two different pH values are most commonly used to purify a peptide completely. The buffering conditions used are dilute phosphate buffer, pH 6.0 or ammonium acetate at pH 5.0–7.0 as a first dimension, followed by 0.1% trifluoroacetic acid (TFA), pH 2.0. This latter step enables collected fractions to be reduced in volume without leaving a residue. Parameters which affect the separation are discussed in Chapter 6, Section 9. When detection is by UV absorbance, the concentrations of buffer and modifier are adjusted to give as level a baseline as possible. It is essential that reagents and solvents are of the highest quality available. With small amounts of sample, it is better to avoid complete drying of the peptide even in coated or plastic tubes.

RP-HPLC of peptides and proteins is ideal as a final stage of purification to desalt samples (e.g. after SDS polyacrylamide gel electrophoresis, SDS-PAGE) to prepare for high sensitivity protein sequencing and mass spectrometry. Peptides are usually detected by UV at 214 nm and/or 280 nm. Detection at 254 nm is also useful for modified residues; for example, pyridethylated cysteines. Diode-array detectors that continuously scan a given range of wavelengths are invaluable in this respect. A fluorescence detector may be useful for tryptophan-containing peptides. On-line radioactivity detectors are also widely used for labeled peptides.

5.1. Prediction of peptide retention times on RP-HPLC

Many of the known post-translational modifications in proteins markedly affect the retention of modified peptides on RP-HPLC. Elution behavior of peptides from RP-HPLC that deviates from the expected position may therefore be useful in detecting previously unknown or novel post-translationally modified peptides. Examples of altered retention times from RP-HPLC of peptides containing many of the principal post-translational modifications are shown in *Table 7*. These peptides may then be selected for further investigation by fast atom bombardment (FAB) mass spectrometry (MS) or on-line HPLC electrospray ionization MS (ESIMS) (see below and Chapter 16).

The development of computer programs to predict the elution position of acetylated and other covalently modified peptides on RP-HPLC has been of great value in the initial purification of these peptides prior to their analysis (21, 23, 24). Peptide retention times on RP-HPLC can be predicted from the hydrophobicity of the constituent residues. A list of retention coefficients on C_{18} RP-HPLC for protein amino acids is given in *Table 8*.

The method for prediction of peptide retention times on RP-HPLC involves calibration of RP C_8 (octyl)- or C_{18} (octadecyl)- silyl HPLC columns with a standard linear (TFA/acetonitrile) gradient using peptides of known sequence and/or composition; for example, trypic digests of a known protein (myoglobin or β-lactoglobulin).

The observed retention times are plotted against $\ln(1+i\Sigma n_i c_i)$, where c_i is the coefficient of amino acid i and n_i is the number of times that it occurs. *Table 8* contains an example of values obtained from a C_{18} μBondapak (A), Hamilton PRP-1 (B) and Vydac C_{18} (C) (5–65%) RP columns with gradients of acetonitrile in 0.1% TFA over 1–2 h at a flow-rate of 1 to 2 ml min^{-1}. Anomalies that may cause column to column variations, including residual ion-exchange effects between basic residues and the column matrix, may also be suppressed in this way and may be reduced by addition of 0.1% triethylamine to the eluting buffers and readjusting the pH appropriately with TFA. In general, however, the method works extremely well and gives good correlation with the relative elution order of peptides. For peptides of up to about 40 residues, there is a high correlation between elution time (or solvent composition) and the sum of the retention coefficients derived from the component amino acids for a given RP system. Peptides of known sequence/composition that elute significantly outside the expected 'windows' calculated from the sum of the retention coefficients are likely to be modified in some way.

Most computer database packages now include a prediction program that will cleave a protein into theoretical peptides, based on the specificity of a wide range of proteolytic and chemical methods, and sort them into their predicted relative elution position; for example, on GCG, one can use 'peptidesort'. These should be used with caution since these programs frequently do not take into account the inability of serine- and many other classes of protease to cleave at proline, nor the fact that free amino acids are rarely released by endopeptidases (see *Table 6*).

Table 7. Altered retention of post-translationally modified peptides on RP-HPLC

Peptide[a]	Modification	Protein	Elution position (percent organic phase)		Column	Ref.
			Actual for modified peptide	Expected or actual[b] for unmodified		
M E T A V N A K	N-acetylated	Fatty acid synthase thioesterase II	29	26	Vydac C$_{18}$	23
G N E	N-myristoylated	Protein phosphatase 2B (B subunit of calcineurin)	57	0	μ bondapak C$_{18}$	25
K I D E P S **T** P Y H S	Phosphothreonine	Phosphatase inhibitor-2	21	23	μ bondapak C$_{18}$	26
D Q I C I G Y H A N **N** S T K	Glycosylated asparagine	Influenza virus hemagglutinin	~22	~23.5	120T-TSK C$_{18}$	27
Q E **P**	Pyroglutamylglutamyl proline amide	Thyrotropin-releasing hormone related peptide	15 and 27	0–4	μ bondapak C$_{18}$ and Spheri-5 RP-18 (1×100 mm)	28
A D **S** G E G D F L A E G G V R	Phosphoserine		19	23[b]	Lichrosorb RP-18 25 mM ammonium acetate, pH 6.0	
A D D **Y** D D E V L P D A R H **Y Y** D D T D E E E R I V S T V D A R	Tyrosine sulfate Mono- and di-tyrosine and sulfate	Fibrinopeptides	10, 23 and 22 respectively	13[b], 25[b]		29
A E I E T D **K** A T I G F E	Lipoyl-lysine	2-oxoacid dehydrogenase	45	28[b]	Aquapore C$_8$ RP-300	Fuller, Perham and Packman (personal communication)
L G L D **S** L M G V E V R	Carboxymethylated, phosphopantetheinyl serine	Acyl carrier domain of fatty acid synthase	33	41	Ultrasphere ODS	30

[a] Modified residues are shown in bold.

[b] Elution position is expressed as apparent concentration of acetonitrile at which the peptides elute. This varies somewhat depending on the flow rate and slope of the gradient since the column effluent is at a lower concentration of organic phase than the mixture entering the top of the column. Mobile phase is water-acetonitrile containing approximately 0.1% TFA (pH~2.0) unless otherwise stated, normally with an increase in acetonitrile of 1 to 2% min⁻¹. Larger shifts in elution position are frequently observed with phospho-, sulfo- and γ-carboxyglutamyl peptides at higher pH where 'ion suppression' is limited and the increased polarity has a greater effect. The elution positions of unmodified peptides marked with an asterisk are actual values.

Protein Chemistry

Table 8. Relative retention coefficients on RP-HPLC

Amino acid or substituent	Retention coefficient (ref.)		
i	A (21)	B (24)	C (23)
Leu	20.0	1.23	1.8
Phe	19.2	1.37	2.5
Trp	16.3	1.79	3.4
Ala	7.3	0.15	0.5
Ile	6.6	1.06	2.5
Tyr	5.9	0.77	2.3
Met	5.6	0.97	1.3
Pro	5.1	0.55	0
Val	3.5	0.73	1.5
Thr	0.8	0.21	0.4
Gln	−0.3	−0.39	0
Gly	−1.2	0.04	0
His	−2.1	−0.12	0.5
Asp	−2.9	−0.08	0
Arg	−3.6	−0.21	0
Lys	−3.7	−0.5	0
Ser	−4.1	−0.16	0
Asn	−5.7	−0.17	0
Glu	−7.1	0.17	0
Cys (CMCys)	−9.2	0.09	0
-CO_2H	2.4		
NH_2-	4.2		
N-acetyl-	10.2		
-amide (C-terminal)	10.3		

6. STRUCTURE ELUCIDATION OF MODIFIED PEPTIDES

6.1. N-terminal modifications of proteins

The majority (perhaps 80–90%) of intact intracellular eukaryotic proteins are 'blocked' at the amino terminus, mainly with acetyl groups. The exact mechanism and the function of this widespread modification is not clear, although studies have suggested a role in stabilizing the protein against degradation, perhaps protecting the protein from ubiquitin-mediated proteolysis (31). Particular proteins with *N*-acetyl groups have been shown to turnover more slowly than the unacetylated form. The half-life of a protein *in vivo* depends on the nature of its N-terminal residue (32). There is a preference for the acetylation of certain amino acids; 35–50% have acetyl-Ser; 27–33% acetyl-Ala; 5–8% acetyl-Gly; 5–6% acetyl-Met; 5–6% acetyl-Thr; 1-3% acetyl-Asp and 2–3% acetyl-Val (33, 34). Although acetyl-Gly- is found in 5–8% of acetylated proteins, many other proteins contain *N*-myristoyl-Gly-, for example. This fatty acyl group also appears to be co-translationally linked at a similar stage in protein synthesis. Acyl peptide hydrolase (35) appears to be a functionally unique serine proteinase which has been employed to remove *N*-acetyl groups. Similar enzymes have been isolated from rabbit muscle (36) and sheep kidney and liver (37). These enzymes also act on *N*-formyl-peptides. Although this enables microsequencing to proceed (37), in practice their substrate specificity is limited to short peptides, less than approximately 20 amino acids. The specificity of the enzyme is also limited to the derivatives of - Met, - Thr, -Ala, -Ser, and (slowly) -Gly.

The amino acid next to the initiator methionine is the most important determinant of the particular mode of N-terminal processing (34) but there is an, as yet, undetermined recognition sequence in the first 40 or so amino acids (38). The 'consensus sequence' that determines whether a mature protein retains the methionine (with or without the acetyl group) or is N-acetylated on the penultimate residue (four possibilities!), is listed in *Table 9*. The requirement for formyl-methionine as the initiation residue, predicts that all prokaryotic proteins must contain an N-terminal formyl- group at some stage of their biosynthesis although the N-formyl-Met- that must be present initially is very rarely found in the mature protein except in recombinant proteins overexpressed in bacteria.

The presence of the acetyl group on the amino terminus of a protein or peptide precludes the use of direct sequencing by the Edman degradation chemistry which requires the primary NH_2 group (or secondary amino in the case of proline) to be unprotonated. Trimethylation of Ala or dimethylation of Pro prevents Edman degradation, but is possible if a lower degree of methylation is present. Formyl groups may be removed by dilute acid (7) and acetyl-Ser and -Thr containing peptides may be 'unblocked' sufficiently by exploiting their ability to undergo N- to O- acyl shift followed by β-elimination (39). MS has again proved invaluable for the detection of the presence of acetyl groups in proteins (23) or pyroglutamyl residues at the amino terminus (28).

6.2. Mass spectrometry

MS is playing a rapidly increasing role in protein chemistry and sequencing and is particularly useful in determining sites of co- and post-translational modification (40–42). This technique can, of course, readily analyze peptide mixtures; therefore it is not always necessary to isolate the

Table 9. N-terminal modifications of proteins

(a) Processing of initial transcript, NH_2-*Met*-Xaa-
(i)   No processing NH_2-*Met*-Arg, -Gln, -His, -Ile, -Leu, -Lys, -Met, -Phe, -Trp, -Tyr
(ii) Initiator Met removed NH_2-Pro, -Val, -Cys
(iii) Initiator Met removed and next residue (Xaa) acetylated   *N-acetyl*-Gly, -Ala, -Ser, -Thr, -Met
(iv) Initiator Met acetylated   *N-acetyl-Met*-Asp, *N-acetyl-Met*-Asn, *N-acetyl-Met*-Glu

(b) Fatty acylation
(i)   N-myristoyl-; myristoeyl (i.e. tetradecaenoyl, C14:1)-; lauryl, C12:0)-; tetradecadienoyl, C14:2) -Gly
(ii) Short-chain alkyl, C_4 (i.e. butyl-); C_6 (hexanoyl-); C_8; C_{10} -Ala

(c) N-methylation
(i) $(CH_3)_3$-Ala-, -Phe-, $(CH_3)_2$-Pro- (also mono- and dimethyl-)

(d) N-pyrrolidone carboxyl- (from Gln-)

(e) Bacteria
(i)   N-formyl-Met- (removed then processed as in a(i)–(iv); often found as partial modification in proteins overexpressed in bacteria)
(ii) N-palmitoyl*, S-(sn-1-dipalmitoyl-glyceryl)-Cys- (from proprotein, -Leu-Leu-Ala-Gly-Cys-Ser-Ser-Asn-) * approximately 65% palmitate + other fatty acids

(f) Others
(i)   N-glucuronyl-
(ii) N-glucosyl (aminoketose)
(iii) N-α-ketoacyl-; pyruvoyl-, malonyl-

constituent peptides. A combination of microsequencing and MS techniques are now commonly employed for complete covalent structure determination. Chromatographic techniques include capillary electrophoresis and HPLC, frequently combined with MS and use of radiolabeled precursors. Nuclear magnetic resonance (NMR) has also been widely used, particularly for the determination of the structure and anomeric configuration involved in glycosylation and prenylation.

Selective detection of modified peptides (e.g. phosphopeptides) is possible on ESIMS. Recent methods of phosphorylation site analysis include: (a) conversion of phosphoserine to S-ethylcysteine, which may be followed by ESIMS as well as sequencing (43); (b) production of phosphate-specific fragment ions of 63 Da (PO_2^-) and 79 Da (PO_3^-) by CID during negative ion liquid chromatography (LC)-ESIMS (44). This latter approach has been extended to include other modifications such as glycosylation, sulfation and acylation (45). ESIMS and FABMS have been used in locating disulfide bonds (46). In FABMS, the use of thioglycerol and mixtures of related compounds in the matrix should be avoided for obvious reasons. It is also important to perform the fragmentation of proteins into peptides at low pH (e.g. with pepsin, Glu-C or cyanogen bromide; see *Table 6*) to prevent disulfide exchange. With peptides of mass greater than about 10 000 Da, the change in mass (2 Da) on disulfide bond formation or breakage is likely to be beyond the resolution of ESIMS. ESIMS lends itself to on-line coupling to chromatographic separations, especially HPLC (47). An on-line trapping system to purify/desalt proteins before introduction to the ESIMS source has been developed (47, 48). Recent developments in ESIMS sources permit on-line microbore HPLC using matrices such as 10 μm Poros resins slurry-packed into columns less than 0.25 mm diameter. Polypeptides can be separated on gradients of 5–75% acetonitrile over 2 min in formic or acetic acid (0.1%) (48).

6.3. Interfering salts and buffers

MS analysis is affected, seriously in some cases, by the presence of particular salts, buffers, detergents etc. Lists of acceptable limits of a large number of interfering substances have been compiled for both ESIMS (47, 49) and MALDITOF (50 and Finnegan LaserMAT manual). The effects of these substances have been found to vary quite widely with the instrument and particularly with the ionization source. In general, nonionic saccharide detergents such as n-dodecyl-β-D-glucopyranoside at 0.01% (w/v) have been found to give the best signals in ESIMS. It would appear that critical micelle concentration is not a good predictor of how well a surfactant will perform (49).

Ladder sequencing by MS involves the generation of a set of nested fragments of a polypeptide chain followed by analysis of the mass of each component. Each component in the ragged polypeptide mixture differs from the next by loss of a mass which is characteristic of the residue mass (which may involve a modified side-chain). In this manner, the sequence of the polypeptide can be read from the masses obtained in MS. The ladder of degraded peptides can be generated by Edman chemistry (51) or by exopeptidase digestion from the N- and C-termini (see *Table 6*). This is essentially a subtractive technique (one looks at the mass of the *remaining* fragment after each cycle); for example, when a phosphoserine residue is encountered, a loss of 167.1 Da is observed in place of 87.1 for a serine residue. This technique therefore avoids one of the major problems of analyzing post-translational modifications. Although the majority are stable during the Edman chemistry, *O*- or *S*-linked esters, for example, (which are very numerous, see *Table 2*) may be lost by β-elimination (e.g. *O*-phosphate) during the cleavage step to form the anilinothioazolidone (52) or undergo *O*- or (*S*- in the case of palmitoylated cysteine) to *N*-acyl shifts which block further Edman degradation (53). *Table 10* lists the elution order of PTH-amino acids from the on-line HPLC of an Applied Biosystems sequencer (52, 54). The elution position is approximate; for example, the basic PTH derivatives of His, Arg and methylated Lys (without the ε-PTC group) will show a large shift in elution position depending on the pH and salt concentration.

The technique of ladder sequencing has particular application in MALDITOF-MS which has high sensitivity and greater ability to analyze mixtures. Exopeptidase digestion may be difficult and the rate of release of amino acid may vary greatly. The use of modified Edman chemistry offers the greatest possibilities (55). The modification consists of carrying out the coupling step with phenylisothiocyanate (PITC) in the presence of a small amount of phenylisocyanate (PIC) which acts as a chain-terminating agent. A development of this technique involves the addition of volatile trifluoroethylisothiocyanate (TFEITC) to the reaction tube to which a fresh aliquot of peptide is added after each cycle. This avoids steps to remove excess reagent and byproducts. This may be combined with subsequent modification of the terminal NH_2 group with quaternary ammonium alkyl N-hydroxysuccinimidyl (NHS) esters which allows increased sensitivity in MALDITOF down to 1 pmol level. Post-source decay (PSD) in MALDITOF instruments results in a cleaner fragmentation spectrum consisting mainly of a and d series ions, where quaternary ammonium alkyl NHS ester modification of the terminal NH_2 group increases the sensitivity down to picomole level. A similar compound based on pyridine analogs is used for the low-energy collisions of MS–MS in a triple quadrupole instrument where complete series of b ions have been demonstrated for peptides.

Table 10. Elution order of PTH-derivatives on HPLC

Cysteic acid	N-ε-acetyl lysine	Pyridylethyl-cysteine
DTT (2 peaks)	Hydroxyproline II	5-hydroxylysine
Asparaginyl-N-acetyl-	1-methyl histidine	S-ethanol cysteine
galactosamine	3-methyl histidine	DPTU
Phosphotyrosine	DTT-dehydroalanine	NO_2.tyrosine
γ-carboxyglutamic acid	(derived from Ser, Cys,	DPU
Aspartic acid	Ser-P, etc.)	Iodotyrosine
Asparagine	Biotinyl lysine	S-ethyl cysteine
O-fucosylthreonine	Arginine	α,γ-diaminobutyric acid
Serine	O-methyl threonine	Tryptophan
S-carboxymethyl cysteine	Cystine	Ornithine
Glutamine	O-methyl glutamic acid	Formyl-Tryptophan
Threonine	Tyrosine	p-NO_2.phenylalanine
Citrulline	DTT adduct of dehydro-α-	Phenylalanine
Homoserine	aminoisobutyric acid	Isoleucine
Glycine	(derived from threonine);	Lysine (PTC)
Tris	2-4 peaks	O-methyltyrosine
Glutamic acid	N-ε-methyl lysine	Lanthionine
DMPTU	N-ε-dimethyl lysine	Leucine
S-carboxamidomethyl cysteine	N-ε-trimethyl lysine	Fluorophenylalanine
Carboxamidomethyl methionine	Canavanine	Norleucine
Methionine sulfone	α-aminobutyric acid	p-chlorophenylalanine
N-ε-acetyl lysine	Methyl arginine	Diiodotyrosine
N-ε-succinyl lysine	S-methyl cysteine	N-ε-methyl lysine
Histidine	Homocystine	(PTC derivative)
5-hydroxylysine derivative	Proline	N-ε-dimethyl lysine
Hydroxyproline I	Methionine	(PTC derivative)
Alanine	Valine	

Protein Chemistry

Some fragmentation may be obtained in ESIMS with capillary-skimmer source fragmentation, when analyzing pure peptides. More sequence information has been generally obtained using tandem mass spectrometry after collision-induced dissociation (56). However, an ion trap quadrupole MS (called 'LCQ') has just been launched by Finnegan-MAT which permits sequence information to be readily obtained. Not only can MS–MS analysis be carried out but due to the high efficiency of each stage, further fragmentation of selected ions may be carried out to MS^n. The charge state of peptide ions is readily determined by a 'zoom-scan' technique which resolves the isotopic envelopes of multiply charged peptide ions. The instrument still allows accurate mass determination to over 100 000 Da at 0.01% mass accuracy.

Recent developments in mass spectrometry, therefore, enable the amino acid sequence and structure of all types of modified peptides to be determined at very high sensitivity.

7. REFERENCES

1. Cattaneo, R. (1992) *Trends Biochem Sci.*, **17**, 4.

2. Colston, M.J. and Davis, E.O. (1994) *Mol. Microbiol.*, **12**, 359.

3. Page, M.J., Aitken, A., Cooper, D.J., Magee, A.I. and Lowe, P.N. (1990) *Methods: A Companion to Methods in Enzymology*, **1**, 221.

4. Uy, R. and Wold, F. (1977) *Science*, **198**, 890.

5. Wold, F. (1981) *Ann. Rev. Biochem.*, **50**, 783.

6. Yan, C.B., Grinnell, B.W. and Wold, F. (1989) *Trends Biochem Sci.*, **14**, 264.

7. Aitken, A. (1990) in *Identification of Protein Consensus Sequences*. Ellis-Horwood, Chichester/Simon and Schuster, New York, p 1.

8. Stadtman, T.C. (1991) *J. Biol. Chem.*, **266**, 16257.

9. Appel, R.D., Bairoch, A. and Hochstrasser, D.F. (1994) *Trends Biochem. Sci.*, **19**, 258.

10. Moore, S. and Stein, W.H. (1963) *Methods Enzymol.*, **6**, 819.

11. Westall, F. and Hesser, H. (1974) *Anal. Biochem.*, **61**, 610.

12. Tsugita, A. and Scheffler, J.-J. (1982) *Eur. J. Biochem.*, **124**, 585.

13. Swadesh, J.K., Thannhauser, T.W. and Scheragha, H.A. (1984) *Anal. Biochem.*, **141**, 397.

14. Matsubara, H. and Sasaki, R.M. (1969) *Biochem. Biophys. Res. Commun.*, **35**, 175.

15. Liu, T-Y. and Chang, Y.H. (1971) *J. Biol. Chem.*, **246**, 2842.

16. Simpson, R.J., Neuberger, M.R. and Liu, T.Y. (1976) *J. Biol. Chem.* **251**, 1936.

17. Penke, B., Ferenczi, R. and Kovacs, K. (1974) *Anal. Biochem.*, **60**, 45.

18. Kemp, B.E. (1980) *FEBS Lett.*, **110**, 308.

19. Martensen, T.M. (1982) *J. Biol. Chem.*, **257**, 9648.

20. Aitken, A. and Stanier, R.Y. (1979) *J. Gen. Microbiol.*, **112**, 219.

21. Aitken, A., Geisow, M.J., Findlay, J.B.C., Holmes, C. and Yarwood, A. (1989) in *Protein Sequencing: A Practical Approach* (M.J. Geisow and J.B.C. Findlay, eds). IRL Press, Oxford, p. 43.

22. Keil, B. and Tong, N.T. (1988) in *Methods in Protein Sequence Analysis* (B. Wittman-Liebold, ed.). Springer-Verlag, Berlin, p. 365.

23. Slabas, A.R., Aitken, A., Howell, S., Welham, K. and Sidebottom, C.M. (1989) *Biochem. Soc. Trans.*, **17**, 886.

24. Walsh, K.A. and Sasagawa, T. (1986) *Methods Enzymol.*, **106**, 22.

25. Aitken, A. and Cohen, P. (1984) *Methods Enzymol.*, **106**, 205.

26. Aitken, A. and Cohen, P. (1982) *FEBS Lett.*, **147**, 54.

27. Deshpande, K.L., Fried,V.A, Ando, M. and Webster, R.G. (1987) *Proc. Natl Acad. Sci. USA*, **84**, 36.

28. Cockle, S.M., Aitken, A., Beg, F. and Smyth, D.G. (1989) *J. Biol. Chem.*, **264**, 7788.

29. Lucas, J. and Henschen, A. (1986) *J. Chromatography*, **369**, 357.

30. McCarthy, A.D., Aitken, A. and Hardie, D.G. (1983) *Eur. J. Biochem.*, **136**, 501.

31. Bachmair, A., Finley, D. and Varshavsky, A. (1986) *Science,* **234**, 179.

32. Bradshaw, R.A. (1989) *Trends Biochem. Sci.*, **14**, 276.

33. Arfin, S.M. and Bradshaw, R.A. (1988) *Biochemistry*, **27**, 7984.

34. Flinta, C., Persson, B., Jornvall, H. and von Heijne, G. (1986) *Eur. J. Biochem.*, **154**, 193.

35. Kobayashi, K. and Smith, J.A. (1987) *J. Biol. Chem.*, **262**, 11435.

36. Radhakrishna, G. and Wold, F. (1989) *J. Biol. Chem.*, **264**, 11076.

37. Farries, T.C., Harris, A., Auffret, A.D. and Aitken, A. (1991) *Eur. J. Biochem.*, **196**, 679.

38. Augen, J. and Wold, W. (1986) *Trends Biochem. Sci.*, **11**, 494.

39. Wellner, D., Panneerselvam, C. and Horecker, B.L. (1990) *Proc. Natl Acad. Sci. USA*, **87**, 1947.

40. Aitken, A. (1989) in *Focus on Laboratory Methodology in Biochemistry* (C. Fini, *et al.*, eds). CRC Press Inc., Boca Raton, FL, Vol. 1, p. 9.

41. Burlingame, A.L., Boyd, R.K. and Gaskell, S.J. (1994) *Anal. Chem.*, **66**, 634R.

42. Mann, M. and Wilm, M. (1995) *Trends Biochem. Sci.*, **20**, 219.

43. Aitken, A., Patel, Y., Martin, H., Jones, D., Robinson, K., Madrazo, J. and Howell, S. (1994) *J. Prot. Chem.*, **13**, 463.

44. Huddleston, M.J., Annan, R.S., Bean, M.F. and Carr, S.A. (1994) in *Techniques in Protein Chemistry V* (J.W. Crabb, ed.). Academic Press, New York, p. 123.

45. Bean, M.F., Annan, R.S., Hemling, M.E., Menter, M., Huddleston, M.J., and Carr, S.A. (1994) in *Techniques in Protein Chemistry VI.* (J.W. Crabb, ed.). Academic Press, New York. p. 107.

46. Aitken, A. (1994) in *Methods in Molecular Biology* (J.M. Walker, ed.). Humana Press, Inc., Totowa, NJ, Vol. 32, p. 351.

47. Kay, I. and Mallet, A.I. (1993) *Rapid Comm. Mass Spec.*, **7**, 744.

48. Aitken, A., Howell, S., Jones, D., Madrazo, J. and Patel, Y. (1995). *J. Biol. Chem.*, **270**, 5706.

49. Loo, R., Dales, N. and Andrews, P.C. (1994) *Protein Sci.,* **3**, 1975.

50. Vorm, Chait, B.T. and Roepstorff, P. (1993) *Proceedings of the 41st ASMS Conference on Mass Spectrometry and Allied Topics.* ASMS, East Lansing, p. 621.

51. Chait, B.T., Wang, R., Beavis, R.C. and Kent, S.B.H. (1993) *Science,* **262**, 89.

52. Geisow, M.J. and Aitken, A. (1989) in *Protein Sequencing: A Practical Approach* (M.J. Geisow and J.B.C. Findlay, eds). IRL Press, Oxford, p. 85.

53. Aitken, A. (1992) in *Lipid Modification of Proteins: A Practical Approach* (N.M. Hooper and A.J. Turner, eds). IRL Press, Oxford, p. 63.

54. Crankshaw, M.W. and Grant, G.A. (1993) *Identification of modified PTH-amino acids in protein sequence analysis (1st Edn).* Compiled for Assocn of Biomol. Resource Facilities.

55. Bartlet-Jones, M., Jeffery, W.A., Hansen, H.F. and Pappin, D.J.C. (1994) *Rapid Comm. Mass Spec.*, **8**, 737.

56. Hunt, D.F., Yates, J.R., Shabanowitz, J., Winston, S. and Hauer, C.R. (1986) *Proc. Natl Acad. Sci. USA*, **83**, 6233.

CHAPTER 24
CHEMICAL MODIFICATION OF AMINO ACID SIDE-CHAINS
R. Lundblad

1. GENERAL ASPECTS OF THE CHEMICAL MODIFICATION OF PROTEINS

The purpose of this chapter is to explore briefly the use of chemical modification to study the relationship between structure and function in proteins. More complete information regarding the use of chemical modification to study structure–function relationships in proteins can be found in several more encyclopedic works (1–8). While it could be argued that the use of chemical modification to study protein structure has been supplanted by protein engineering (9–14), it is clear that the combined use of the two techniques can be extremely powerful (15–26).

This chapter will focus on the use of site-specific chemical modification to study protein function. Site-specific chemical modification is defined as the process by which a single class of chemical groups is modified in a protein with a chemical reagent. Such modification can be absolutely site-specific, as is the situation with affinity labels (27–29; see also Chapter 25), or a measure of the unique reactivity of a particular functional group, as is the situation with the reaction of active-site histidine residues with α-halo acids (30–33). Modification within a single class of residues can be somewhat less specific and even random (34, 35).

While it is rare that a chemical reagent will react with a unique group in a protein, even with multiple reactive sites, information regarding the relative importance of specific amino acid residues and/or functional domains in proteins can be obtained through the careful use of the reagents described in this chapter.

2. CHARACTERIZATION OF SITE-SPECIFIC MODIFIED PROTEINS

2.1. Establishment of reaction stoichiometry

The determination of the stoichiometry of the modification of a protein by a chemical reagent can be either very easy or impossible, depending on the reagent or amino acid modified. If the reagent has a property such as a unique spectral absorption or radioactivity, then the extent of modification can be determined by characterizing the modified protein. Examples of reagents with spectral properties which can be used to establish reaction stoichiometry include 5,5′-dithiobis(2–nitrobenzoate) (36–38), tetranitromethane (39, 40), 2-hydroxy-5-nitrobenzyl bromide (41, 42), ninhydrin (43), 2,4,6-trinitrobenzenesulfonic acid (44–45) and diethylpyrocarbonate (46). Radiolabels have been used extensively and frequently a radiolabeled moiety is attached to a base chemical reagent. Examples of this approach include iodinated compounds attached to base chemical reagents such as N-ethylmaleimide (47–49) or the α-halo acids (50–52). On occasion, the extent of modification can be determined by the loss of the amino acid modified on amino acid analysis as in the case of photooxidation (53–55). Amino acid analysis of the modified proteins is strongly encouraged as this permits an approach to the confirmation of reaction specificity. There are situations where the modified amino acid has properties which can be used for quantification. An example of this approach is the spectral analysis of tryptophan modification by *N*-bromosuccinimide (56–58).

In addition to the determination of the concentration of the modified residue(s), it is necessary also to have an accurate estimate of the concentration of the modified proteins. Absorbance at 280 nm can frequently be used for this purpose provided the modification reaction does not involve tryptophan or tyrosine (while both phenylalanine and cystine contribute to absorbance at 280 nm, the contribution is negligible). In situations where the modification does involve tryptophan or tyrosine, a colorimetric protein assay such as a dye-binding assay (59, 60), the Lowry assay (61) or related bicinchoninic acid assay (62) or the biuret reaction (63) should be used. Care should be taken to use an appropriate standard because the various assays can have qualitative differences (64). The biuret reaction shows the least variation between different proteins but is by far the least sensitive of the assays available. The reader is referred to Chapter 4 for a further discussion of this topic.

2.2. Correlation of the extent of modification with concomitant changes in biological activity

Correlation of the extent of chemical modification with concomitant changes in biological activity can be either an easy task or somewhat more difficult, depending on the functional role of the amino acid or amino acids modified. The modification of a residue that is actually involved in the catalytic process generally results in a total loss of enzymatic activity, while the modification of a residue that is involved in substrate binding or an allosteric effect generally does not result in a complete loss of enzymatic activity. In fact, in some situations, site-specific chemical modification might result in either no observed change in activity or an increase in activity.

The most frequent situation is where there are multiple sites of modification. While analysis might be tedious, it is possible to obtain satisfactory information from such experiments (see Chapter 1 in ref. 2). There are examples of multiple sites of modification in a preparation where the presence of multiple sites of modification reflects the presence of unique modified species rather than multiple sites of modification on a single protein. The classic example is provided by the reaction of α-haloalkanoic acids with bovine pancreatic ribonuclease A (32) where it is possible to separate the carboxymethyl-His[119] derivative from the carboxymethyl-His[12] derivative (65). Where there are multiple sites of modification on the same protein, rigorous assignment of changes in activity to a single site requires the ability to quantify modification at each site and then correlate such modification with changes in biological activity (66, 67).

Modifications of enzymes where there is incomplete loss of catalytic activity on stoichiometric modification require additional analysis. Residual activity must be accounted for in such analysis. This is fairly straightforward as advanced by Levy *et al.*, (68) and by Tsou (69). These approaches have been used by a number of other investigators (70–74).

2.3. Identification of residue(s) modified

This involves the structural analysis of the protein. The general approach involves the fragmentation of the protein followed by identification of the modified residue. The usual fragmentation scheme will involve cyanogen bromide cleavage followed by proteolytic enzyme digestion (75, 76). The resulting peptides are then resolved by high-performance liquid chromatography (HPLC) and the modified residue(s) identified (see Chapter 23). In classic technology, the identification involved a spectral probe (77) or the incorporation of radiolabel (78) or an altered chromatographic pattern where the modified residue was identified by amino acid analysis (79). More recent experimental approaches have used mass spectrometry for identification of the modified residue(s). The reader is referred to Chapter 16 and other recent works in this area (76, 80–84).

2.4. Comparison of the conformation of the modified protein to that of the native protein

One of the most frequent criticisms of the use of site-specific chemical modification of proteins has been that such modification causes major conformational change in the protein and that changes in biological activity are a reflection of such changes rather than the modification of a specific amino acid residue. It should be noted that such criticism could also be made regarding protein engineering.

There are several approaches to this issue. The first is to utilize a reversible modification such as N-acetylimidazole (85) or citraconic anhydride (86). This approach, however, will be the exception as opposed to the rule. A more satisfactory approach is to use a sensitive affinity procedure to purify the modified protein. Examples would include affinity chromatography (87) or immunoaffinity chromatography (88). More sophisticated approaches might include the application of nuclear magnetic resonance (NMR; see Chapter 19 and refs 89–93). The application of techniques such as 'trace labeling' (94–98) to such conformational analysis would also prove useful, as would be the use of conformational probes (99–102) or circular dichroism (CD; see Chapter 18 and ref. 103).

3. MODIFICATION OF SPECIFIC AMINO ACID RESIDUES

3.1. Cysteine

Cysteine is potentially the most powerful nucleophile in a protein and, as a result, is frequently the easiest to modify with a variety of reagents. The most commonly used reagents are listed in *Table 1*. The chemistry of cysteine has been reviewed and the reader should consult these reviews for further detail (8, 104). The unique reactivity of cysteine has prompted investigators to use site-specific mutagenesis to place cysteine at particular points in a protein for the subsequent attachment of structural probes (105, 106). Dithiobis(2-nitrobenzoic acid) has been used frequently, as the extent of reaction can be easily determined by spectral measurement (107–109).

Table 1. Reagents for the modification of cysteine in proteins

Reagent	Other AA[c] modified	Ref.
Iodoacetate[a]	Histidine, lysine, methionine	114
N-Ethylmaleimide	Lysine	115
5,5'-dithiobis(2-nitrobenzoic acid)	–	116
p-Hydroxymercuribenzoate[b]	–	117

[a] This includes related α-haloketo compounds such as bromoalkanoic acids, chloroalkanoic acids and the related amides.
[b] Included are related organic mercurial derivatives including mercurinitrophenol derivatives and mercuriphenylsulfonate derivatives.
[c] AA = amino acids.

The base compounds listed in *Table 1* have been used as means for attaching structural probes. Examples include α-haloalkanoic acids (110, 111) and *N*-alkylmaleimides (112, 113).

3.2. Cystine

Cystine is responsible for connecting portions of the protein together. Modification of cystine is accomplished by using one of the various reagents listed in *Table 2* to reduce the disulfide bond resulting in the formation of cysteine. The use of oxidation to break the disulfide bonds (118) is not included in this list. The reduction of cystine is generally used as a mechanism to denature the protein fully prior to structural analysis (119).

Table 2. Reagents for the modification of cystine in proteins

Reagent	Other AA modified	Ref.
2-Mercaptoethanol	–	120
Dithiothreitol (DTT)	–	121
Tris–n–butylphosphine	–	122
Tris–n–(2-carbo-ethoxy)phosphine	–	123
Sulfite	–	124

3.3. Methionine

The modification of methionine in a protein (see *Table 3*) is generally accomplished with considerable difficulty. This is a reflection of the fact that, as a relatively hydrophobic residue, methionine is frequently buried in a protein (125). As a reflection of the relatively drastic solvent conditions which are generally required for the modification of methionine, group-specific modification is difficult to obtain. Cleavage at methionine with cyanogen bromide is frequently used during the structural analysis of a protein (126) as, due to its relatively infrequent occurrence (127), large fragments are obtained.

Table 3. Reagents for the modification of methionine in proteins

Reagent	Other AA modified	Ref.
Chloramine T	Cysteine, cystine, histidine, tyrosine, tryptophan	128
Hydrogen peroxide	Cysteine, cystine, histidine	129
Iodoacetate	Cysteine, histidine, lysine	130
N-Chloro-succinimide	–	131

3.4. Histidine

Modification of histidine residues in enzymes has been a popular practice reflecting its importance in the catalytic process (see *Table 4*). However, until the introduction of diethyl pyrocarbonate(ethoxyformic anhydride) (132), modification of this residue in proteins was often

Table 4. Reagents for the modification of histidine in proteins

Reagent	Other AA modified	Ref.
Photooxidation[a]	Cysteine, methionine, cystine, tryptophan	135
Diethyl pyrocarbonate (ethoxyformic anhydride)	Tyrosine, lysine	136
Iodoacetate[b]	Cysteine, methionine	137

[a] Photooxidation is a relatively nonspecific reaction where a photosensitive compound such as methylene blue (138) is used to capture the energy of radiation which is then used to oxidize available sensitive residues (139).
[b] Included in this class of compounds are other α-haloalkanoic acids such as bromoacetate and chloroacetate. The corresponding amide derivatives such as iodoacetamide are also included.

more fortuitous than not. An example is provided by the study of Liu with streptococcal proteinase (133), where it was necessary to block a reactive sulfydryl group prior to modification of the histidine residue. The reactivity of histidine with α-haloketones has been utilized for the development of peptide chloromethylketones used extensively with serine proteinases (134).

3.5. Lysine

The modification of lysine is relatively easy to obtain (see *Table 5*). It is, however, somewhat more difficult to obtain site-specific modification. Pyridoxal-5′-phosphate has been used to obtain site-specific modification as a reflection of the charge on the modifying reagent to target the modification to basic regions of the protein. A particularly good example is provided by the

Table 5. Reagents for the modification of lysine in proteins

Reagent	Other AA modified	Ref.
2,4-Dinitrofluorobenzene	α-amino groups, histidine, cysteine	149
Acetic anhydride	α-amino groups, tyrosine[a]	150
Methyl acetyl phosphate	α-amino groups	151
Citraconic anhydride (methylmaleic anhydride)	α-amino groups, tyrosine	152
Cyanate	α-amino groups[b], cysteine	153
Imidoesters	–	154
Pyridoxal-5′-phosphate	–	155
Reductive alkylation[c]	–	156
2,4,6-Trinitrobenzenesulfonic acid[d]	α-amino groups	157

[a] Reaction at tyrosine residues yields a derivative which is unstable. Incubation of the modified protein at alkaline pH results in the facile hydrolysis of these derivatives. This is also true of succinylation. Reaction with maleic anhydride generally results in an extremely transient derivative of tyrosine and the formation of amino derivatives can be readily reversed at low pH.
[b] Reaction at amino groups occurs more readily than reaction at the ε-amino groups of lysine. Since cyanate can occur as a result of the dismutation of urea (158), blockage of α-amino groups can occur during the use of urea as a denaturing agent. As a result, the use of freshly recrystallized urea is recommended, as is the inclusion of a chelating agent such as ethylenediaminetetraacetic acid (EDTA).
[c] The reaction of glucose with proteins can readily occur (159) and this can result in the formation of a variety of products due to the Amadori rearrangement (160, 161).
[d] Titration with 2,4,6-trinitrobenzoic acid (TNBS) is used to estimate the free amino groups in a protein (162, 163).

Chemical Modification

modification of thrombin, where reaction occurs in the anion-binding exosite (140). Lysine residues must be unprotonated to function as satisfactory nucleophiles, so alkaline pH conditions are required.

Reaction at lysine has been used to attach a variety of structural probes (141–145). In addition, the technique of trace labeling has been used to explore conformational changes in proteins (146–148).

3.6. Arginine

The site-specific modification of arginine is almost impossible to achieve with currently available reagents (see *Table 6*). While not directly involved in catalysis, arginine residues are considered to function as general anion recognition sites in proteins (164). It is, however, relatively easy to obtain group-specific modification. The discovery of the accelerating effect of bicarbonate on the reaction (165) greatly facilitated the study of the chemistry of this reaction. 2,3-Butanedione has been the most frequently used reagent.

Table 6. Reagents for the modification of arginine in proteins

Reagent	Other AA modified	Ref.
Phenylglyoxal	–	166
2,3-Butanedione	–	167
1,2-Cyclohexanedione	–	168

3.7. Tryptophan

As with methionine, the facile modification of tryptophanyl residues in proteins has been difficult to obtain. In general, it is possible to obtain group-specific modification by reaction at low pH (low pH is required to obviate reaction at other nucleophiles in the protein) with reagents such as 2-hydroxy-5-nitrobenzyl bromide (see *Table 7*). Specificity with this class of reagents has been obtained by the use of derivatives such as alkoxy derivatives which are hydrolyzed by the enzyme (i.e. chymotrypsin) to release the active reagent (169, 170). *N*-Bromosuccinimide has been used more extensively than other reagents, but pH is an important consideration to obtain reaction specificity. Titration with *N*-bromosuccinimide has been used to quantify tryptophanyl residues in proteins but this has been largely supplanted by the development of hydrolysis conditions that do not result in the loss of tryptophanyl residues (171–173). Cleavage at tryptophan with *N*-bromosuccinimide has also been observed but the use of *N*-chlorosuccinimide is preferred for this purpose (174, 175).

Table 7. Reagents for the modification of tryptophan in proteins

Reagent	Other AA modified	Ref.
N-Bromosuccinimide	Histidine, cysteine, methionine	176
2-Hydroxy-5-nitrobenzyl bromide		177
2-Nitrophenylsulfenyl chloride		178

3.8. Tyrosine

The group-specific modification of tyrosine residues is relatively easy to achieve in most proteins (see *Table 8*). Side-reactions, which are possible, are generally not a problem. Notable among the side-reactions are the acetylation of amino groups with *N*-acetylimidazole (these are easy to differentiate from the modification of tyrosine as the the O-acetylation of tyrosine is base labile) (179, 180), the modification of tryptophan (181, 182) or crosslinking of peptide chains (183, 184) with tetranitromethane.

Table 8. Reagents for the modification of tyrosine in proteins

Reagent	Other AA modified	Ref.
Arsanilic acid (diazotized)[a]	Lysine, histidine	185
N-Acetylimidazole	Lysine, histidine	186
Tetranitromethane	Tryptophan, cross-linking, cysteine	187

[a] Included in this category are various diazotized aromatic compounds.

3.9. Carboxyl groups

It is relatively easy to obtain group-specific modification of carboxyl groups in proteins (see *Table 9*). It is much more difficult to obtain site-specific modification, or even differentiation between aspartic and glutamic acid residues. As a reflection of this specificity difficulty and the necessity to work at acid pH, there has been limited effort in this area. Carbodiimide-mediated activation of carboxyl groups has been used for protein cross-linking (188–190).

Table 9. Reagents for the modification of carboxyl groups in proteins

Reagent	Other AA modified	Ref.
Alkyl/aryl isoxazolium salts	–	191
Carbodiimides	Tyrosine	192

4. REFERENCES

1. Lundblad, R.L. (1991) in *Chemical Reagents for Protein Modification (2nd Edn)*. CRC Press, Boca Raton, FL.

2. Lundblad, R.L. and Noyes, C.M. (1984), *Chemical Reagents for the Modification of Proteins*, Vols. I, II. CRC Press, Boca Raton, FL.

3. Means, G.R. and Feeney, R.E. (1971) *Chemical Modification of Proteins*. Holden-Day, San Francisco, CA.

4. Feeney, R.E. (1987) *Int. J. Pept. Prot. Res.*, **29**, 145.

5. Hirs, C.H.W. and Timasheff, S.N. (1977), *Methods Enzymol.*, **47**, 3.

6. Kaiser, E.T., Lawrence, D.S. and Rokita, S.E. (1985) *Annu. Rev. Biochem.*, **54**, 565.

7. Hirs, C.H.W. and Timasheff, S.N. (1983) *Methods Enzymol.*, **91**, 2.

8. Lundblad, R.L. (1994) *Techniques in Protein Modification*. CRC Press, Boca Raton, FL.

9. Wells, J.A. and Estell, D.A. (1988) *Trends Biochem. Sci.*, **13**, 291.

Chemical Modification

10. Graham, L.D., Haggett, K.D., Jennings, P.A., Le Brocque, D.S. and Whittaker, R.G. (1993) *Biochemistry*, **32**, 6250.

11. Ohnishi, H., Matsumoto, H., Sakai, H. and Ohta, T. (1994) *J. Biol. Chem.*, **269**, 3503.

12. Sheffield, W.P. and Blajchman, M.A. (1994) *FEBS Lett.*, **339**, 147.

13. Richardson, M.A., Gerlitz, B. and Grinnell, B.W. (1994) *Protein Sci.*, **3**, 711.

14. Perona, J.J. and Craik, C.S. (1995) *Protein Sci.*, **4**, 337.

15. Nureki, O., Kohno, T., Sakamoto, K., Miyazawa, T. and Yokoyama, S. (1993) *J. Biol. Chem.*, **268**, 15368.

16. Miyazaki, K., Kadono, S., Sakurai, M., Moriyama, H., Tanaka, N. and Oshima, T. (1994) *Protein Engineering*, **7**, 99.

17. Ohnishi, H., Matsumoto, H., Sakai, H. and Ohta, T. (1994) *J. Biol. Chem.*, **269**, 3503.

18. Ghose-Dastidar, J., Green, R. and Ross, J.B.A. (1994) *J. Steroid Biochem. Mol. Biol.*, **48**, 139.

19. Corbier, C., Michels, S., Wonacott, A.J. and Branlant, G. (1994) *Biochemistry*, **33**, 3260.

20. Kim, D.-W., Yoshimura, T., Esaki, N., Satoh, E. and Soda, K. (1994) *J. Biochem. (Tokyo)*, **115**, 93.

21. Gross, M., Furter-Graves, E.M., Walliman, T., Eppenberger, H.M. and Furter, R. (1994) *Protein Sci.*, **3**, 1058.

22. Jung, H., Jung, K. and Kaback, H.R. (1994) *Protein Sci.*, **3**, 1052.

23. Brown, N.F., Anderson, R.C., Caplan, S.L., Foster, D.W. and McGarry, J.D. (1994) *J. Biol. Chem.*, **269**, 19157.

24. Charng, Y., Iglesias, A.A. and Preiss, J. (1994) *J. Biol. Chem.*, **269**, 24107.

25. Kataoka, K., Tanizawa, K., Fukui, T., Ueno, H., Yoshimura, T., Esaki, N. and Soda, K. (1994) *J. Biochem. (Tokyo)*, **116**, 1370.

26. Czupryn, M.J., McCoy, J.M. and Scoble, H.A. (1995) *J. Biol. Chem.*, **270**, 978.

27. Chuan, H. and Wang, J.H. (1988) *J. Biol. Chem.*, **263**, 13003.

28. Cooperman, B.S. (1988) *Methods Enzymol.*, **164**, 341.

29. Plapp, B.V. (1982) *Methods Enzymol.*, **87**, 469.

30. Stark, G.R., Stein, W.H. and Moore, S. (1961) *J. Biol. Chem.*, **236**, 436.

31. Heinrikson, R.L., Stein, W.H., Crestfield, A.M. and Moore, S. (1965) *J. Biol. Chem.*, **240**, 2921.

32. Fruchter, R.G. and Crestfield, A.M. (1967) *J. Biol. Chem.*, **242**, 5807.

33. Lin, M.C., Stein, W.H. and Moore, S. (1968) *J. Biol. Chem.*, **243**, 6167.

34. Ueno, H, Popowizc, A.M. and Manning, J.M. (1993) *J. Prot. Chem.*, **12**, 561.

35. Bazaes, S., Silva, R., Goldie, H., Cardemil, E. and Jabalquinto, A.M. (1993) *J. Prot. Chem.*, **12**, 571.

36. Sheikh, S. and Katiyar, S.S. (1993) *Biochim. Biophys. Acta*, **1202**, 251.

37. Chang, G.-G., Satterlee, J. and Hsu, R.Y. (1993) *J. Prot. Chem.*, **12**, 7.

38. Nakayama, T., Tanabe, H., Deyashiki, Y., Shinoda, M., Hara, A. and Sawada, H. (1992) *Biochim. Biophys. Acta.*, **1120**, 144.

39. Caruso, C., Cacace, M.G. and Di Prisco, G. (1987) *Eur. J. Biochem.*, **166**, 547.

40. Skawinski, W.J., Adebodun, F., Cheng, J.T., Jordan, F. and Mendelsohn, R. (1993) *Biochim. Biophys. Acta*, **1162**, 297.

41. Kirtley, M.E. and Koshland, D.E.,Jr. (1966) *Biochem. Biophys. Res. Commun.*, **23**, 810.

42. Loudon, G.M. and Koshland, D.E.,Jr. (1970) *J. Biol. Chem.*, **245**, 2247.

43. Takahashi, K. (1976) *J. Biochem. (Tokyo)*, **80**, 1173.

44. Haniu, M., Yuan, H., Chen, S., Iyanagi, T., Lee, T.D. and Shively, J.E. (1988) *Biochemistry*, **27**, 6877.

45. Xia, C., Meyer, D.J., Chen, H., Reinemer, P., Huber, R. and Ketterer, B. (1993) *Biochem. J.*, **293**, 357.

46. Altman, J., Lipka, J.J., Kuntz, I. and Waskell, L. (1989) *Biochemistry*, **28**, 7516.

47. Brown, R.D. and Matthews, K.S. (1979) *J. Biol. Chem.*, **254**, 5128.

48. Brown, R.D. and Matthews, K.S. (1979) *J. Biol. Chem.*, **254**, 5135.

49. Marquez, J., Iriarte, A. and Martinez-Carrion, M. (1989) *Biochemistry*, **28**, 7433.

50. Kaslow, H.R., Schlotterback, J.D., Mar, V.L. and Burnette, W.N. (1989) *J. Biol. Chem.*, **264**, 6386.

51. Makinen, A.L. and Nowak, T. (1989) *J. Biol. Chem.*, **264**, 12148.

52. First, E.A. and Taylor, S.S. (1989) *Biochemistry*, **28**, 3598.

53. Ray, W.J., Jr. and Koshland, D.E., Jr. (1962) *J. Biol. Chem.*, **237**, 2493.

54. Nakamura, S. and Kaziro, Y. (1981) *J. Biochem. (Tokyo)*, **90**, 1117.

55. Funakoshi, T., Abe, M., Sakata, M., Shoji, S. and Kubota, Y. (1990) *Biochem. Biophys. Res. Commun.*, **168**, 125.

56. O'Gorman, R.B. and Matthews, K.S. (1977) *J. Biol. Chem.*, **252**, 3572.

57. Spande, T.F. and Witkop, B. (1967) *Methods Enzymol.*, **11**, 498.

58. Ohnishi, M., Kawagishi, T. and Hiromi, K. (1989) *Arch. Biochem. Biophys.*, **272**, 46.

59. Bradford, M.M. (1976) *Anal. Biochem.*, **72**, 248.

60. Sedmak, J.J. and Grossberg, S.E. (1977) *Anal. Biochem.*, **79**, 544.

61. Lowry, O.H., Rosebrough, N.J, Farr, A.L. and Randall, R.J. (1951) *J. Biol. Chem.*, **193**, 265.

62. Smith, P.K., Krohn, R.I., Hermanson, G.T., Mallia, A.K., Gartner, F.H., Provesano, M.D., Fujimoto, E.K., Goeke, N.M., Olson, B.J. and Klenk, D.C. (1985) *Anal. Biochem.*, **150**, 76.

63. Gornall, A.G., Bardawill, C.J. and David, M.M. (1949) *J. Biol. Chem.*, **177**, 751.

64. Jenzano, J.W., Hogan, S.L., Noyes, C.M., Featherstone, G.L. and Lundblad, R.L. (1986) *Anal. Biochem.*, **159**, 370.

65. Crestfield, A.M., Stein, W.H. and Moore, S. (1963) *J. Biol. Chem.*, **238**, 2413.

66. Ray, W.J. and Koshland, D.E.,Jr. (1961) *J. Biol. Chem.*, **236**, 1973.

67. Horike, K. and McCormick, D.B. (1979) *J. Theoret. Biol.*, **79**, 403.

68. Levy, H.M., Leber, P.D. and Ryan, E.M. (1963) *J. Biol. Chem.*, **238**, 3654.

69. Tsou, C.-L. (1962) *Scientia Sinica,* **11**, 1535.

70. Dunham, K.R. and Selman, B.R. (1981) *J. Biol. Chem.*, **256**, 212.

71. Chang, G.-G. and Huang, T.-M. (1981) *Biochim. Biophys. Acta*, **660**, 341.

72. Bhagwat, A.S. and Ramakrishna, J. (1981) *Biochim. Biophys. Acta*, **662**, 181.

73. Carrillo, N., Arana, J.L. and Vallejos, R.H. (1981) *J. Biol. Chem.*, **256**, 6823.

74. Lundblad, R.L., Noyes, C.M., Featherstone, G.L., Harrison, J.H. and Jenzano, J.W. (1988) *J. Biol. Chem.*, **263**, 3729.

75. Shively, J. (ed.) (1986) *Methods of Protein Microcharacterization.* Humana Press, Inc., Clifton, NJ.

76. Shively, J.E. (1994) *Methods*, **6**, 207.

77. Kobayashi, R., Kanatani, A., Yoshimoto, T. and Tsuru, D. (1989) *J. Biochem. (Tokyo)*, **106**, 1110.

78. Paine, L.J., Perry, N., Popplewell, A.G., Gore, A.G. and Atkinson, T. (1993) *Biochim. Biophys. Acta*, **1202**, 235.

79. Lin, S.-R. and Chang, C.-C. (1992) *Biochim. Biophys. Acta*, **1159**, 255.

80. Smith, J.B., Thevenon-Emeric, G., Smith, D.L. and Green, B. (1991) *Anal. Biochem.*, **193**, 118.

81. Carr, S.A., Hemling, M.E., Bean, M.F. and Roberts, G.D. (1991) *Anal. Chem.*, **63**, 2802.

82. Tuinman, A.A., Thomas, D.A., Cook, K.D., Xue, C.-B., Naider, F. and Becker, J.M. (1991) *Anal. Biochem.*, **193**, 173.

83. Rosenfeld, R., Philo, J.S., Haniu, M., Stoney, K., Rohde, M.F., Wu, G.-M., Narhi, L.O., Wong, C., Boone, T., Hawkins, N.N., Miller, J.M. and Arakawa, T. (1993) *Protein Sci.*, **2**, 1664.

84. de Almeida Oliveira, M.G., Rogana, E., Rosa, J.C., Reinhold, B.B., Andrade, M.H., Greene, L.J. and Mares-Guia, M. (1993) *J. Biol. Chem.*, **268**, 26893.

85. Arguello, J.M. and Kaplan, J.H. (1990) *Biochemistry*, **29**, 5775.

86. Atassi, M.Z. and Habeeb, A.F.S.A. (1972) *Methods Enzymol.*, **25**, 546.

87. Lundblad, R.L., Tsai, J., Wu, H., Jenzano, J.W., White, G.C., II and Connolly, T.M. (1993) *Arch. Biochem. Biophys.*, **302**, 109.

88. Murphy, R.F., Imam, A., Hughes, A.E., McGucken, M.J., Buchanan, K.D., Conlon, J.M. and Elmore, D.T. (1976) *Biochim. Biophys. Acta*, **420**, 87.

89. Bagert, U. and Rohm, K.-H. (1989) *Biochim. Biophys. Acta*, **999**, 36.

90. Jordan, P.M., Warren, M.J., Williams, H.J., Stolowich, N.J., Roessner, C.A., Grant, S.K. and Scott, A.I. (1988) *FEBS Lett.*, **235**, 189.

91. Auchus, R.J., Covey, D.F., Bork, V. and Schaefer, J. (1988) *J. Biol. Chem.*, **263**, 11640.

92. Altman, J., Lipka, J.J., Kuntz, I. and Waskell, L. (1989) *Biochemistry*, **28**, 7516.

93. Liu, F. and Fromm, H.J. (1989) *J. Biol. Chem.*, **264**, 18320.

94. Bosshard, H.R., Koch, G.L.E. and Hartley, B.S. (1978) *J. Mol. Biol.*, **119**, 377.

95. Richardson, R.H. and Brew, K. (1980) *J. Biol. Chem.*, **255**, 3377.

96. Winkler, M.A., Fried, V.A., Merat, D.L. and Cheung, W.Y. (1987) *J. Biol. Chem.*, **262**, 15466.

97. Suckau, D., Mak, M. and Przybylski, M. (1992) *Proc. Natl Acad. Sci. USA*, **89**, 5630.

98. Hitchcock-De Gregori, S.E., Lewis, S.F. and Mistrik, M. (1988) *Arch. Biochem. Biophys.*, **264**, 410.

99. Sackett, D.L. and Wolff, J. (1987) *Anal. Biochem.*, **167**, 228.

100. Weber, L.D., Tulinsky, A., Johnson, J.D. and El-Boyoumi, M.A. (1979) *Biochemistry*, **18**, 1297.

101. Mock, D.M., Lankford, G. and Horowitz, P. (1988) *Biochim. Biophys. Acta*, **956**, 23.

102. Musci, G., Metz, G.D., Tsunematsu, H. and Berliner, L.J. (1985) *Biochemistry*, **24**, 2034.

103. Hlavica, P., Kellerman, J., Golly, I. and Lehnerer, M. (1994) *Eur. J. Biochem.*, **224**, 1039.

104. Liu, T.-Y. (1979) in *The Proteins (3rd Edn)* (H. Neurath and R. B.Hill, eds). Academic Press, New York, p. 239.

105. Bech, L.M. and Breddam, K. (1988) *Carlsberg Res. Commun.*, **53**, 381.

106. Jung, K., Jung, H., Wu, J., Privé, G.G. and Kaback, H.R. (1993) *Biochemistry*, **32**, 12273.

107. Ellman, G.L. (1959) *Arch. Biochem. Biophys.*, **82**, 70.

108. Habeeb, A.F.S.A. (1972) *Methods Enzymol.*, **25**, 457.

109. Narasimhan, C., Lai, C.-S., Haas, A. and McCarthy, J. (1988) *Biochemistry*, **27**, 1988.

110. Seifreid, S.E., Wang, Y. and von Hippel, P.H. (1988) *J. Biol. Chem.*, **263**, 13511.

111. First, E.A. and Taylor, S.S. (1989) *Biochemistry*, **28**, 3598.

112. Pardo, J.P. and Slayman, C.W. (1989) *J. Biol. Chem.*, **264**, 9373.

113. Esmann, M., Sar, P.C., Hideg, K. and Marsh, D. (1993) *Anal. Biochem.*, **213**, 336.

114. Kaslow, H.R., Schlotterbeck, J.D., Mar, V.L., and Burnette, W.N. (1989) *J. Biol. Chem.*, **264**, 6386.

115. May, J.M. (1989) *Biochem. J.*, **263**, 875.

116. Okonjo, K.O. and Adejor, I.A. (1993) *J. Protein Chem.*, **12**, 33.

117. Bai, Y. and Hayashi, R. (1979) *J. Biol. Chem.*, **254**, 8473.

118. Drozdz, R., Naskalski, J.W. and Sznajd, J. (1988) *Biochim. Biophys. Acta*, **957**, 47.

119. Barkholt, V. and Jensen, A.L. (1989) *Anal. Biochem.*, **177**, 318.

120. Liu, T.-Y. (1979). in *The Proteins (3rd Edn)* (H. Neurath and R.B. Hill, eds). Academic Press, New York, p. 239.

121. Cleland, W.W. (1964) *Biochemistry*, **3**, 480.

122. Ruegg, U.T. and Rudinger, J. (1977) *Methods Enzymol.*, **47**, 111.

123. Burns, J.A., Butler, J.C., Moran, J. and Whitesides, G.M. (1991) *J. Org. Chem.*, **56**, 2648.

124. Kella, N.K.D. and Kinsella, J.E. (1985) *J. Biochem. Biophys. Meth.*, **11**, 251.

125. Lundblad, R.L. (1991) in *Chemical Reagents for Protein Modification (2nd Edn)*. CRC Press, Boca Raton, FL, p. 99.

126. Goodlett, D.R., Armstrong, F. B., Creech, R.J. and van Breemen, R.B. (1990) *Anal. Biochem.*, **186**, 116.

127. Tristram, G.R. and Smith, R.H. (1963) in *The Proteins* (H. Neurath ed.). Academic Press, New York, p. 45.

128. Oda, T. and Tokushige, M. (1988) *J. Biochem. (Tokyo)*, **104**, 178.

129. Drozdz, R., Naskalski, J.W. and Sznajd, J. (1988) *Biochim. Biophys. Acta*, **957**, 47.

130. Kleanthous, C., Campbell, D.G. and Coggins, J.R. (1990) *J. Biol. Chem.*, **265**, 10929.

131. Padrines, M., Rabuad, M. and Bieth, J.G. (1992) *Biochim. Biophys. Acta*, **1118**, 174.

132. Miles, E.W. (1977) *Methods Enzymol.*, **47**, 431.

133. Liu, T.-Y. (1967) *J. Biol. Chem.*, **242**, 4029.

134. Kettner, C. and Shaw, E. (1981) *Methods Enzymol.*, **80**, 826.

135. Funakoshi, T., Abe, M., Sakata, M., Shoji, S. and Kubota, Y. (1990) *Biochem. Biophys. Res. Commun.*, **168**, 125.

136. Dumas, D. P. and Raushel, F.M. (1990) *J. Biol. Chem.*, **265**, 21498.

137. Inagami, T. and Hatano, H., (1969) *J. Biol. Chem.*, **244**, 1176.

138. Weil, L., James, S. and Buchert, A.R. (1953) *Arch. Biochem. Biophys.*, **46**, 266.

139. Ray, W.J., Jr., and Koshland, D.E., Jr. (1962) *J. Biol. Chem.*, **237**, 2493.

140. White, G.C., II, Lundblad, R.L. and Griffith, M.J. (1981) *J. Biol. Chem.*, **256**, 1763.

141. Tshabalala, M.A. and Latz, H.W. (1981) *Anal. Biochem.*, **111**, 343.

142. Yang, C.-C. and Chang, L.-S. (1989) *Biochem. J.*, **262**, 855.

143. Dick, L.R., Geraldes, C.F.G.C., Sherry, A.D., Gray, C.W. and Gray, D.M. (1989) *Biochemistry*, **28**, 7896.

144. Tuls, J., Geren, L. and Millett, F. (1989) *J. Biol. Chem.*, **264**, 16421.

145. Miki, M. (1989) *J. Biochem. (Tokyo)*, **106**, 651.

146. Kaplan, H., Stevenson, K.J. and Hartley, B.S. (1971) *Biochem. J.*, **124**, 289.

147. Rieder, R. and Bosshard, H.R. (1978) *J. Biol. Chem.*, **253**, 6045.

148. Wei, Q., Jackson, A.E., Pervaiz, S., Carraway, K.L., III, Lee, E.Y.C., Puett, D. and Brew, K. (1988) *J. Biol. Chem.*, **263**, 19541.

149. Bünning, P., Kleemann, S.G. and Riordan, J.F. (1990) *Biochemistry*, **29**, 10488.

150. Riordan, J.F. and Vallee, B.L. (1967) *Methods Enzymol.*, **11**, 565.

151. Ueno, H., Pospischil, M.A., Manning, J.M. and Kluger, R. (1986) *Arch. Biochem. Biophys.*, **244**, 795.

152. Shetty, J.K. and Kinsella, J.E. (1980) *Biochem. J.*, **191**, 269.

153. Stark, G.F. (1972) *Methods Enzymol.*, **25**, 579.

154. Plapp, B.V., Moore, S. and Stein, W.H. (1971) *J. Biol. Chem.*, **246**, 939.

155. Miller, A.D., Packman, L.C., Hart, G.J., Alefounder, P.R., Abell, C. and Battersby, A.R. (1989) *Biochem. J.*, **262**, 119.

156. Means, G.E. and Feeney, R.E. (1995) *Anal. Biochem.*, **224**, 1.

157. Goldfarb, A.R. (1966) *Biochemistry*, **5**, 2574.

158. Stark, G.R., Stein, W.H. and Moore, S. (1960) *J. Biol. Chem.*, **235**, 3177.

159. Bunn, H.F. and Higgins, P.J. (1981) *Science*, **213**, 222.

160. Acharya, A.S., Cho, Y.J. and Manjula, B.N. (1988) *Biochemistry*, **27**, 4522.

161. Dunn, J.A., Patrick, J. S., Thorpe, S.R. and Baynes, J.W. (1989) *Biochemistry*, **28**, 9464.

162. Habeeb, A.F.S.A. (1966) *Anal. Biochem.*, **14**, 328.

163. Fields, R. (1972) *Methods Enzymol.*, **25**, 464.

164. Riordan, J.F., McElvany, K.D. and Borders, C.L., Jr. (1977) *Science*, **195**, 884.

165. Cheung, S.-T. and Fonda, M.L. (1979) *Biochem. Biophys. Res. Commun.*, **90**, 940.

166. Krell, T., Pitt, A.R. and Coggins, J.R. (1995) *FEBS Lett.*, **360**, 93.

167. Epperly, B.R. and Dekker, E.E. (1989) *J. Biol. Chem.*, **264**, 18296.

168. Suckau, D., Mak, M. and Przybylski, M. (1992) *Proc. Natl Acad. Sci. USA.*, **89**, 5630.

Chemical Modification

169. Horton, H.R. and Young, G. (1969) *Biochim. Biophys. Acta*, **194**, 272.

170. Uhteg, L.C. and Lundblad, R.L. (1977) *Biochim. Biophys. Acta*, **491**, 551.

171. Simpson, R.J., Neuberger, M.R. and Liu, T.-Y. (1976) *J. Biol. Chem.*, **251**, 1936.

172. Yokoto, Y., Arai, K.M. and Akahane, K. (1986) *Anal. Biochem.*, **152**, 245.

173. Nakazawa, M. and Manabe, K. (1992) *Anal. Biochem.*, **206**, 105.

174. Schechter, Y., Patchornki, A. and Burstein, Y. (1976) *Biochemistry*, **15**, 5071.

175. Lischwe, M.A. and Sung, M.T. (1977) *J. Biol. Chem.*, **252**, 4976.

176. O'Gorman, R.B. and Matthews, K.S. (1977) *J. Biol. Chem.*, **252**, 3572.

177. Horowitz, J. and Heller, J. (1974) *J. Biol. Chem.*, **249**, 7181.

178. Fontana, A. and Scoffone, E. (1972) *Methods Enzymol.*, **25**, 482.

179. Perlmann, G.E. (1966) *J. Biol. Chem.*, **241**, 153.

180. Connellan, J.W. and Shaw, D.C. (1970) *J. Biol. Chem.*, **245**, 2845.

181. Cuatrecasas, P., Fuchs, S. and Anfinsen, C.B. (1968) *J. Biol. Chem.*, **243**, 4787.

182. Riggle, W.L., Long, J.A. and Borders, C.L., Jr. (1973) *Can. J. Biochem.*, **51**, 1433.

183. Bristow, A.F. and Virden, R. (1978) *Biochem. J.*, **169**, 381.

184. Hugli, T.E. and Stein, W.H. (1971) *J. Biol. Chem.*, **246**, 7191.

185. Riordan, J.F. and Vallee, B.L. (1972) *Methods Enzymol.*, **25**, 521.

186. Riordan, J.F. and Vallee, B.L. (1972) *Methods Enzymol.*, **25**, 500.

187. Riordan, J.F. and Vallee, B.L. (1972) *Methods Enzymol.*, **25**, 515.

188. Yamamoto, K., Sekine, T. and Sutoh, K. (1988) *J. Biochem. (Tokyo)*, **104**, 251.

189. Deen, C., Claassen, E., Gerritse, K., Zegers, N.D. and Boersma, W.J.A. (1990) *J. Immunol. Methods*, **129**, 119.

190. Grabarek, Z. and Geregly, J. (1990) *Anal. Biochem.*, **185**, 131.

191. Paterson, A.K. and Knowles, J.R. (1972) *Eur. J. Biochem.*, **31**, 510.

192. Pennington, R.M. and Fisher, R.R. (1981) *J. Biol. Chem.*, **256**, 8963.

CHAPTER 25
AFFINITY-BASED COVALENT MODIFICATION
K. Brocklehurst

1. INTRODUCTION

1.1. Scope and general principles of the technique

In its broadest sense affinity-based covalent modification encompasses a variety of processes whereby one molecule (usually a protein) undergoes chemical reaction at a particular site, directed in some way by specific binding interactions. A volume of *Methods in Enzymology* (1) is devoted to this topic and is a valuable source of detailed information (some individual chapters are cited below). The phenomenon obviously includes classical affinity labeling (reviewed in refs 2–5), in which the modifying reagent contains both a chemically reactive group and binding features analogous to those of the substrate. It also includes, however, more complex processes such as 'k_{cat} inhibition' (mechanism-based, enzyme-activated or suicide inhibition) (reviewed in refs 6–9) and 'syncatalytic inhibition' (reviewed in ref. 10). In the former a relatively unreactive compound with structural similarity to the substrate for a particular enzyme is converted by the enzyme, via its normal catalytic mechanism, into a molecular species that reacts chemically with the enzyme without prior release from the active center. In the latter, uniquely high reactivity of a single amino acid side-chain is induced by the presence of the substrate undergoing turnover and thus the inhibition reaction occurs synchronously with catalysis. Affinity-based modification, therefore, is distinct from covalent modification by supposedly group-specific reagents (see Chapter 24 and review in ref. 11) where site selectivity, if it exists, derives from uniquely high reactivity produced by the protein microenvironment. Striking examples of the latter are provided by the specific reactions of (Cys)-S$^-$/(His)-Im$^+$H ion-pairs (notably in the catalytic sites of cysteine proteinases) with nonsubstrate-derived 2-pyridyl disulfides at low pH, which provide the basis for the use of these compounds as active center titrants and reactivity probes (reviewed in refs 12, 13) and in covalent chromatography (reviewed in Chapter 7 and ref. 14). Specific reaction derives from the coexistence at pH 3.0–4.0 of the nucleophilic thiolate component of the ion-pair and the activated disulfide hydronated (protonated) on the pyridyl N atom. In this pH region other thiol groups exist almost entirely in the undissociated, nonnucleophilic SH form.

1.2. Applications: an overview

Important functions in biology depend upon molecular recognition, that is the specific binding of a ligand as in antibody–antigen, hormone–receptor and enzyme–substrate interactions. Early studies on enzyme binding sites by separate experiments, using reversible competitive inhibitors on the one hand and more or less group-specific covalent (irreversible) inhibitors on the other, progressed via differential labeling studies (15) to classical affinity labeling (active-site-directed covalent modification) with the goal of achieving highly site-specific covalent modification. Early examples of the technique developed by studies on both enzymes (16–18) and antibodies (19) demonstrated its great potential in protein structure-function studies and in the related area of rational drug design. Affinity labeling and variations and extensions of the technique continue to provide valuable structural information on binding site topology of a growing number of protein systems prior to crystallographic characterization and also contribute to the study of dynamic aspects of reaction mechanism, for example, through the use of substrate-derived time-dependent inhibitors as reactivity probes (see Section 8).

2. CLASSICAL AFFINITY LABELS

Classical affinity labels are also known as active-site-directed reagents or K_s reagents.

The reaction involves two distinct steps: specific binding of the inhibitor (I) at a site on the protein (P) followed by covalent bond formation (P–I) within the adsorptive complex (P.I):

$$P + I \underset{k_{-1}}{\overset{k_{+1}}{\rightleftharpoons}} P.I. \overset{k_{+2}}{\rightarrow} P-I. \qquad \text{Equation 1}$$

Binding features are usually based on those of known specific substrates, with due attention being given to the appropriate spatial relationship within the inhibitor of binding features and the reactive chemical group targeted on a particular site in the protein. The two major types of affinity label in this respect are the endo and exo reagents (20,21). Endo reagents (20) react covalently relatively close to the catalytic site. The larger exo reagents provide covalent modification further away from the catalytic site in regions that may still contribute to substrate specificity (see e.g. refs 21, 22 for a discussion of benzamidine derivatives as exo reagents).

Experimental criteria for recognition of affinity labeling are:
(i)   the rate saturation effect (see below);
(ii) stoichiometric reaction of one reagent molecule per binding site;
(iii) protection against reaction by substrate or a competitive inhibitor.

Saturation kinetics are predicted by the model of Equation 1 (see e.g. refs 23–27). Thus when the modification reaction is carried out under pseudo first-order conditions with I in excess, the observed first order-rate constant, k_{obs} is given by Equation 2:

$$k_{obs} = k_{+2}[I]/(K_I + [I]), \qquad \text{Equation 2}$$

where K_I is either the steady state assembly, $(k_{-1} + k_{+2})/k_{+1}$, or the dissociation equilibrium constant of P.I, k_{-1}/k_{+1}, if, as is probably the case for many modification reactions, quasi-equilibrium around PI may be assumed, as was done by Kitz and Wilson (23). The conventional condition for quasi-equilibrium $(k_{+2} \ll k_{-1})$ necessarily requires $k_{obs}/[I]$ to be much less than k_{+1} (see ref. 27 and note that the denominator of Equation 2 of that paper is wrongly printed as $k_{-1}k_{+2}$ instead of $k_{-1} + k_{+2}$). Equation 2 above is analogous to the Michaelis–Menten and Briggs–Haldane equations for hyperbolic enzyme catalysis. When both k_{-1} and k_{+2} are large it may not be possible to use concentrations of I that are comparable to the value of K_I and then k_{obs} may appear to be a linear function of [I]. The slope of the linear plot of k_{obs} versus [I] provides the value of k_{+2}/K_I which is analogous to the specificity constant (k_{cat}/K_m) for catalysis and is the apparent second-order rate constant for the overall reaction of P and I to form product P–I. It is a matter of preference whether the second-order rate constant, thus obtained, is regarded as k_{+2}/K_I of Equation 1, where the large value of K_I precludes observation of the saturation effect, or as the second-order rate constant for the reaction of E with I in a direct encounter mechanism. The former recognizes the possibility of transient complex formation and may aid more thoughtful interpretation if the (undetected) binding process is considered.

Most of the experimental protocols for the study of mechanism-based enzyme inactivation (see ref. 8, Vol. I) apply also to the use of classical affinity labels.

The range of affinity labels currently described in the literature is very large indeed. Selected examples are described in ref. 11, Vol. II, and a particularly large range in ref. 1. Early examples of the use of affinity labels include the well-known studies on chymotrypsin and trypsin using the chloromethyl ketones Tos-PheCH$_2$Cl (TPCK) and Tos-LysCH$_2$Cl) (TLCK) (reviewed in ref. 28). The difference in site specificity exhibited by halomethyl ketones in their reactions with serine proteinases (His) and cysteine proteinases (Cys) is discussed in ref. 29. In both cases the catalytic site nucleophile (–OH or –S$^-$) probably attacks the carbonyl group of the inhibitor to form a tetrahedral adduct (hemiketal) prior to the alkylation reaction. Whereas alkylation of the catalytic site His imidazole group occurs in the serine proteinases, the Cys S atom is the site of alkylation in the cysteine proteinases. Alkylation on S may involve reaction via a 3-membered cyclic transition state. Systems for which affinity labeling is discussed in ref. 1 include proteolytic enzymes, proteins with binding sites for nucleotides and nucleic acids, carbohydrates, amino acids and steroids, antibodies, receptors, transporters and a variety of other enzymes and binding proteins.

The recent literature contains many studies that make use of affinity labeling. Examples of the use of well-established reagent types such as chloromethylketones and diazomethylketones include studies on a multicatalytic endopeptidase complex (proteasome) (30), calpain in intact human platelets (31), cathepsin S (32), and unfolding of a proteinase domain of urokinase-type plasminogen activator (33). Examples of more recently developed affinity labels include 8-(4-bromo-2,3-dioxobutylthio) nicotinamide adenine dinucleotide (NAD) for studies on the allosteric NAD-dependent isocitrate dehydrogenase (34), 17β-[(bromoacetyl)oxy]-5α-androstan-3-one as a label for the androgen receptor in rat prostate cytosol (35), and 10-(3-propionyloxysuccinimide)-2-(trifluoromethyl) phenothiazine as a specific label for the hydrophobic drug-binding domains of calmodulin (36).

3. QUIESCENT AFFINITY LABELS

These reagents (37) contain an electrophilic carbon atom and, although they are unreactive towards low M_r nucleophiles, they can react with catalytic site nucleophiles involved in catalysis. Examples of such reagents where high reactivity results from assistance from binding include the cysteine proteinase inhibitors, E-64 and acyloxymethylketones and 3-(fluoromethyl)-3-butenyldiphosphate, an inhibitor of isopentenyldiphosphate isomerase (37).

4. PHOTOAFFINITY LABELS

Photoaffinity labels (reviewed in refs 38, 39) are reagents containing appropriate binding features and, instead of the chemically reactive group of a classical affinity label, a chemically inert group that is transformed into a highly reactive and chemically nonselective group by irradiation with visible or near ultraviolet (UV) light.

The activatable groups are usually diazoacyl compounds, which produce carbenes, arylazides, which produce nitrenes, or, less commonly, free radicals or triplet states.

Covalent Modification

Major advantages of photoaffinity reagents for some applications include:

(i) the fact that the highly reactive carbenes or nitrenes will attack many structural features of a protein molecule and thus do not require a specific electrophilic center to become aligned with a nucleophilic center in the bound reagent;

(ii) the reagent is activated by irradiation only when it should have reached its target site, which decreases the possibility of nonspecific labeling.

A useful collection of various types of photoaffinity label, their syntheses and applications is available in ref. 39. Recent examples include:

(i) S-(2-nitro-4-azidophenyl)glutathione in studies on the hydrophobic region of the active centers of glutathione transferases (40);

(ii) the use of a photoaffinity label (41) to distinguish the anti-chaperone activity of protein disulfide isomerase from its disulfide isomerase and chaperone activity (42).

5. TRANSITION STATE AFFINITY ($K_s^{\ddagger}$) LABELS

The principles involved in the design and use of transition state analogs as highly effective inhibitors, which can provide useful mechanistic insights, and the possibility of combining transition state inhibition chemistry with specific binding features in the inhibitor are discussed in ref. 43. Transition state and multisubstrate analog inhibitors are listed and discussed in ref. 44.

6. MECHANISM-BASED IRREVERSIBLE INHIBITORS

Mechanism-based irreversible inhibitors (referred to also as k_{cat} inhibitors, enzyme-activated inhibitors and suicide substrates or inhibitors) are reviewed extensively in refs 6–9.

The inhibitor is produced by the catalytic action of the enzyme on a specifically designed analog substrate (Equation 3):

$$E + S \rightleftharpoons ES \xrightarrow{k_{cat}} E.I \xrightarrow{k_i} E-I. \qquad \text{Equation 3}$$

Catalytic action produces E.I from ES in the step with rate constant k_{cat} and inhibition to produce E–I occurs in the step with rate constant k_i (45).

One class of such inhibitors is the acetylenic inhibitors, reviewed in ref. 46. Conjugated allenes, which are powerful Michael acceptors, are produced enzymically from acetylenes (Equation 4, where X=O, N or S).

$$R-C\equiv C-CH(R')-C(=X)R'' \xrightarrow{E} R-CH=C=C(R')-C(=X)R''$$
$$\downarrow E \qquad \text{Equation 4}$$
$$R-CH=C(E)-CH(R')-C(=X)R''.$$

Enzymes that produce a carbanion or carbanion-like intermediate adjacent to an acetylenic moiety are candidates for inhibition by acetylenic inhibitors. Members of this class of inhibitor have been designed to inhibit isomerization, redox, elimination and transamination reactions. Interesting examples include the inhibition of β-hydroxy-decanoylthioesterdehydrase by 3-decanoyl-N-acetylcysteamine (47) and the inhibition of $\Delta^{3,5}$-ketosteroid isomerase by γ-acetylenic analogs of substrate (Δ^5 steroids) (48).

A recent chapter in *Methods in Enzymology* (9) describes methodologies and approaches for studies using mechanism-based inhibitors and provides examples involving inhibition of γ-aminobutyric acid aminotransferase and monoamine oxidase.

k_{cat} inhibitors generally exert a higher specificity than K_s inhibitors because they are chemically unreactive towards other biomolecules that do not specifically convert them to the reactive forms. For this reason they are the reagents of choice for studies on crude extracts or on *in vivo* systems. Another advantage is that because the inhibitor form is produced within the active center, highly reactive chemical species can be used to provide for covalent bonding without risk of non-specific bonding at other sites.

7. SUBSTRATE-DEPENDENT NONAFFINITY LABELS

Techniques that may be considered to be related to affinity labeling and involve the use of mixtures of a substrate or substrate analog and the nonaffinity labeling reagent include the following.

Ligand-induced modification makes use of the fact that whereas substrate usually protects against chemical modification, in some cases binding enhances reactivity towards modifying reagents, presumably by ligand-induced conformational change (49).

Syncatalytic modification (10) makes use of the fact that increases in reactivity of some groups occurs, not on adsorptive complex formation, but rather during subsequent covalent phases of the catalysis. A striking example is the large variation in the reactivity of the thiol group of Cys 390 of aspartate aminotransferase as the pyridoxal form is transformed via various covalent intermediates to the pyridoxamine form during the catalytic cycle (see ref. 50 and references therein).

Paracatalytic modification (51) involves a direct chemical reaction between the enzyme-activated substrate and the modifying reagent, as can occur during the oxidation of enzyme–substrate carbanion intermediates produced, for example, in the reaction catalyzed by fructose-1,6-bisphosphate aldolase (52).

8. SUBSTRATE-DERIVED SITE-SPECIFIC REACTIVITY PROBES

Kinetic analysis of substrate-derived covalent modification reactions can provide valuable information about active center chemistry.

To understand molecular recognition in enzyme–substrate and enzyme–time-dependent inhibitor systems it is necessary to consider the interdependence of binding interaction and the covalency changes that have a central role in catalytic-site chemistry. This dynamic aspect of molecular recognition is one of the least well understood aspects of enzyme chemistry, and one of the most sensitive and direct ways of investigating catalytic site geometry and its modulation by ligand binding in the study of chemical reactivity characteristics. Transition-state geometry is the fundamental characteristic on which catalytic ability depends, and the reactivity of a catalytic site nucleophile that plays a central role in the catalytic act is a prime target for investigation of factors that contribute to the control of transition state and catalytic site geometry.

Covalent Modification

Substrate-derived 2-pyridyl disulfides such as Ac-[Phe]-NH-(CH$_2$)$_n$-S-S-2-Py, Ac-[Phe]-O-(CH$_2$)$_n$-S-S-2-Py, Ac-NH-(CH$_2$)$_n$-S-S-2-Py and Ac-O-(CH$_2$)$_n$-S-S-2-Py (which are chromogenic, thiol-specific, two-protonic-state electrophiles (11)) have been used to explore how the reactivity of the catalytic site of the cysteine proteinase, papain, responds to various combinations of hydrogen bonding and hydrophobic effects in its extended binding site (see e.g. refs 53, 54 and references therein).

Stopped-flow kinetic studies of the pH-dependence characteristics of these reactions permit an assessment of the contributions of the (P$_1$)-NH$\cdots$O=C$<$(Asp158) and (P$_2$)$>$C=O$\cdots$HN-(Gly66) hydrogen bonds and enantiomeric P$_2$–S$_2$ hydrophobic contacts in papain–ligand association in two manifestations of dynamic molecular recognition:

(i) signaling to the catalytic-site region to provide for an Im$^+$H-assisted transition state;
(ii) the dependence of P$_2$–S$_2$ stereochemical selectivity on hydrogen bonding interactions outside the S$_2$ subsite.

Activation of the 2-mercaptopyridine leaving group of these reagents may occur by formal protonation at low pH or, if binding interactions provide for association of the catalytic-site (His159)-Im+H with the pyridyl N atom simultaneously with reaction of the catalytic-site (Cys25)-S$^-$ at the distal S atom, by this hydrogen bonding interaction. These two circumstances are readily distinguished because the former results in a rate maximum at pH 3.0–4.0, whereas the latter provides for a rate maximum (of much greater reactivity) at pH 6.0–7.0. This marked change in the form of the pH-k profile provides opportunities to define the combinations of binding interactions that control transition state geometries.

9. ACKNOWLEDGMENTS

I thank Hasu Patel for searching the literature and SERC, BBSRC and AFRC for project grants and Earmarked Studentships.

10. REFERENCES

1. Jakoby, W.B. and Wilchek, M. (1977) *Methods Enzymol.*, **46**.

2. Shaw, E. (1970) in *The Enzymes (3rd Edn)*, Vol. 1 (P.D. Boyer, ed.). Academic Press, New York, p. 91.

3. Wold, F. (1977) *Methods Enzymol.*, **46**, 3.

4. Plapp, B.V. (1982) *Methods Enzymol.*, **87**, 469.

5. Lundblad, R.L. and Noyes, C.M. (1985) in *Chemical Reagents for Protein Modification*, Vol. II. CRC Press Inc., Boca Raton, FL, p. 141.

6. Rando, R.R. (1977) *Methods Enzymol.*, **46**, 28.

7. Seiler, N., Jung, M.J. and Koch-Weser, J. (1978) *Enzyme-activated Irreversible Inhibitors.* Elsevier, Amsterdam.

8. Silverman, R.B. (1988) *Mechanism-Based Enzyme Inactivation: Chemistry and Enzymology*, Vols I and II. CRC Press Inc., Boca Raton, FL.

9. Silverman, R.B. (1995) *Methods Enzymol.*, **249**, 240.

10. Birchmeier, W. and Christen, P. (1977) *Methods Enzymol.*, **46**, 41.

11. Lundblad, R.L. and Noyes, C.M. (1985) *Chemical Reagents for Protein Modification*, Vols. I and II. CRC Press Inc., Boca Raton, FL.

12. Brocklehurst, K. (1982) *Methods Enzymol.*, **87**, 427.

13. Brocklehurst, K. (1995) in *Enzymology Labfax* (P. Engel, ed.). BIOS Scientific Publishers, Oxford/Academic Press, San Diego, p. 59.

14. Brocklehurst, K., Carlsson, J. and Kierstan, M.P.J. (1985) *Top. Enzyme Ferment. Biotechnol.*, **10**, 146.

15. Phillips, A.T. (1977) *Methods Enzymol.*, **46**, 59.

16. Baker, B.R., Lee, W.W., Tong, E. and Ross, L.O. (1961) *J. Am. Chem. Soc.*, **83**, 3713.

17. Schoellmann, G. and Shaw, E. (1962) *Biochem. Biophys. Res. Commun.*, **7**, 36.

18. Lawson, W.B. and Schramm, H.J. (1962) *J. Am. Chem. Soc.*, **84**, 2017.

19. Wofsy, L., Metzger, H. and Singer, S.J. (1962) *Biochemistry*, **1**, 1031.

20. Baker, B.R. (1964) *J. Pharm. Sci.*, **53**, 347.

21. Cory, M., Andrews, J.J. and Bring, D.H. (1977) *Methods Enzymol.*, **46**, 115.

22. Baker, B.R. and Cory, M. (1971) *J. Med. Chem.*, **14**, 305.

23. Kitz, R. and Wilson, I.B. (1962) *J. Biol. Chem.*, **237**, 3245.

24. Malcolm, A.D.B. and Radda, G.K. (1970) *Eur. J. Biochem.*, **15**, 555.

25. Meloche, H.P. (1967) *Biochemistry*, **6**, 2273.

26. Cornish-Bowden, A. (1979) *Eur. J. Biochem.*, **93**, 383.

27. Brocklehurst, K. (1979) *Biochem. J.*, **181**, 775.

28. Powers, J.C. (1977) *Methods Enzymol.*, **46**, 197.

29. Brocklehurst, K. and Malthouse, J.P.G. (1978) *Biochem. J.*, **175**, 761.

30. Savory, P.J., Djaballah, H., Angliko, H., Shaw, E. and Rivett, A.J. (1993) *Biochem. J.*, **296**, 601.

31. Anagli, J., Hagmann, J. and Shaw, E. (1993) *Biochem. J.*, **289**, 93.

32. Shaw, E., Mohanty, S., Colic, A., Stoka, V. and Turk, V. (1993) *FEBS Lett.*, **334**, 340.

33. Nowak, U.K., Cooper, A., Saunders, D., Smith, R.A. and Dobson, C.M. (1994) *Biochemistry*, **33**, 2951.

34. Kumar, A. and Colman, R.F. (1994) *Arch. Biochem. Biophys.*, **308**, 357.

35. Chang, C.H., Lobl, T.J., Rowley, D.R. Tindall, D.J. (1984) *Biochemistry*, **23**, 2527.

36. Jarrett, H.W. (1984) *J. Biol. Chem.*, **259**, 10136.

37. Krantz, A. (1992) *BioMed. Chem. Lett.*, **2**, 1327.

38. Knowles, J.R. (1972) *Acc. Chem. Res.*, **5**, 155.

39. Bayley, H. and Knowles, J.R. (1977) *Methods Enzymol.*, **46**, 69.

40. Cooke, R.J., Bjornestedt, R., Douglas, K.T., McKie, J.H., King, M.D., Coles, B., Ketterer, B. and Mannervik, B. (1994) *Biochem. J.*, **302**, 383.

41. Noiva, R., Freedman, R.B. and Lennarz, W.J. (1993) *J. Biol. Chem.*, **268**, 19210.

42. Puig, A., Lyles, M.M., Noiva, R. and Gilbert, H.F. (1994) *J. Biol. Chem.*, **269**, 19128.

43. Wolfenden, R. (1977) *Methods Enzymol.*, **46**, 15.

44. Radzicka, A. and Wolfenden, R. (1995) *Methods Enzymol.*, **249**, 284.

45. Rando, R.R. (1974) *Science*, **185**, 320.

46. Rando, R.R. (1977) *Methods Enzymol.*, **46**, 158.

47. Bloch, K. (1969) *Acc. Chem. Res.*, **2**, 193.

48. Batzold, F.H. and Robinson, C.H. (1975) *J. Am. Chem. Soc.*, **97**, 2376.

49. Citri, N. (1973) *Adv. Enzymol.*, **37**, 397.

50. Wilson, K.J., Birchmeier, W. and Christen, P. (1974) *Eur. J. Biochem.*, **41**, 471.

51. Christen, P. (1977) *Methods Enzymol.*, **46**, 48.

52. Christen, P., Cogoli-Greuter, M., Healy, M.J. and Lubin, D. (1976) *Eur. J. Biochem.*, **63**, 223.

53. Brocklehurst, K., Brocklehurst, S.M., Kowlessur, D., O'Driscoll, M., Patel, G., Salih, E., Templeton, W., Thomas, E.W., Topham, C.M. and Willenbrock, F. (1988) *Biochem. J.*, **256**, 543.

54. Patel, M., Kayani, I.S., Templeton, W., Mellor, G.W., Thomas, E.W. and Brocklehurst, K. (1992) *Biochem. J.*, **287**, 881.

CHAPTER 26
CROSS-LINKING REAGENTS FOR PROTEINS
J.R. Coggins

1. INTRODUCTION

The purpose of this chapter is first to introduce the reader to the versatility of protein cross-linking reagents as tools for investigating many aspects of protein structure and function, and second to provide practical advice on the choice of reagent for particular types of experiment.

Cross-linking reagents were first used to study the structure of proteins in solution in the 1950s (1) and initially the major interest was to obtain information about the arrangement of side-chains on the surface of monomeric proteins (2–4). This was achieved by forming intramolecular cross-links between particular pairs of amino acid side-chains with bifunctional reagents of defined length. Although, in the absence of a physically determined three-dimensional (3D) structure, intramolecular cross-linking remains a useful method for studying the tertiary structure of monomeric proteins, the major application of protein cross-linking reagents is now in the study of oligomeric proteins (5), multi-enzyme complexes (6,7) and large assemblies of proteins such as those found in the ribosome (8, 9), in chromatin (10) or in membranes (11). Protein cross-linkers are also being used increasingly to detect protein–protein interactions (12), to study conformational changes in proteins (13–15), to study protein–ligand interactions (16), to monitor changes in quaternary structure (16) and to prepare protein conjugates, for example to link enzymes to antibodies (17) or to prepare immunotoxins (17). The subject of protein cross-linking has recently been comprehensively reviewed in a book by Wong (17) and readers are recommended to consult this volume for further examples.

2. REAGENTS

Only a minority of the amino acid side-chains found in proteins are chemically reactive and therefore suitable as targets for chemical modification (see Chapter 24). The two most reactive side-chains are the thiol group of cysteine and the amino group of lysine and these are the side-chains usually targeted in protein cross-linking studies. Thiol groups are the most reactive but generally amino groups are very much more abundant in proteins and the most commonly used cross-linking reagents are amino group specific reagents.

Cross-linking reagents are made by connecting two group-specific chemical modification reagents together with a simple carbon chain. The length of this chain or linker can be varied to facilitate distance measurements and topology studies. In most cross-linkers the linker is a simple, inert chain and the cross-linking reaction is irreversible. However reversible cross-linking reagents are available containing linkers that can be specifically cleaved, for example by oxidation (6), reduction (11), acid treatment (18) or photochemically (19). These cleavable reagents are particularly useful for studying complex protein assemblies when the analysis of the cross-linked species is greatly simplified if the reaction can be reversed to facilitate the identification of the cross-linked components (6, 8, 9, 11). Cross-linking reagents may be homobifunctional (containing two identical reactive groups) or heterobifunctional (containing

Table 1. Some amino group-directed homobifunctional cross-linking reagents

Reagent	Structure
Dimethyl suberimidate . 2HCl (DMS)	$H_3C-O-\overset{\overset{\displaystyle N^+H_2Cl^-}{\|}}{C}-(CH_2)_6-\overset{\overset{\displaystyle N^+H_2Cl^-}{\|}}{C}-O-CH_3$
Dimethyl 4,4'-dithiobisbutyrimidate . 2HCl (DTBB)	$H_3C-O-\overset{\overset{\displaystyle N^+H_2Cl^-}{\|}}{C}-(CH_2)_3-S-S-(CH_2)_3-\overset{\overset{\displaystyle N^+H_2Cl^-}{\|}}{C}-O-CH_3$
N, N'-bis(2-carboximido-methyl)tartarimide dimethyl ester. 2HCl (CMTD)	$H_3C-O-\overset{\overset{\displaystyle N^+H_2Cl^-}{\|}}{C}-CH_2.NH.CO.\overset{\overset{\displaystyle OH}{\|}}{C}H-\overset{\overset{\displaystyle OH}{\|}}{C}H.CO.NH.CH_2-\overset{\overset{\displaystyle N^+H_2Cl^-}{\|}}{C}-O-CH_3$
Disuccinimidylsuberate (DSS); N-hydroxysuccinimidylsuberate (NHS-SA)	N—O.CO.(CH₂)₆.CO.O—N
Bis(sulfosuccinimidyl)-suberate (BSSS)	N—O.CO.(CH₂)₆.CO.O—N
3,3'-dithiobis-(succinimidylpropionate) (DTSP, DSP); Lomant's reagent	N—O.CO.(CH₂)₂—S—S—(CH₂)₂.CO.O—N

DTBB and DTSP cross-links are cleavable by treatment with dithiothreitol (DTT) and CMTD cross-links by treatment with periodate. BSSS is membrane impermeable.

two different reactive groups). Examples of some commonly used homobifunctional reagents directed against amino groups are shown in *Table 1*, while *Table 2* shows some reagents directed against thiol groups. *Table 3* some heterobifunctional reagents. Most of the cross-linking reagents mentioned in the tables are commercially available from Pierce (3747 N. Meridian Road, P.O. Box 117, Rockford IL 61105, USA; Tel. (815) 968 0747). The tables are intended to illustrate the type and range of reagents available. Comprehensive lists including more than 300 reagents, most of which are not commercially available, are given in ref. 17.

2.1. Control of pH
Most of the protein modification reactions utilized for cross-linking are nucleophilic reactions

Table 2. Some sulfhydryl group-directed homobifunctional cross-linking reagents

Reagent	Structure
N,N'-Hexamethylenebismaleimide; *bis*(*N*-maleimido)-1,6-hexane (BMH)	
4,4'-Dimaleimidostilbene	
4,4'-Dimaleimidylstilbene-2,2'-disulfonic acid (DMSDS)	
Maleimidomethyl-3-maleimido-propionate (MMP)	
Bis(*N*-maleimido)-4,4'-bibenzyl (BMB)	
N,N'-Bis (α-iodoacetyl)-2,2'-dithiobisethylamine (DIDBE)	

MMP cross-links are cleavable by treatment with base and DIDBE cross-links by treatment with dithiothreitol (DTT).

involving the alkylation or acylation of cysteine or lysine side-chains. The nucleophilicity of the thiol group (as thiolate) is significantly greater than that of the amino group and with most reagents the thiol groups of proteins will react preferentially. Because protonation affects nucleophilicity the pH of the medium affects the rate of nucleophilic reactions. It is therefore essential to use a buffer to maintain the pH at an appropriate value. Buffers containing reactive groups, such as Tris, which has a primary amino group, must be avoided (a good replacement for Tris is triethanolamine which buffers in a similar pH range but contains an unreactive tertiary amino group).

CROSS-LINKING REAGENTS FOR PROTEINS

Table 3. Some representative heterobifunctional cross-linking reagents

Reagent	Structure
N-Succinimidyl 3-(2-pyridyldithio) propionate (SPDP)	
N-Succinimidyl 6-maleimidocaproate; *N*-succinimidyl 6-maleimidyl-hexanoate (SMH)	
N-Succinimidyl 4-(*N*-maleimido-methyl)cyclohex-ane-1-carboxylate (SMCC)	
N-Sulfosuccinimidyl 4-(*N*-maleimido-methyl)cyclohexane -1-carboxylate (sulfo-SMCC)	
N-Succinimidyl-6-(4′-azido-2′-nitro phenylamino)hexanoate (SANPAH); (Loman's reagent II)	
Sulfosuccinimidyl-2-(*p*-azidosalicyl-amino)ethyl-1,3′-dithiopropionate (SASD)	

The position marked with an asterisk in SASD can be radiolabeled by iodination.

2.2. Lysine-directed cross-linking reagents

The abundance and accessibility of lysine side-chains on the surface of proteins means that the most frequently used cross-linking reagents are lysine-directed. Acylating reagents such as active esters, imidoesters, acid azides, acid chlorides, isocyanates and isothiocyanates have all been used. The currently favored reagents are the *bis*-*N*-hydroxysuccinimidyl esters and the *bis*-imidoesters (*Table 1*). *N*-Hydroxysuccinimidyl esters are easily synthesized from the corresponding acids, are highly activated and react with amino groups under mild conditions (pH 7.0–9.0) (17). The imidoesters can also be readily synthesized from the corresponding nitriles, for example by the Pinner method (5), they are very soluble in water, react under mild conditions (pH 8.0) and show a very high degree of specificity for amino groups (20). The maximum distance between the two imidoester groups (i.e. the maximum cross-linking distance which can be spanned) can be varied by changing the number (n) of methylene groups. Thus when n is 2, 4, 6 or 8, the distance is 0.6, 0.9, 1.1 or 1.4 nm respectively. Another lysine-directed reagent widely used to cross-link proteins is the dialdehyde glutaraldehyde (16, 17). The chemistry of glutaraldehyde cross-linking is complex and not well understood (17) but the reagent is simple to use, it reacts rapidly under mild conditions (pH 7.0) and the cross-linking process is irreversible.

2.3. Cysteine-directed cross-linking reagents

Two types of cross-linking reagent are most frequently used to react with cysteine residues: alkylating reagents, particularly N-substituted *bis*-maleimides, and disulfide forming reagents (*Table 2*). In principle, alkylating reagents will react with any nucleophilic side-chain but in practice at pH 7.0 maleimides react 1000 times faster with cysteines than with lysines and the reaction at other nucleophilic side-chains is negligible (17). The *bis*-maleimides are relatively easy to synthesize from the appropriate amine and maleic anhydride. Another useful class of readily synthesized alkylating reagents which are cysteine-specific contain the haloacetyl group (17).

Two characteristic properties of cysteine residues, which are exploited in cross-linking experiments, are their ability to form mixed disulfides and their ability to form disulfide bridges following mild oxidation. The thiol–disulfide interchange reaction is frequently used to achieve cross-linking, for example pyridyl disulfides, which contain the excellent leaving group thiopyridine, are used in many heterobifunctional reagents (17, see *Table 3*). The resulting mixed disulfide linkage can subsequently be cleaved by reduction with, for example, dithiothreitol (DTT). Because of the relative rarity of cysteine residues in proteins, methods have been developed to introduce additional cysteine residues to provide more sites for cross-linking. Among the most commonly used reagents for this purpose is methyl 4-mercaptobutyrimidate which is generally used in its cyclic form (2-iminothiolane, Traut's reagent) for thiolation of proteins (22, see *Table 3*). The added thiol groups can be used to produce disulfide cross-links following mild oxidation; these cystine cross-links can subsequently be cleaved by reduction (22).

2.4. Other types of cross-linking reagent

Cross-linking reagents directed towards other types of amino acid side-chain, for example carboxylate directed, guanidino-directed, phenolate-directed and imidazole-directed, have also been described (17) but they tend to have specialized applications. There are also a number of photoactivatable cross-linking reagents which are based on the chemistry developed for the photo-affinity labeling of proteins (21). On irradiation these compounds give rise to highly reactive intermediates such as nitrenes and carbenes which are described as 'nondiscriminatory', since they can react with most kinds of amino acid side-chain. The aryl azides, which on irradiation give nitrenes, have proved to be the most useful reagents in this class (17).

3. APPLICATIONS

This section describes some typical applications of protein cross-linking reagents to the study of protein structure and function. Some of the practical aspects of the methodology are discussed with particular reference to the use of the simple *bis*-imidoesters to study the quaternary structure of oligomeric proteins. The *bis*-imidoesters have the virtue of being among the simplest and most widely available of the cross-linking reagents and the precautions required in their use serve to illustrate some of the potential difficulties and limitations of the cross-linking technique.

Cross-linking Reagents

3.1. Estimation of the native molecular mass of oligomeric proteins

Cross-linking, combined with polyacrylamide gel electrophoresis (PAGE) in the presence of sodium dodecyl sulfate (SDS), provides one of the simplest methods of estimating the native molecular mass of oligomeric proteins (5). By paying careful attention to reaction conditions (see below) it is possible to ensure that cross-linking proceeds efficiently only within the oligomer. A relatively long *bis*-imidoester such as dimethylsuberimidate (DMS, see *Table 1*) when added to a tetrameric protein will give a mixture of species in which two, three or four subunits have been cross-linked as well as material lacking any inter-subunit cross-links; this mixture will give a characteristic four-band ladder pattern on SDS–PAGE (23). For multisubunit proteins containing different polypeptide chains, measurement of the molecular mass of the individual subunits and of the largest cross-linked species can also be used to estimate the numbers of polypeptide chains in the assembly (6). Cross-linking studies, especially with reagents of different length, can also be used to provide information on the arrangement of subunits. Thus aldolase and a number of other tetrameric proteins have been shown to consist of a dimer of dimers (24, 25).

In cross-linking experiments that are intended to explore quaternary structure, it is important to keep the protein concentration low (usually 0.5–2.0 mg ml^{-1}) to avoid inter-oligomer cross-links being formed. Alternatively, the ionic strength can be raised to minimize aggregation of oligomers (6).

Cross-linking with *bis*-imidoesters is generally carried out with 1–10 mM reagent for 1 h at 20°C in 0.1 M triethanolamine hydrochloride buffer at pH 8.0 or 8.5. It is important when making up the solutions of imidoesters: (a) to add the exact amount of alkali to neutralize the dihydrochloride (6), and (b) to use freshly prepared solutions since they are rather unstable (the half-life of dimethylsuberimidate at pH 8.5 and 25°C is 46 min (25)). At the end of the reaction the reagent is quenched with excess ammonium bicarbonate and the cross-linked samples boiled in the presence of SDS and 2-mercaptoethanol (5). PAGE in the presence of SDS is carried out in sodium phosphate buffer, pH 6.5 (6, 23) or borate/sodium acetate buffer, pH 8.5 (5). The cross-linked bands are separated satisfactorily by this technique and good straight line plots of mobility against log(relative molecular mass) are obtained (5, 26). The cross-linking process does not adversely affect the linearity of the plots. If the higher resolution Laemmli technique is used (27), the cross-linked bands frequently appear as multiplets making interpretation difficult and the molecular mass calibration plots are less satisfactory.

It should be noted that there is at least one example, *Escherichia coli* β-galactosidase, of an oligomeric protein which is not cross-linked by *bis*-imidoesters (24). Presumably this tetrameric protein contains no suitably placed pairs of lysine residue to allow intersubunit cross-linking.

3.2. Nearest neighbor analysis in oligomeric proteins

By using a number of cross-linking agents of different types and lengths it is possible to examine the subunit arrangement of complex assemblies. Reversible reagents are particularly valuable for identifying the subunits which have been linked.

The technique has been used to study multienzyme complexes such as the five subunit *E. coli* RNA polymerase (6) and the nine subunit *E. coli* F$_1$-adenosine triphosphatase (ATP) (7). A wide variety of reagents including both cleavable and noncleavable *bis*-imidoesters and *bis-N*-hydroxysuccinimidyl esters were used. However the most impressive application has been to the study of ribosome topology. Complete maps of the 30S and 50S ribosomal subunits have been obtained (9, 17). *Bis*-imidates, *bis*-maleimides and 2-iminothiolane have figured prominently in these studies but many other reagents, both cleavable and noncleavable have been used. The most difficult problem has been to identify which subunits have been cross-linked. This has been solved either by using reversible reagents or by using immunoblotting techniques (28).

3.3. Studies on membrane proteins

Cross-linking reagents have been very extensively used to study membrane proteins (11, 17). A wide variety of reagents have been employed and the technical difficulties encountered have provided a strong impetus to develop many novel reagents. Both hydrophobic and hydrophilic reagents have been produced to probe different regions of the membrane (17). Arguably the most important studies have been aimed at identifying membrane surface receptors using the method of macromolecular affinity labeling (29). This technique involves cross-linking a macromolecular ligand to its receptor using a photosensitive heterobifunctional reagent. The reagent is first bound to the ligand through a conventional group-specific reaction, such as that between an amino group and an imidoester or an active ester. After purification the labeled ligand is then incubated with a membrane preparation containing its receptor. Cross-linking is achieved by photoactivation of the photosensitive group. The cross-linked products are isolated and the receptor protein is identified. The ideal reagent is heterobifunctional with an imidoester or N-hydroxysuccinimidyl ester at one end and an aryl azide at the other. A good example is the reagent SANPAH (*Table 3*) which has been used to identify the dopamine receptor (30). Another example is SASD (*Table 3*), which has the added advantages that it can be radiolabeled at the receptor end by iodination (it contains phenolic group) after the cross-linking reaction and then the ligand moiety can be cleaved off by reduction. This ingenious reagent has been used to characterize the phytohemagglutinin cell surface receptor (31). The simpler homobifunctional reagent disuccinimidyl suberate (DSS) (*Table 1*) has proved to be a particularly valuable reagent for receptor identification (17), for example in the identification of the interleukin receptor (32). Another flourishing area of membrane research where cross-linking is extensively used is the study of the structural organization of membrane protein complexes (17). The experimental approaches closely follow those used for nearest neighbor analysis of soluble protein assemblies.

3.4. Detection of conformational change

Conformational changes of proteins that result from ligand binding can often be detected by observing changes in the cross-linking patterns; for example structural changes in glycogen phosphorylase caused by various activators, inhibitors and substrates have been studied by observing the changes in cross-linking pattern produced by a series of *bis*-imidoesters of increasing chain length (14). Ligand-induced changes in the enzyme phosphoenolpyruvate carboxylase (13) have also been detected using *bis*-imidoesters.

3.5. Detection of subunit association/dissociation

Cross-linking provides a very simple method for examining the effect of ligands and denaturing agents such as guanidinium hydrochloride on the state of aggregation of oligomeric proteins. A good example of this is the use of glutaraldehyde cross-linking to investigate the cofactor induced dissociation of a mutant form of phosphoglycerate mutase (16). Glutaraldehyde cross-links oligomeric proteins very rapidly and the reagent can therefore be used very simply to monitor changes in quaternary structure.

3.6. Conjugation of proteins

Protein cross-linking reagents are used for the linking of enzymes to antibodies or antigens for use in enzyme immunoassays (17). For the conjugation of enzymes to antibodies it is usual first to introduce thiol groups into the immunoglobulin, either by reducing the native cystine residues or by chemically introducing thiol groups, for example with 2-iminothiolane (17). Conjugation to enzymes is then achieved using heterobifunctional reagents selective towards amino and thiol groups such as SMCC and SMH (*Table 3*).

CROSS-LINKING REAGENTS FOR PROTEINS

Cross-linking Reagents

Heterobifunctional reagents are also used for the preparation of immunotoxins; for example the reagent SPDP (*Table 3*) is first used to introduce a thiol group into an immunoglobulin fragment by amino group modification followed by cleavage of the thiopyridyl group with DTT. Then SPDP is used in a second reaction to link to an amino group of the toxin. Finally a mixed disulfide is formed between the the new thiol group of the immunoglobulin and the toxin–SPDP adduct with the release of another thiopyridyl group (17).

4. REFERENCES

1. Zahn, H. (1955) *Angew. Chem.*, **67**, 56.

2. Fasold, H., Klappenberger, J., Keyer, C. and Remold, H. (1971) *Angew. Chem. Internat. Edit*, **10**, 795.

3. Wold, F. (1967) *Methods Enzymol.*, **11**, 617.

4. Wold, F. (1972) *Methods Enzymol.*, **25**, 623.

5. Davies, G.E. and Stark, G. (1970) *Proc. Natl Acad. Sci. U.S.A.*, **66**, 651.

6. Coggins, J.R., Lumsden, J. and Malcolm, A.D.B. (1978) *Biochemistry*, **16**, 1111.

7. Bragg, P.D. and Hou, C. (1980) *Eur. J. Biochem.*, **106**, 495.

8. Bickle, T., Hershey, J.W.B. and Traut, R.R. (1972) *Proc. Natl Acad. Sci. U.S.A.*, **69**, 1327.

9. Lambert, J.M., Boileau, G., Cover, J.A. and Traut, R.R. (1983) *Biochemistry*, **22**, 3923.

10. Thomas, J.O. and Kornberg, R.D. (1975) *Proc. Natl Acad. Sci. U.S.A.*, **72**, 2626.

11. Wang, K. and Richards, F.M. (1974) *J. Biol. Chem.*, **249**, 8005.

12. Harrison, J.K., Lawton, R.G. and Gnegy, M.E. (1989) *Biochemistry*, **28**, 6023.

13. Jones, R., Wilkins, M.B., Coggins, J.R., Fewson, C.A. and Malcolm, A.D.B. (1978) *Biochem. J.*, **175**, 391.

14. Hajdu, J., Dombradi, V., Bot, G. and Friedrich, P. (1980) *Biochemistry*, **18**, 4037.

15. Schenker, E. and Kohanski, R.A. (1988) *Biochem. Biophys. Res. Commun.*, **157**, 140

16. White, M.F., Fothergill-Gilmore, L.A., Kelly, S.M. and Price, N.C. (1993) *Biochem. J.*, **291**, 479.

17. Wong, S.S. (1993) *Chemistry of Protein Conjugation and Cross-Linking*. CRC Press, Boca Raton, FL.

18. Blattler, W.A., Kuenzi, B.S., Lambert, J.M. and Senter, P.D. (1985) *Biochemistry*, **24**, 1517.

19. Senter, P.D., Tansey, M.J., Lambert, J.M. and Blattler, W.A. (1985) *Photochem. Photobiol.*, **42**, 231.

20. Hunter, M.J. and Ludwig, M.L. (1972) *Methods Enzymol.*, **25**, 595.

21. Bayley, H. and Knowles, J.R. (1977). *Methods Enzymol.*, **46**, 69.

22. Traut, R.R., Bollen, A., Sun, T.T., Hershey, J.W.B., Sundberg, J. and Pierce, L.R. (1973) *Biochemistry*, **12**, 3266.

23. Shapiro, A.L. and Maizel, J.V. (1969) *Anal. Biochem.*, **29**, 505.

24. Hucho, F., Mullner, H. and Sund, H. (1975) *Eur. J. Biochem.*, **59**, 79.

25. Hajdu, J., Bartha, F. and Friedrich, P. (1975) *Eur. J. Biochem.*, **68**, 373.

26. Carpenter, F.H. and Harrington, K.T. (1972) *J. Biol. Chem.*, **247**, 5580.

27. Laemmli, U.K. (1970) *Nature*, **227**, 680.

28. Redl, B., Walleczek, J., Stoffler-Meilicke, M. and Stoffler, G. (1989) *Eur. J. Biochem.*, **181**, 351.

29. Ji, T.H. (1977) *J. Biol. Chem.*, **252**, 1566.

30. Amlaiky, N., Berger, J.G., Chang, W., McQuade, R.J., and Caron, M.G. (1987) *Mol. Pharmacol.*, **31**, 129.

31. Shephard, E.G., de Beer, Fl.C., von Holt, C. and Hopgood, J.P. (1988) *Anal. Biochem.*, **168**, 306.

32. Murthy, S.C., Mui, A.L. and Krystal, G. (1990) *Exp. Hematol.*, **18**, 11.

INDEX

activity staining   8, 9
affinity labels
    examples   301
    experimental criteria   300, 301
    principles   299, 300
affinity chromatography
    activation of matrix   44
    affinity elution   46–48
    operation of   45–46
    purification of fusion proteins   46
    theory of   43–44
    use to assess purity   81
affinity-based covalent modification of proteins
    299–304
amino acid analysis
    for protein determination   28
    practical aspects 270–272
    use to assess purity   82
amino acids
    abbreviations for   181, 270
    masses of   181
    post-translational modifications of   262–268
ammonium sulfate
    precipitation of proteins by   31
    solubility tables   32
    use in protein crystallization   158
analytical ultracentrifuge   161–174
arginine, modification of   292
aspartic acid, modification of   293
assay method, choice of   1–11

bacterial cells, disruption of   16
bicinchoninic acid (BCA) method for protein
    determination   29
bile acid derivative detergents   104
binding assays   83
biuret method for protein determination   28
block diagram of protein purification   132–133
blotting
    blocking solutions   94
    membranes   94
    protein stains   95
    transfer buffers   95
Brookhaven National Laboratory (BNL)   241, 242
buffers
    preparation of   24–25
    table of   22–23
    temperature dependence of   25

capillary electrophoresis   77–80
cell disruption, large scale   133–134
centrifugation, large scale   134
chemical modification of proteins   287–293
    group-selective reagents   289–293
    identification of modified residues   288–289

stoichiometry of modification   287–288
chromatofocusing
    operation of   39–40
    principles of   39
    use to assess purity   81
chromatographic methods
    affinity chromatography   43–47
    chromatofocusing   39, 40
    dye–ligand chromatography   47–52
    gel filtration   53–55
    hydrophobic interaction chromatography
        40–42
    immobilized metal affinity chromatography
        52–53
    ion-exchange chromatography   35–38
    materials for   35
    reverse phase chromatography   42–43
    theory of   33–35
circular dichroism (CD)
    far UV circular dichroism   195–197
    near UV circular dichroism   195–197
circular dichroism, applications of
    binding of ligands   198–199
    kinetics of structural changes   199
    protein denaturation   197
    secondary structure of proteins   195–197
    tertiary structure of proteins   195–197
circular dichroism, practical aspects of
    amounts of sample   200–201
    choice of solvent   201
    concentration of protein   201
    sources of errors   201–203
cleavage of proteins
    chemical   275–276
    enzymic   272–274, 277
CLUSTAL program   251
computer-based analysis of protein structure
    241–252
concentrating proteins
    adsorption   56–57
    centrifugation   57
    gas pressure   57–58
    osmotic pressure   58
    precipitation   56
    removal of water   57
    water pressure   58
conjugation of proteins to antibodies   313
consensus sequences, table of   254–261
continuous assays   2
Coomassie blue method for protein
    determination   29
covalent chromatography   65–70
criteria of purity   73–83
critical micelle concentration (CMC)   102–104
cross-linking reagents

Index

cysteine-directed 309, 311
lysine-directed 308, 310–311
other types 311
crystallization of proteins
  factors affecting crystal growth 156
  hanging drop crystallization 158, 159
  methods 155–159
  special methods for membrane proteins 105, 106
  suppliers of equipment and reagents 158
cysteine, modification of 289–290
cysteine proteinases, purification of 66, 69, 70
cystine, modification of 290

denaturation of proteins 233–238
detection methods for PAGE
  activity staining 91
  Coomassie blue staining 92
  dye staining 92
  fluorophore labeling 93
  radiolabelled proteins 95, 96
detergents, table of 102–104
Dictionary of Secondary Structure information in Proteins (DSSP) 246, 247
discontinuous assays 2
disruption of tissues and cells 13–16
disulfide bond formation 125–126
DSSP database 246, 247
dye–ligand chromatography
  choice of buffers 50–52
  differential chromatography 50
  selection of dye adsorbent 48–49

electron paramagnetic resonance (EPR)
  characteristics of spectra 221–222
  electron–electron interactions 222–223
  EPR-detectable species in proteins 218
  hyperfine interactions 222
  principles 217, 219, 220
electron paramagnetic resonance, applications of
  free radicals 227
  mobility of proteins 229
  molecular interactions 230
  paramagnetic metal ions 227
  redox proteins 225, 227
electrospray ionization mass spectrometry (ESIMS)
  advantages and disadvantages 175
  applications 81–82, 177, 281–282
enzyme-linked immunosorbent assay (ELISA) 10
epitope tag 111
European Biomolecular Institute (EBI) 241, 242

far UV absorbance for protein determination 28
filtration, large scale 134, 135
flow sheet of protein purification 132–133
formulation of industrial scale protein products 136
FPLC 80
free solution capillary electrophoresis 77, 79, 80
frictional ratio — see Perrin ratio
fusion proteins
  C-terminal fusions 109
  detection of 113–116

induction of expression 111–112
N-terminal fusions 109
proteolytic cleavage of 112
purification of 112–113
sandwich fusions 109

β-galactosidase tag 109–111
gel filtration
  media for gel filtration 55
  separation of proteins 53–55
  use to assess purity 80
gene fusion vectors 111
Genetics Computer Group 249, 250
glutamic acid, modification of 294
glutathione-S-transferase tag 109–111
glycosylations
  mass changes 183
gravimetric method for protein determination 28
guanidine hydrochloride 233, 235

heterobifunctional reagents for cross-linking 310
His-tag 109–111
histidine, modification of 290–291
Homology-derived Secondary Structure of Proteins (HSSP) 246, 248
HPLC — see Reverse-phase HPLC
HSSP database 246, 248
hydrodynamic methods 161–174
hydrolysis of proteins 270–271
hydrophobic interaction chromatography (HIC) 41–42

immobilized metal affinity chromatography (IMAC) 52–53
immunoblotting 9
immunoelectrophoresis
  reagents for 88
  staining methods 92
immunological assays 9–10
inclusion bodies
  factors affecting formation 120
  isolation of 122
  solubilization of protein in 122–123
  use in protein purification 119–122, 126, 127
industrial scale protein purification 131–136
infrared (IR) spectroscopy — see vibrational spectroscopy
ion-exchange chromatography
  anion exchangers 36
  buffers for 37
  cation exchangers 36
  operation of 37–39
  use to assess purity 80
ionic detergents 102
isoelectric focusing
  electrolytes for 87
  immobilized pH gradients 90
  marker proteins for 89–90
  mixtures for 87
  staining methods for 91
  use to assess purity 77
isotopic masses

amino acids 181
C-terminal groups 182
elements in proteins 181
N-terminal groups 182

J-coupling 205, 209

liposomes 105
liquid chromatography, large scale 134–136
Lowry method for protein determination 29
lysine, modification of 291–292

maltose binding protein tag 109–111
mammalian tissues, disruption of 14
marker proteins
for isoelectric focusing 89–90
for SDS–PAGE 89–90
mass spectrometry (MS) 81–82, 174–183, 281–282
matrix-assisted laser desorption ionization time of flight MS (MALDITOF-MS)
advantages and disadvantages 175
applications 81–82, 177, 281–282
MaxEnt software 178
mechanism-based irreversible inhibitors 302–303
membrane proteins
assay 105
cross-linking 313
crystallization 105, 106
purification 101, 105
reconstitution 105
solubilization 101–104
metal ion buffers 25–26
methionine, modification of 290
mobility of proteins
probed by EPR 229
probed by NMR 215
molecular dipsticks 227–228
molecular mass
calculation of 183
determination of 161–183
multi-dimensional NMR techniques 208

N-terminal modification of proteins 280–281
N-terminal sequence analysis 82, 280–281
national facilities for ultracentrifugation 173
nearest neighbor analysis by cross-linking 312
NMR of proteins 205–215
NMR resonances, assignment of
heteronuclear techniques 208–209
homonuclear techniques 207–208
nonionic detergents 102–103
nuclear magnetic resonance — see NMR
nuclear Overhauser effect (NOE) 205, 209
nuclei, magnetic properties of 205–206
nucleic acids, presence in protein samples 73
number-average molecular mass 164

o-phthalaldehyde (OPA) method for protein determination 29
oxygen electrode 7

partial specific volume 167–168
PDB file 242, 243
peptide synthesis — see solid-phase peptide synthesis 139–152
Perrin ratio 166
pH-stat 7–8
photoaffinity labels 301–302
PILEUP program 251
plant tissues, disruption of 15
polishing steps in industrial scale purification 136
polyacrylamide gel electrophoresis (PAGE)
detection methods 92–96
immunoelectrophoresis 90
isoelectric focusing (IEF) 87
manufacturers and suppliers 96–97
marker proteins 89–90
nondenaturing gels 86
SDS–PAGE 85–86
staining methods 91, 92
two-dimensional electrophoresis 90
use to assess purity 76–77
polyethylene glycol (PEG) in protein crystallization 159
post-translational modifications
consensus sequences for 262–268, 269
list of 254–261, 269
mass changes 182, 254–261
PROSITE database 250, 251
protein A tag 109–111
protein concentration, determination of 27–30
protein databank (PDB) 241–243
protein motions database 249
protein stability 233–238
proteinase inhibitors 17, 18
proteolysis, control of 17
purity, criteria of 73–83
2-pyridyl disulfide 66

quiescent affinity labels 301

radiometric assays 4, 5–6
Raman spectroscopy — see vibrational spectroscopy
RASMOL molecular graphics program 243–246
renaturation of denatured protein 123–126
retention times of peptides on reverse-phase HPLC 278–280
reverse-phase chromatography — see reverse-phase HPLC
reverse-phase HPLC 6–7, 42–43, 58–62, 80, 277–280
reverse phase HPLC-based assays 6–7
root mean squared (RMS) fit 246, 249

S-tag 109–111
s value — see sedimentation coefficient
salting out 31–32
scale down of industrial processes 132
scale up of processes 131–132
SCOP database 249
SDS–PAGE
marker proteins for 89–90

mixtures for 86
use to assess purity, 76
secondary structure of proteins
    by circular dichroism 195–197
    by infrared or Raman spectroscopy 192, 193
sedimentation coefficient 165–166
sedimentation equilibrium 162–164, 170–172
sedimentation velocity 164–166, 171–173
separation of cells 14
separation of peptides 277–280
SEQNET 241, 242
sequencing of peptides 281–284
small peptides, isolation of 58–62
small scale dialysis in protein crystallization 159
solid-phase peptide synthesis
    activation of carboxyl groups 143–146
    analysis of product 146–148, 150
    deprotection 144
    monitoring and control 146–149
    peptide–resin linkages 141–143
    protection of amino groups 141–144
    supports for 140–141
specific activity 76
spectrophotometric assays 2–5
spin labels 227, 229
spin probes 227
Strep-tag 109–111
structural classification of proteins (SCOP) 249
structural information
    from circular dichroism 195–197
    from infrared or Raman spectroscopy 192, 193
structure determination by NMR
    distance geometry algorithms 211
    quality of structures 211, 212
    restrained molecular dynamics 211
    table of structures 212–214
substrate-dependent nonaffinity labels 303
substrate-derived site-specific reactivity probes 303–304

synthetic peptides for antibody production 149–152

thermal denaturation of proteins 237–238
thiol groups, protection of 19
thiol–disulfide interchange 65–70
thioredoxin tag 109–111
transition state affinity labels 302
tryptophan, modification of 292
two state model of protein unfolding 235
two-dimensional electrophoresis
    mixtures for 88, 90
    staining methods 91
two-dimensional NMR techniques 207
tyrosine, modification of 293

unfolding of proteins 233–238
urea 233, 235
US National Center for Biotechnology Information (NCBI) 241, 242
UV absorbance for protein determination 28

vibrational spectroscopy
    applications 192–194
    assignment of bands 190–191
    Fourier transform infrared spectroscopy (FTIR) 188
    Fourier transform Raman spectroscopy 188
    infrared (IR) spectroscopy 187–194
    raman spectroscopy 187–194
    sampling methods 189

weight-average molecular mass 164
western blotting — see immunoblotting
world wide web (WWW) 241

yeasts, disruption of 15–16

z–average molecular mass 164
Zwitterionic detergents 103